SHAPES OF TIME

signale
modern german letters, cultures, and thought

Series editor: Paul A. Fleming, Cornell University
Peter Uwe Hohendahl, Founding Editor

Signale: Modern German Letters, Cultures, and Thought publishes new English language books in literary studies, criticism, cultural studies, and intellectual history pertaining to the German-speaking world, as well as translations of important German-language works. *Signale* construes "modern" in the broadest terms: the series covers topics ranging from the early modern period to the present. *Signale* books are published under a joint imprint of Cornell University Press and Cornell University Library in electronic and print formats. Please see http://signale.cornell.edu/.

SHAPES OF TIME

*History and Eschatology
in the Modernist Imagination*

MICHAEL MCGILLEN

A Signale Book

CORNELL UNIVERSITY PRESS AND CORNELL UNIVERSITY LIBRARY
ITHACA AND LONDON

Cornell University Press and Cornell University Library gratefully acknowledge the College of Arts & Sciences, Cornell University, for support of the Signale series.

Copyright © 2023 by Cornell University

All rights reserved. Except for brief quotations in a review, this book, or parts thereof, must not be reproduced in any form without permission in writing from the publisher. For information, address Cornell University Press, Sage House, 512 East State Street, Ithaca, New York 14850.

First published 2023 by Cornell University Press
and Cornell University Library

Library of Congress Cataloging-in-Publication Data

Names: McGillen, Michael, 1982– author.
Title: Shapes of time : history and eschatology in the modernist imagination / Michael McGillen.
Description: Ithaca [New York] : Cornell University Press and Cornell University Library, 2023. | Series: Signale: modern German letters, cultures, and thought | Includes bibliographical references and index.
Identifiers: LCCN 2023001990 (print) | LCCN 2023001991 (ebook) | ISBN 9781501772818 (hardcover) | ISBN 9781501772825 (paperback) | ISBN 9781501772849 (pdf) | ISBN 9781501772832 (epub)
Subjects: LCSH: Time in literature. | Mathematics in literature. | Modernism (Literature)—Europe, German-speaking. | Barth, Karl, 1886-1968—Criticism and interpretation. | Rosenzweig, Franz, 1886-1929—Criticism and interpretation. | Kracauer, Siegfried, 1889-1966—Criticism and interpretation. | Musil, Robert, 1880-1942—Criticism and interpretation.
Classification: LCC PN56.T5 M34 2023 (print) | LCC PN56.T5 (ebook) | DDC 809/.9336—dc23/eng/20230505
LC record available at https://lccn.loc.gov/2023001990
LC ebook record available at https://lccn.loc.gov/2023001991

Contents

Acknowledgments	vii
Note on Translations and Abbreviations	ix
Introduction	1
1. Constructivism, Mathematics, and the Space of History	33
2. The Eschatological Limit: Spatial Form in Karl Barth's Dialectical Theology	81
3. The Arc of History: Franz Rosenzweig's Figures of Time and Eternity	129
4. Temporal Exterritoriality: Siegfried Kracauer and the Shape of History	182
5. Images without End: Robert Musil's Narrative Ruptures	234
Epilogue: The Ends of Modernism	289
Bibliography	301
Index	319

Acknowledgments

This book is the culmination of many years of research and thinking, and I am indebted to a great number of colleagues, teachers, and friends who have inspired me in ways large and small. I owe a special debt of gratitude to Michael W. Jennings, whose thoughts and commentary helped to shape the project in its formative stages, and whose seminars on Musil and Kracauer provided the germ cells for my approaches to their work. I would also like to acknowledge Klaus Mladek, Brigid Doherty, Jane O. Newman, and David Kleinberg-Levin, who imparted critical impulses at key junctures in my work on the book. My colleagues in the Department of German Studies at Dartmouth have given me essential support and guidance that have made this book possible. I feel fortunate to have found a home for the book at Cornell University Press, and Kizer S. Walker and Mahinder S. Kingra have been supportive, responsive, and a pleasure to work with. I would especially like to thank Paul Fleming, the Signale Editorial Board, and two anonymous reviewers for their feedback and suggestions, which have been pivotal in giving the book its final form.

My research for the book has been supported by the German Academic Exchange Service (DAAD), by a Charlotte Elizabeth Proctor Honorific Fellowship (Princeton University), and by the Office of the Dean of the Faculty at Dartmouth College. An earlier version of parts of chapter 2 was published as "Theology's Weimar Moment: History before the Eschatological Limit," in *The Weimar Moment: Liberalism, Political Theology, and Law*, ed. Leonard Kaplan and Rudy Koshar (Lanham, MD: Lexington Books, 2012), 267–87, and appears here with permission from Rowman & Littlefield. I would like to thank Allison Van Deventer for editing the manuscript and sharpening its prose. Above all, I am grateful to Petra McGillen, who has been an exceptional first reader and an unwavering source of support over the years.

Note on Translations and Abbreviations

Unless otherwise noted, all translations are the author's. On occasion, the following published translations were consulted:

Karl Barth, *The Epistle to the Romans*, trans. Edwyn C. Hoskyns (London: Oxford University Press, 1933).
Franz Rosenzweig, *The Star of Redemption*, trans. Barbara E. Galli (Madison: University of Wisconsin Press, 2005).
Siegfried Kracauer, *The Mass Ornament: Weimar Essays*, trans. Thomas Y. Levin (Cambridge, MA: Harvard University Press, 1995).
Robert Musil, *The Man without Qualities*, trans. Sophie Wilkins (New York: Alfred A. Knopf, 1995).

The following abbreviations are used for in-text references to frequently cited primary sources:

R   Karl Barth, *Der Römerbrief (zweite Fassung), 1922*, 17th ed. (Zurich: Theologischer Verlag Zürich, 2005).

S   Franz Rosenzweig, *Der Stern der Erlösung* (Frankfurt am Main: Suhrkamp, 1988).

B   Franz Rosenzweig, *Briefe*, ed. Ernst Simon and Edith Rosenzweig (Berlin: Schocken, 1935).

L   Siegfried Kracauer, *History: The Last Things before the Last* (New York: Oxford University Press, 1969).

O   Siegfried Kracauer, *Das Ornament der Masse: Essays* (Frankfurt am Main: Suhrkamp, 1963).

M   Robert Musil, *Der Mann ohne Eigenschaften*, in *Gesammelte Werke*, vol. 1, ed. Adolf Frisé (Reinbek bei Hamburg: Rowohlt, 1978).

P   Robert Musil, *Prosa und Stücke, Kleine Prosa, Aphorismen, Autobiographisches, Essays und Reden, Kritik*, in *Gesammelte Werke*, vol. 2, ed. Adolf Frisé (Reinbek bei Hamburg: Rowohlt, 1978).

Shapes of Time

Introduction

The Shapes of Time

In *History: The Last Things before the Last* (1969), Siegfried Kracauer proposes that beyond the "empty vessel" of chronological time stands a variety of "shaped times" or "time curves" (*L* 201, 147, 144). The impetus for this bold assertion is the imagination of time beyond its classical historicist attributes. What if time, Kracauer ponders, does not flow like a current, steadily and continuously, ticking off second after second with the regularity of a clock? Suppose that history is not a linear continuum defined by steady progress and stages of development, and that the movement of time cannot be fully captured by the image of a present moment slipping ineluctably into the past as the future becomes present. How might we imagine a more complex fabric of time and history, one with manifold shapes, curves, textures, and openings?

Philosophers have long considered space and time to be the basic categories of our perceptual apparatus. Immanuel Kant famously described space and time as the pure forms of our intuition: it is in time and space, on a preconceptual level, that we receive sensations.[1] Yet space and time, though equally foundational, are generally conceived as separate and opposite categories that operate on fundamentally different levels. Whereas space is characterized by extension in three dimensions, time is one-dimensional and is determined by duration. Before the twentieth century, there was little cross-pollination of these categories. With the rise of modernism in art, literature, philosophy, and the sciences, however, new possibilities emerged for imagining the spatiality of time.

Over the past few decades, critics have pointed to a range of factors that contributed to new perspectives on time and space around 1900 and to the various literary, philosophical, and aesthetic manifestations of these perspectives. In his pathbreaking *The Condition of Postmodernity* (1989), David Harvey interpreted modernism as a response to a crisis in the experience of space and time resulting from what he calls "time–space compression"—an acceleration of the pace of life in modern capitalism, via new modes of transportation and technologies of communication, that overcomes spatial barriers and collapses time horizons.[2] In *Raumgeschichten* (2007), Oliver Simons showed how the perceptual inversions of *Kippfiguren* (multistable figures), as developed by Gestalt psychology and in the philosophical reflections of Ludwig Wittgenstein, induced new strategies for representing space in modernist literature.[3] Andreas Huyssen argues in *Miniature Metropolis* (2015) for the fundamental spatiality of the "metropolitan miniature"—a short form of urban prose in Baudelaire, Rilke, Kafka, Kracauer, Benjamin, Musil, and

---

1. In the *Critique of Pure Reason*, Kant claims that there are "two pure forms of sensuous intuition [*sinnlicher Anschauung*]" that serve as "a priori principles of knowledge"—namely, "space and time" (*Raum und Zeit*) (Immanuel Kant, *Kritik der reinen Vernunft*, ed. Jens Timmermann [Hamburg: Meiner, 1998], 96).

2. David Harvey, *The Condition of Postmodernity: An Enquiry into the Origins of Cultural Change* (Oxford: Basil Blackwell, 1989), 240, 265.

3. Oliver Simons, *Raumgeschichten: Topographien der Moderne in Philosophie, Wissenschaft und Literatur* (Munich: Fink, 2007).

Adorno—which he describes as "one of the few genuinely innovative modes of spatialized writing created by modernism."[4] More recently, in *Precarious Times* (2019) Anne Fuchs shows that within literary and aesthetic modernism one can find forms of resistance to the narrative of modernity as defined by the acceleration of time, and a range of alternative temporalities that run counter to the figure of the "arrow of time" and the idea of a "linear progression through history."[5]

Whereas it is commonplace in social theory to claim that modernity has privileged time over space, foregrounding the ways the temporal processes of modernization transform spatial orders, Fuchs and Harvey have shown how the "aesthetic thrust of modernism" uses strategies of spatialization to imagine alternative timescapes.[6] Building on these investigations of modernism's spatial turn, this book shows how modernism's imagination of new shapes of time was conditioned by a hitherto unrecognized encounter between mathematics and religious thought. The language and imagery of modernist mathematics, especially the imagination of nonintuitive and higher-dimensional spaces in non-Euclidean geometry, provided an interface for describing history in relation to an unrepresentable "time of the now." Complicating the picture of a "linear and irreversible 'time's arrow'" that "flows through all events and makes them chronologically measurable,"[7] modernism conceived of the

---

4. Andreas Huyssen, *Miniature Metropolis: Literature in an Age of Photography and Film* (Cambridge, MA: Harvard University Press, 2015), 2.

5. Anne Fuchs, *Precarious Times: Temporality and History in Modern German Culture* (Ithaca, NY: Cornell University Press, 2019), 12.

6. See Fuchs, *Precarious Times*, 51–52; and Harvey, *The Condition of Postmodernity*, 205–6. Harvey refers to the traditions of social thought emanating from Marx, Weber, Adam Smith, and Marshall as having privileged time over space (205). An early statement of the priority of time over space in modernity, dating to 1843, can be found in an oft-cited reflection by Heinrich Heine on the experience of train travel: "Sogar die Elementarbegriffe von Zeit und Raum sind schwankend geworden. Durch die Eisenbahnen wird der Raum getödtet, und es bleibt uns nur noch die Zeit übrig" (Heinrich Heine, "Lutezia II," in *Historisch-kritische Gesamtausgabe der Werke*, ed. Manfred Windfuhr [Hamburg: Hoffmann und Campe, 1990], 9–98, at 58).

7. See Aleida Assmann, *Is Time out of Joint? On the Rise and Fall of the Modern Time Regime*, trans. Sarah Clift (Ithaca, NY: Cornell University Press, 2020), 14.

shape of time as a curved space in which apparently parallel lines can intersect, a space that folds back on itself and produces a contemporaneity of temporally distant moments, a space defined by a liminal relation to the end of history.

The imagination of history in terms of shaped times was especially fruitful for a range of modernist writers and thinkers who sought to reconceive the end of history not as the goal of history but as a figure of discontinuity and rupture inherent in each moment in time. Reflections on eschatology—traditionally understood as the doctrine of the last things—took on a renewed significance: no longer a parochial theological concern, it became a field of inquiry that engaged writers across a spectrum between religion and the secular, between Judaism and Christianity.[8] With recourse to mathematical figures and spatial forms, the key writers explored in this book—Karl Barth, Franz Rosenzweig, Siegfried Kracauer, and Robert Musil—detached the end of history from a narrative framework and experimented with shaped times that reconfigured the relation of history and eschatology. At stake is an afterlife of religious thought in modernism that fundamentally reimagined the shape of history beyond its historicist form.

In his essay "Das Ende aller Dinge" ("The End of All Things") (1794), Kant proposes that the idea of "an end of all time" (*ein Ende aller Zeit*) is simultaneously "terrible and sublime" (*furchtbarerhaben*). We cannot form a clear concept of the end of time because it is utterly incommensurable with our temporal existence, yet the very thought proves irresistible; it takes us to the "edge of an abyss" (*an den Rand eines Abgrunds*), yet we cannot help but turn our gaze toward it again and again, however we may recoil at the thought.[9] Although the end of history appears to resist repre-

---

8. As Benjamin Lazier notes in his study of gnosticism and pantheism in the work of Hans Jonas, Leo Strauss, and Gershom Scholem, theology in Weimar Germany was not a mere "parochial pursuit" but rather "a cultural and intellectual practice" that served as "a vehicle for commentary on the political, aesthetic, and philosophical present common to us all" (*God Interrupted: Heresy and the European Imagination between the World Wars* [Princeton, NJ: Princeton University Press, 2008], 3).

9. Immanuel Kant, "Das Ende aller Dinge," in *Was ist Aufklärung? Ausgewählte kleine Schriften*, ed. Horst D. Brandt (Hamburg: Meiner, 1999), 62–76, at 62.

sentation, its very obscurity stimulates the imagination. Traditionally, the end of time is imagined as "the last day" (*de[r] jüngste[] Tag*), a day that belongs to time and passes judgment over the entirety of time, hence a "judgment day" (*Gerichtstag*).[10] As such, the eschatological realm of the last things, which stands on the threshold of time and eternity, is understood as a moment in time—namely, as the last day that marks the end of time as its cessation.

In the first decades of the twentieth century, however, the imagination of eschatology as the temporal end of history was supplanted by spatial representations of the relationship between time and eternity. The problem of how to represent what exceeds representation—the sublime moment of the end of time—was approached through a transposition into the realm of space. For Barth, the most prominent figure in a nascent Dialectical Theology, the end of history is conceived in terms of metaphors of spatial proximity and distance, such as limits, thresholds, and hollow spaces. The relationship of history and eschatology becomes legible in a mathematical language of planes, tangent lines, and points without extension. These spatial forms allow Barth to represent a moment that stands outside historical and narrative trajectories. Similarly, in the work of Musil, a key figure in literary modernism, time appears to come to a standstill in the moment of the "other condition"—an end of time that emerges, absent temporal closure, in a space of images. In the famous scene of hovering blossoms in *The Man without Qualities*, Musil provides an image of how we might picture the suspension of time.

The reimagination of the end of history in spatial terms provides a means of representing the unrepresentable without making recourse to a temporal concept of the end, as Kant felt compelled to do. Conceived in temporal terms, the end is the final element in a series or the culmination of a teleological movement toward a goal. By contrast, when conceived spatially, the end emerges as a boundary or threshold that inheres in each moment. This reimagination of the end had far-reaching consequences not only for theological debates but also for the concept of history at large. This is because a temporal understanding of eschatology, in secularized form, provided a foundation

---

10. Kant, "Das Ende aller Dinge," 63.

for the idealist and historicist concept of history that emerged in the work of Johann Gottfried Herder, Friedrich Schiller, and Georg Wilhelm Friedrich Hegel. In their work, the moment of divine judgment is temporalized and brought within the contours of a history of development. Schiller's famous dictum "World history is the world's court of judgment" (*Die Weltgeschichte ist das Weltgericht*) is emblematic of this process of historicization. A formerly transcendent source of meaning emerges here in secular form, its unfolding now coinciding with the movement of history itself. As Reinhart Koselleck comments in a gloss on Schiller's epigram: "The moral of history was temporalized as process.... The renunciation of retributive justice in the beyond leads to its temporalization [*Verzeitlichung*]. History in the here and now attains an inescapable quality."[11]

The concept of history that began to circulate in the eighteenth and nineteenth centuries was thus the result of a temporalization of thought in which history emerged as a process of development. According to Koselleck, the notion of historical development in German idealism depended on the historicization of eschatology as *Weltgericht*: "The path was thus clear, in the wake of idealistic philosophies of history, to dissolve Christian eschatology in processual terms.... The last judgment—the crisis—is effectively extended to *the developmental sequence of history* [*auf die geschichtliche Entwicklungsreihe ausgedehnt*]."[12] In this historicization of eschatology, the negativity of the end is temporalized and located in the future as the *telos* of history. In Hegel's philosophy of history, an inner-historical dialectic mediates a progressive and processual emergence of the ideal within the immanence of history. As Karl Löwith concludes, the "philosophy of history originates with the Hebrew and Christian faith in a fulfilment and ... ends with the secularization of its eschatological pattern [*Säkularisierung ihres eschatologischen Vorbildes*]."[13]

---

11. Reinhart Koselleck, "Geschichte," in *Geschichtliche Grundbegriffe: Historisches Lexikon zur politisch-sozialen Sprache in Deutschland*, vol. 2 (Stuttgart: Ernst Klett, 1975), 593–718, at 667.

12. Koselleck, "Geschichte," 685.

13. Karl Löwith, *Meaning in History* (Chicago: University of Chicago Press, 1949), 2; Löwith, *Weltgeschichte und Heilsgeschehen: Die theologischen Voraussetzungen der Geschichtsphilosophie* (Stuttgart: Kohlhammer, 1953), 11–12.

Meanwhile, within the discipline of history, historical research blossomed in the nineteenth century. Historians such as Leopold von Ranke and Johann Gustav Droysen—whose approach to history has come to be known as "historicism"—produced comprehensive historiographies of past epochs, amassing and objectively evaluating historical data in order to produce a picture of the past "as it really was." Their historiographies carried forward the idea of a universal history of human development of which individual cultures, peoples, and nations are inextricably a part. The aim of their work was to reconstruct the historical causalities that account for historical development. In historicism, a belief in the progressive movement of history, undeterred by epochs of decline, was coupled with a desire to justify and give meaning to history through historiographical reconstruction.

The emergence of a discursive structure of historical thought was one of the most decisive transformations of the eighteenth and nineteenth centuries, and it depended in no small measure on the secularization and immanentization of the end of history. By the early twentieth century, however, the historicization of thought that Schiller and Hegel had set in motion had begun to tremble. History itself—its constitution, its representability, the way it determines the logic and grammar of thought, and its experience—emerged as a pressing cultural problem. In the intellectual historical formation of modernism, the concepts of time and history underwent a profound transformation.[14] The result was a proliferation of dual-aspect theories of history that recognized how different senses of time can exist side by side, as chronological time and shaped time do in Kracauer's work. For example, Rosenzweig conceived of the "today" both as the present moment and as a springboard to eternity, while Barth considered the eschatological now to be both coextensive and incommensurable with historical time. Complicating the idea of history as a narrative, Musil pictured history as a space of entropic dispersion that lacks distinct beginnings and ends.

---

14. As Fuchs notes, modernism and contemporary art and literature bring forth "alternative temporal trajectories and experiences" that are "pluritemporal" (*Precarious Times*, 4–5).

Whereas the historicist project, as it emerged in Hegel and Schiller, conceived of the eschatological moment of judgment as immanent to the unfolding of history, modernism gave rise to a more complex afterlife of eschatological thought. In modernism, eschatology is decoupled from a teleological model of historical development: the end of history no longer stands for an apocalyptic narrative of catastrophe in which time and history come to an end in the moment of the last judgment or for a moment of utopian fulfillment that is deferred into a more or less imminent future. The writers considered in this book deflect the temporalization of eschatology in which the end of history is imagined as a terminal point that concludes a temporal sequence. Instead, they conceive of eschatology in spatial terms as a limit phenomenon. As in the mathematical concept of the limit—in which a variable approaches a fixed value yet maintains an infinitesimal difference—modernism constructs the eschatological moment in a relation of spatial proximity to each moment of history. Barth, Rosenzweig, Kracauer, and Musil attend to a liminal relation of time and eternity, history and eschatology, in which history stands at the threshold of its incommensurable other, yet remains at a distance from it. The spatial dimension of the eschatological moment opens up a field of tension at each moment in time rather than succeeding history as its cessation. This staging of an encounter of history and eschatology produced a novel concept of time that complicated the historicist notion of time's arrow. In short, eschatological discourse made manifest what Kracauer calls the antinomy of time—namely, that "time not only conforms to the conventional image of a flow but must also be imagined as being not such a flow" (*L* 199).

Such alternative models of historiography make use of key modernist aesthetic strategies to reimagine the contours of history. Indeed, the spatial imagination of time resonates with new literary forms that emphasized discrete series of images rather than narratives of development. In place of the teleological structure of the *Bildungsroman*, Rainer Maria Rilke's *Die Aufzeichnungen des Malte Laurids Brigge* (*The Notebooks of Malte Laurids Brigge*) (1910) and Walter Benjamin's *Berliner Kindheit um Neuzehnhundert* (*Berlin Childhood around 1900*) (1938), to name just two prominent examples, produced disjunctive images of the subject, drawing on literary strategies that emphasize the spatial relations of constellations of images rather than the temporal lineage of an embryonic moment that comes to fruition in the course of a narrative. Moreover, these new models of history have an affinity with the constructivist turn in historiography and the visual arts. In the first decades of the twentieth century, theorists of history such as Karl Mannheim and Ernst Troeltsch claimed that history is constructed from the vantage point of the present, while Karl Heussi and Oswald Spengler argued that the historian imposes a structure on historical figures that is mobile and constantly changing shape. In the Bauhaus and Russian Constructivism, meanwhile, the primacy of geometrical forms emerged as a specifically modernist aesthetic mode.

But a crucial and largely unrecognized impetus for spatializing time and conceiving of history in terms of shaped times can be found in the new theory of space that arose in non-Euclidean geometry and modernist mathematics. Whereas in Euclidean geometry the constitution of space is fixed in a uniform three-dimensionality, the work of Carl Friedrich Gauss, Bernhard Riemann, and Nikolai Lobachevsky showed that space can have different degrees of curvature, resulting in spaces in which parallel lines can intersect and in which figures can be deformed. The curved spaces of non-Euclidean geometry, while mathematically consistent, stretch the limits of what can be visualized intuitively. As Oliver Simons notes, non-Euclidean spaces deviate from empirical experience and result in a "loss of intuitiveness" (*Verlust der Anschaulichkeit*) that has ripple effects well beyond the discipline of mathematics.[15] In *The Fourth Dimension and Non-Euclidean Geometry in Modern Art* (2013), Linda D. Henderson shows how the non-Euclidean revolution spurred the radical innovations in modernist art in the early twentieth century—from Cubism, Futurism, and Surrealism to Suprematism and Russian Constructivism.[16] But non-Euclidean curved

---

15. Simons, *Raumgeschichten*, 14.
16. Linda D. Henderson, *The Fourth Dimension and Non-Euclidean Geometry in Modern Art* (Cambridge, MA: MIT Press, 2013).

spaces also provided an impetus for rethinking the shape of time in new approaches to history and eschatology that push beyond the framework of historicism. The instability of figures in non-Euclidean spaces provided a model for the changing shapes of figures in history. In adapting insights from the geometry of curved space to conceptualize new shapes of time, Barth, Rosenzweig, Kracauer, and Musil find in the language of spatial forms a set of metaphors well suited to the nonintuitive and liminal relations of time and eternity, history and eschatology.

Mathematics and Modernism

The non-Euclidean revolution could provide a paradigm for thinking about history and eschatology in new spatial terms because it fundamentally transformed the way we think about space. As historians of mathematics such as Herbert Mehrtens and Jeremy Gray have shown, over the course of the nineteenth and early twentieth centuries, a sea change took place in the way mathematicians conceived of their objects of study. In its modernist guise, mathematics was concerned with representing not the physical world but rather objects and spaces that it creates through the development of a language of mathematical signs.[17] According to Gray, between 1890 and 1930 mathematics underwent a modernist transformation in which it came to define "an autonomous body of ideas, having little or no outward reference, placing considerable emphasis on formal aspects of the work."[18] The germ cell of mathematical modernism can be traced to the work of Gauss and Riemann and to the discov-

---

17. See Herbert Mehrtens, *Moderne—Sprache—Mathematik: Eine Geschichte des Streits um die Grundlagen der Disziplin und des Subjekts formaler Systeme* (Frankfurt am Main: Suhrkamp, 1990), 9.

18. Jeremy Gray, *Plato's Ghost: The Modernist Transformation of Mathematics* (Princeton, NJ: Princeton University Press, 2008), 1. See also Nina Engelhardt, *Modernism, Fiction and Mathematics* (Edinburgh: Edinburgh University Press, 2018), 7: "The notion that maths does not have a direct representational relationship with nature is a main feature of its modern development, and the release from the constraints of realist representation had enormous implications for the discipline and beyond."

ery of non-Euclidean geometry by Lobachevsky and János Bolyai. These figures ushered in an epochal break that shattered the idea that mathematics represents and reproduces a natural reality.[19]

Mathematics is an integral part of modernism because it posits that the world is not given but rather a construction. Comparing modernist mathematics to the figure of Baron Münchhausen, who pulls himself out of a swamp by his own hair, Mehrtens argues that in modernism it becomes apparent that mathematics has no ground outside itself.[20] In this respect, modernist mathematics has a strong affinity with modernism in music, art, and literature, which likewise abandon the representation of a given reality and undermine the foundation of a space ordered through a central Renaissance perspective.[21] Just as Cubism redefines the relationship between surface and space, producing new modes of visualizing the world, non-Euclidean geometry reveals the possibility of curved spaces in which the basic premises of the three-dimensional space of our empirical experience do not hold. Mathematics thus participated in a larger context of cultural modernism that grappled with the loss of traditional unities: just as the ego could no longer claim sovereignty over itself and national identity could no longer serve as the guarantor of political unity, so too, as Mehrtens notes, did geometry no longer describe a single space: many geometries now dealt with many possible spaces.[22] As a kind of thinking that creates or constructs space, mathematics contributed to the constructivist turn in modernism.

The construction of new and multiple forms of space in the wake of the non-Euclidean revolution had a formative impact on discourses on eschatology in the early twentieth century. In particular, modernist writers interested in the relation of history and eschatology drew inspiration from non-Euclidean spaces that appear to be nonintuitive and paradoxical, pushing the limits of what we can visualize. For example, the image of intersecting parallel lines—an apparent paradox that becomes possible in the curved spaces of

---

19. See Mehrtens, *Moderne—Sprache—Mathematik*, 29.
20. Mehrtens, *Moderne—Sprache—Mathematik*, 519–20.
21. See Mehrtens, *Moderne—Sprache—Mathematik*, 546.
22. Mehrtens, *Moderne—Sprache—Mathematik*, 564.

non-Euclidean geometry—appears again and again as a metaphor for the liminal relation of time and eternity, or history and eschatology. Similarly, the nonintuitive character of non-Euclidean space frequently serves as a proxy for the unrepresentability of God or the eschatological now. The paradoxical possibilities of non-Euclidean space provided a stimulus for reflections on the nonintuitive contours and curvatures of time.

Mathematics, non-Euclidean geometry, and their constructions of space and spatial figures gave Barth, Rosenzweig, Kracauer, and Musil a reservoir of metaphors and images for rethinking the end of history and the shape of time. In their work, modernist mathematics provides a language for constructing history in a nonintuitive relationship to the eschatological now, such that for Barth and Rosenzweig the end of history is always present, and for conceiving of shaped times in which endings are impossible to visualize because they are multidimensional, as evident in the absence of narrative endings in Musil's *The Man without Qualities* and in the ever-changing shapes of time in Kracauer's topological imagination of history. To be sure, as these writers transpose mathematical insights about curved and nonintuitive spaces into their concepts of history, they produce spatialized images of time and history that are not always strictly mathematical. Nevertheless, the modernist turn in mathematics provided a crucial impetus for new concepts of the end. For example, Barth and Rosenzweig constructed curved spaces of history in which the end of history is folded back on itself, as in spherical geometries, resulting in a coincidence of beginning and end, *Urgeschichte* and *Endgeschichte*, and the proximity of the end to each present. In contrast, Kracauer's and Musil's mobile spaces of history can be compared to a bundle of lines that never intersect, as in pseudospherical geometries, implying the impossibility of an end of history.

Shapes of Time contributes to a nascent body of scholarship that explores the mathematical dimension of modernism's spatial turn and the affinities and exchanges between mathematics, literature, intellectual history, and the arts. As Nina Engelhardt notes in *Modernism, Fiction and Mathematics* (2018), while scholarship on the history of science and its relationship to cultural modernism has roots that extend back to the 1980s, mathematics has received far less attention.[23] Similarly, critics such as Henri Lefebvre, Edward W. Soja, Fredric Jameson, and David Harvey explore the phenomenon of space in its geographical, sociological, cultural historical, and political dimensions,[24] but have not considered how mathematics—especially the destabilization of the idea of a stable three-dimensional container space in non-Euclidean geometries—informed new theories of space in twentieth-century discourses in the humanities and social sciences.[25]

In addition to the contributions of Mehrtens, Gray, and Henderson on the proximity of mathematics to cultural modernism, especially in the visual arts, recent scholarship has shown how mathematics has informed the thinking and poetics of major figures in the German-speaking context. Engelhardt points to the importance of mathematics for Hermann Broch's and Robert Musil's poetics,[26] while John H. Smith demonstrates the key role played by infinitesimal calculus in the work of Hermann Cohen, Franz Rosenzweig, Gershom Scholem, and Karl Barth.[27] In *The Mathematical Imagination* (2019), Matthew Handelman uncovers the significance of mathematical logic, calculus, and geometry (in its Euclidean and Cartesian forms) for Scholem's,

---

23. Engelhardt, *Modernism, Fiction and Mathematics*, 2. For an overview of the extensive scholarship on the history of science, especially physics, and its relationship to literature, see Robert Bud and Morag Shiach, "Being Modern: Introduction," in *Being Modern: The Cultural Impact of Science in the Early Twentieth Century*, ed. Robert Bud, Paul Greenhalgh, Frank James, and Morag Shiach (London: University College of London Press, 2018), 1–19, at 4.

24. On the contributions of Lefebvre, Soja, Jameson, Harvey, and others to the "spatial turn," see Jörg Döring and Tristan Thielmann, "Einleitung: Was lesen wir im Raume? Der Spatial Turn und das geheime Wissen der Geographen," in *Spatial Turn: Das Raumparadigma in den Kultur- und Sozialwissenschaften*, ed. Jörg Döring and Tristan Thielmann (Bielefeld: Transcript, 2008), 7–45.

25. On the ripple effects of mathematical concepts of space—along with those of physics—in the humanities and social sciences, see Stephan Günzel, "Physik und Metaphysik des Raums: Einleitung," in *Raumtheorie: Grundlagentexte aus Philosophie und Kulturwissenschaften*, ed. Jörg Dünne and Stephan Günzel (Frankfurt am Main: Suhrkamp, 2006), 19–43, at 11, 41.

26. Engelhardt, *Modernism, Fiction and Mathematics*, 59–125.

27. John H. Smith, "The Infinitesimal as Theological Principle: Representing the Paradoxes of God and Nothing in Cohen, Rosenzweig, Scholem, and Barth," *MLN* 127 (2012): 562–588.

Rosenzweig's, and Kracauer's contributions to critical theory.[28] By contrast, this book turns to the construction of space in non-Euclidean geometry as a model for new concepts of history in the work of Barth, Rosenzweig, Kracauer, and Musil.

Religion and Modernism

These pages tell a story about how modernist mathematics informs new concepts of history and eschatology. At the same time, they provide a new account of the multiple afterlives of religion in modernity. In the past decades there has been a burgeoning interest in new ways of understanding the status of religious modes of thought in twentieth-century modernism. Calling into question the classical theory of secularization in writers such as Max Weber, Carl Schmitt, and Karl Löwith, scholars have drawn attention to the persistence of religion in post-secular societies.[29] As early as the 1960s, in a secularization debate directed against Schmitt and Löwith, Hans Blumenberg criticized both a quantitative model of secularization—as a "subtraction thesis" that provides evidence of a "decline of religion"—and a qualitative or transitive model summed up in the formula "B is the secularized A."[30] Rejecting the binary choice between modernity as an emancipation from religion and as purely

---

28. Matthew Handelman, *The Mathematical Imagination: On the Origins and Promise of Critical Theory* (New York: Fordham University Press, 2019).

29. In the past two decades, scholars from a wide range of disciplines have given renewed attention to the persistence of religion in a post-secular age. The turn to religion has become prominent in theoretical discourses, philosophy, sociology, cultural studies, and literary studies. For an overview of these discussions, see Hent de Vries and Lawrence E. Sullivan, eds., *Political Theologies: Public Religions in a Post-Secular World* (New York: Fordham University Press, 2006); Martin Treml and Daniel Weidner, eds., *Nachleben der Religionen: Kulturwissenschaftliche Untersuchungen zur Dialektik der Säkularisierung* (Munich: Fink, 2007); and Judith Butler et al., *The Power of Religion in the Public Sphere*, ed. Eduardo Mendieta and Jonathan Van Antwerpen (New York: Columbia University Press, 2011).

30. Hans Blumenberg, *The Legitimacy of the Modern Age*, trans. Robert M. Wallace (Cambridge, MA: MIT Press, 1983), 3–11; Hans Blumenberg, *Die Legitimität der Neuzeit* (Frankfurt am Main: Suhrkamp, 1997), 11–19.

derivative of its theological forerunners, Blumenberg proposes a model of "reoccupation" that allows for a more complex account of the inheritance and persistence of religion in modernity, involving both continuity and discontinuity.[31]

In the wake of Blumenberg's work, new paradigms have explored the afterlives of religion in modernism as phenomena that cannot be reduced to the translation of religion into secular modes of thought.[32] As Martin Treml and Daniel Weidner note, the concept of the "afterlife" (*Nachleben*) of religion provides an alternative to the binary of the disappearance of religion and its return, since an afterlife fluctuates between "survival" (*Überleben*) and "bestowal" (*Hinterlassen*), implying neither "uninterrupted continued existence" (*ununterbrochenes Fortleben*) nor a "clearly defined discontinuation" (*klar bestimmter Abbruch*).[33] To imagine a complex afterlife of religion in modernity, it is necessary to dispense with a theory of secularization conceived as a zero-sum game in which the transference of meaning or concepts from the realm of transcendence to that of immanence is a one-way and irreversible process. In place of the narrative of a disenchanted modernity that has left behind the illusions of religion, the scholarship has proposed a more nuanced relation of religion and modernity under the banner of a "dialectic of secularization"—that is, a mutual interaction between religion and the secular beyond any binary opposition.[34] The idea of a dialectic of secularization allows for both continuities and discontinuities between religious ways of ordering the world and their secular counterparts. To speak of the afterlife of religion reflects the ambivalence with which a secular

---

31. Blumenberg, *Legitimacy of the Modern Age*, 65; Blumenberg, *Legitimität der Neuzeit*, 75.

32. On the concept of the "afterlife of religion," see Martin Treml and Daniel Weidner, "Zur Aktualität der Religionen: Einleitung," in *Nachleben der Religionen: Kulturwissenschaftliche Untersuchungen zur Dialektik der Säkularisierung* (Munich: Fink, 2007), 7–22.

33. Treml and Weidner, "Zur Aktualität der Religionen," 11.

34. See Daniel Weidner, "Einleitung: Walter Benjamin, die Religion und die Gegenwart," in *Profanes Leben: Walter Benjamins Dialektik der Säkularisierung*, ed. Daniel Weidner (Berlin: Suhrkamp, 2010), 7–35, esp. 8–9; and Treml and Weidner, "Zur Aktualität der Religionen," 11.

modernity both displaces religion and marks its preservation and spectral presence.[35]

In the spirit of these theoretical approaches, the interface between eschatological thought and the concept of history is explored here not only in the work of deeply theological thinkers such as Barth and Rosenzweig, but also in the work of writers like Kracauer and Musil, whose writing is saturated by religious thought in much less obvious ways. An account of history and eschatology in the modernist imagination thus adds to a substantial body of scholarship in a range of disciplines that explores the role of religion and theology in key figures in European intellectual history, philosophy, and literature. As part of a "turn to religion," this research has explored the religious and theological subtext of the work of Emmanuel Levinas, Jean-Luc Marion, Jacques Derrida, Walter Benjamin, Franz Kafka, Paul Celan, Theodor W. Adorno, Hannah Arendt, and Erich Auerbach, to name just a few of the most prominent examples.[36]

Similarly, the fruitful and reciprocal interaction of theology and modernism in eschatological discourse suggests that the boundary between religion and the secular, or between theology and the profane, was highly porous. Kracauer's notion of an encounter with "theology in the profane" and Musil's concept of the other condi-

---

35. On the ambiguities and complexities of the concept of secularization, see Daniel Weidner, "Thinking beyond Secularization: Walter Benjamin, the 'Religious Turn,' and the Poetics of Theory," *New German Critique* 37, no. 3 (2010): 131–148, esp. 134.

36. On Levinas, Marion, and Derrida, see Hent de Vries, *Philosophy and the Turn to Religion* (Baltimore: Johns Hopkins University Press, 1999); on Benjamin, see Colby Dickinson and Stéphane Symons, eds., *Walter Benjamin and Theology* (New York: Fordham University Press, 2016); on Kafka, see Manfred Engel and Ritchie Robertson, eds., *Kafka, Religion, and Modernity* (Würzburg: Königshausen & Neumann, 2014); on Auerbach, see Michael McGillen, "Erich Auerbach and the Seriality of the Figure," *New German Critique* 45, no. 1 (2018): 111–154; and Jane O. Newman, "Auerbach's Dante: Poetical Theology as a Point of Departure for a Philology of World Literature," in *Approaches to World Literature*, ed. Joachim Küpper (Berlin: Akademie, 2013), 39–58. Readings of the status of religion in a range of figures can be found in Anna Glazova and Paul North, eds., *Messianic Thought outside Theology* (New York: Fordham University Press, 2014); and Leonard V. Kaplan and Rudy Koshar, eds., *The Weimar Moment: Liberalism, Political Theology, and Law* (Lanham, MD: Lexington Books, 2012).

tion as a form of "profane religiosity" give a sense of the fluidity of these boundaries, while Barth's and Rosenzweig's uses of principles of spatial construction drawn from modernist aesthetics and mathematics show how even explicitly religious and theological discourses participated in modernist culture.[37] Opening up new avenues for thinking about the relationship between modernism and religion, *Shapes of Time* demonstrates that a key afterlife of religious thought in modernity consists in its resistance to the narrative closure of teleological models of history and its use of spatial forms to represent the aporias and antinomies of time.

The excavation of a modernist concept of eschatology that challenges the historicist model of history revises the prevailing scholarly view that the legacy of eschatology in the modern period consists in its secularization and historicization. Löwith's work on the secularization of eschatology in the philosophy of history, for example, explored how from the late eighteenth century onward a previously transcendent eschatological moment of judgment and fulfillment was mapped onto history as an inner-historical *telos*.[38] This explains how Schiller and Hegel could grasp the *Weltgericht*, a formerly theological concept, as identical with the movement and development of *Weltgeschichte* itself. This process of historicization and secularization resulted in what Eric Voegelin, in *The New Science of Politics* (1952), called an "immanentization of the eschaton"—that is, a Hegelian or gnostic form of eschatology that entails "a this-worldly, immanent re-creation of the world."[39] Similarly, in *Meaning in History* (*Weltgeschichte und Heilsgeschehen*) (1949), Löwith argues that in writers such as Marx, Hegel, and Schelling, "historical

---

37. On Kracauer's idea of "theology in the profane," see his letter to Ernst Bloch dated May 27, 1926, in Ernst Bloch, *Briefe, 1903–1975*, vol. 1, ed. Karola Bloch et al. (Frankfurt am Main: Suhrkamp, 1985), 274. On Musil's concept of "profane religiosity," see Musil, "Der deutsche Mensch als Symptom," *P* 1398.

38. Löwith, *Meaning in History*; Löwith, *Weltgeschichte und Heilsgeschehen*.

39. See Brian G. Mattson, "Bavinck's 'Revelation and the Future': A Centennial Retrospective," *Kuyper Center Review* 2 (2011): 126–56, at 144–45. On Voegelin's phrase "to immanentize the eschaton" as a description of the secularization of gnostic thought, see Stanley Corngold, "Kafka (with Nietzsche) as Neo-Gnostic Thinkers," in *Franz Kafka: The Ghosts in the Machine*, ed. Stanley Corngold and Benno Wagner (Evanston, IL: Northwestern University Press, 2011), 151–78, at 153.

consciousness" has as its foundation the eschatological motif of an end of history:

> The future is the "true" focus of history, provided that the truth abides in the religious foundation of the Christian Occident, whose historical consciousness is, indeed, determined by an eschatological motivation [*eschatologische[s] Motiv*], from Isaiah to Marx, from Augustine to Hegel, and from Joachim to Schelling. The significance of this vision of an ultimate end [*ein letztes Ende*], as both *finis* and *telos*, is that it provides a scheme of progressive order and meaning [*ein Schema fortschreitender Ordnung und Sinnhaftigkeit*], a scheme which has been capable of overcoming the ancient fear of fate and fortune. Not only does the *eschaton* delimit the process of history by an end, it also articulates and fulfils it by a definite goal [*ein bestimmtes Ziel*]. . . . Comparable to the compass which gives us orientation in space [*der uns im Raum Orientierung gibt*], and thus enables us to conquer it, the eschatological compass gives orientation in time [*Orientierung in der Zeit*] by pointing to the Kingdom of God as the ultimate end and purpose [*das letzte Ziel und Ende*].[40]

In his account of the end of history as *finis* and *telos*, Löwith applies a historicist framework to understand the significance of eschatology for historical consciousness. By identifying *Ende* and *Ziel*, he conceives of eschatology in terms of a teleological model of history. Accordingly, the thesis of a secularization of eschatology in Marx, Hegel, and Schelling is itself the product of a specific set of assumptions about the constitution of history. By contrast, modernism calls into question the narrative of secularization by reimagining the interface of eschatology and history in nonteleological terms.

Other influential accounts of eschatology, such as Jacob Taubes's *Abendländische Eschatologie (Occidental Eschatology)* (1947), likewise fail to recognize the new concept of eschatology that arose in twentieth-century modernism.[41] Taubes lays important groundwork for recognizing how eschatology exceeds and disrupts the narrative of chronological history, especially in his readings of Marx and Kierkegaard, but he neglects religious thinkers like Barth and Rosen-

---

40. Löwith, *Meaning in History*, 18; *Weltgeschichte und Heilsgeschehen*, 25–26.
41. Jacob Taubes, *Abendländische Eschatologie* (Bern: A. Francke, 1947).

zweig whose work defined the years of his youth.[42] More recently, in *The Time That Remains* (2005), Giorgio Agamben argues that Walter Benjamin's messianic concept of "now-time" (*Jetztzeit*) has its origin in Paul's phrase *ho nun kairos*, which he translates as "the time of the now."[43] Like Taubes, Agamben disregards the importance of Barth's work in twentieth-century intellectual history,[44] overlooking the fact that in the second edition of *Der Römerbrief* (*The Epistle to the Romans*) (1922), Barth had translated Paul's phrase, in contrast to Luther, as "the time of the now" (*die Zeit des Jetzt*).[45] These curious omissions are remedied in the story told here of eschatology as it was reinvented in new spatial terms in modernism.

On a deeper level, Agamben is predisposed not to recognize the significance of eschatological discourse in the early twentieth century

---

42. Even in his posthumously published *Die politische Theologie des Paulus* (*The Political Theology of Paul*) (1987), Barth is relegated to a minor role and dismissed as a "zealot of the absolute." See Jacob Taubes, *Die politische Theologie des Paulus: Vorträge, gehalten an der Forschungsstätte der evangelischen Studiengemeinschaft in Heidelberg, 23.–27. Februar 1987*, ed. Aleida Assmann and Jan Assmann (Munich: Fink, 1993), 86–97. Dieter Schellong points out that Barth is mentioned only twice in Taubes's *Abendländische Eschatologie* and that *Der Römerbrief* is strangely absent from the bibliography, despite the clear relevance of Barth's work for the problem of eschatology ("Jacob Taubes zu Karl Barth," in *Abendländische Eschatologie: Ad Jacob Taubes*, ed. Richard Faber, Eveline Goodman-Thau, and Thomas H. Macho [Würzburg: Königshausen & Neumann, 2001], 385–405, at 385ff).

43. Giorgio Agamben, *The Time That Remains: A Commentary on the Letter to the Romans* (Stanford, CA: Stanford University Press, 2005).

44. Similarly, Weidner notes that Agamben conceals the role of key twentieth-century discourses, such of those of Albert Schweitzer and Rudolf Bultmann, in his reading of Paul's Epistle to the Romans: "der Text . . . ist tatsächlich wohl in nicht geringerem Maße von anderen Diskursen abhängig, die er verschweigt: von der Rückführung des Christentums auf die Parusieverzögerung bei Albert Schweitzer . . . und von der präsentistischen Eschatologie Rudolf Bultmanns, die Agambens radikaler Aktualisierung des Messianismus auffällig ähnelt" (Weidner, "Benjamin, die Religion und die Gegenwart," 33). According to Michael W. Jennings, it is highly likely that Benjamin was familiar with Barth's *Epistle to the Romans* and the theological debates of the 1920s ("Towards Eschatology: The Development of Walter Benjamin's Theological Politics in the Early 1920s," in *Walter Benjamins anthropologisches Denken*, ed. Carolin Duttlinger, Ben Morgan, and Anthony Phelan [Freiburg: Rombach, 2012], 41–57, at 55n31).

45. Barth's translation and interpretation of this phrase is discussed in the context of Romans 8:18ff in chapter 2.

because he places it in strict opposition to the messianic tradition. "It is of utmost importance," he writes, "that we rectify the frequent misunderstanding that occurs when messianic time is flatly identified with eschatological time," a declaration that sets up a familiar opposition between Jewish and Christian understandings of redemption.[46] Yet Agamben's distinction between the *eschaton* as the "end of time" and the messianic as "the time of the end" had already been undercut by Barth and his contemporaries. Agamben's devaluation of eschatology vis-à-vis messianism depends on a teleological understanding of eschatology that was no longer tenable in the early twentieth century. The confluence of messianic and eschatological (or Jewish and Christian) concepts of history in modernism already pointed to a more nuanced understanding of the end.

Finally, Walter Benjamin's "Über den Begriff der Geschichte" ("On the Concept of History") (1940) is without a doubt a key text for understanding the role of religion in the critique and reformulation of the concept of history in the twentieth century. Through the lineage of Scholem, Adorno, Taubes, and Agamben, Benjamin's work has come to define the nexus of religion and the concept of history. The notion of history as a construction whose place is defined not by chronological time but by its relation to a "now time" (*Jetztzeit*), for example, resonates with reflections on the relation of history and eschatology developed in the first decades of the twentieth century.[47] Yet an examination of its prehistory shows that Benjamin's text arose neither in isolation nor exclusively in the context of German-Jewish messianic thought. Rather, the contestation of history was a general cultural problem in modernism, one that was broached across the boundaries between Judaism and Christianity and between religion and the secular. Benjamin's "On the Concept of History" is thus part of a larger field of reflection on the status of history in modernism.

---

46. Agamben, *The Time That Remains*, 62.
47. As Benjamin writes, "Die Geschichte ist Gegenstand einer Konstruktion, deren Ort nicht die homogene und leere Zeit sondern die von Jetztzeit erfüllte bildet" ("Über den Begriff der Geschichte," in *Gesammelte Schriften*, vol. 1, ed. Rolf Tiedemann and Hermann Schweppenhäuser [Frankfurt am Main: Suhrkamp, 1980], 691–704, at 701).

Showing how mathematical and geometrical models of space offered new ways of thinking about history and eschatology, *Shapes of Time* offers a unique account of the imbrication of religion and modernism in early twentieth-century thought.[48] Far from providing substitutes for religion in a process of secularization, modernist mathematics and its spatial constructions gave new life to religious thought. The emergence of new shapes of time in this context was the result of a conjunction of religious concepts of the end of history and modernist forms of spatiality that reimagined endings in nonteleological terms as thresholds, limits, and borderline concepts.

Historicism and the Teleology of History

The contours of the modernist concept of history explored here can be thrown into sharper relief by a comparison with the models of history used in nineteenth-century philosophies of history and historicism. The nineteenth-century concept of history is often understood in straightforward terms as embodying notions of progress and development. But it is not a monolithic concept—it is a multifaceted set of ideas about the teleological orientation of history and how history can be reconstructed. The modernist concept of history responds to the tensions and nuances of these diverse historicist premises.

Schiller and Hegel provide a key starting point for this discussion because they each articulated teleological concepts of history in which the eschatological end was brought within the immanent frame of history. In his poem "Resignation" (1786), Schiller famously wrote,

---

48. Other important contributions to this field of research include David N. Myers, *Resisting History: Historicism and Its Discontents in German-Jewish Thought* (Princeton, NJ: Princeton University Press, 2003); Zachary Braiterman, *The Shape of Revelation: Aesthetics and Modern Jewish Thought* (Stanford, CA: Stanford University Press, 2007); and Peter Fenves, *The Messianic Reduction: Walter Benjamin and the Shape of Time* (Stanford, CA: Stanford University Press, 2011). Myers analyzes the challenge to historicism in key twentieth-century German-Jewish thinkers; Braitermann explores the influence of the aesthetic theories of Wassily Kandinsky, Paul Klee, and Franz Marc on the German-Jewish philosophy of Martin Buber and Franz Rosenzweig; and Fenves uncovers the importance of Gershom Scholem's early studies of mathematics for Walter Benjamin's thinking about time.

"World history is the world's court of judgment" (*Die Weltgeschichte ist das Weltgericht*),[49] and Hegel cited this dictum at the conclusion of his *Grundlinien der Philosophie des Rechts* (*Elements of the Philosophy of Right*) (1820).[50] Whereas in Jewish and Christian apocalyptic literature the "judgment of the world" is conceived as a moment of divine judgment at the end of time, for Schiller and Hegel world history is not the object of this judgment but rather its execution.[51] Schiller's claim that "world history is the world's court of judgment" amounts to a secularization of eschatology that conceives of the highest moral authority in inner-worldly terms.[52]

While Schiller's secularization of eschatology collapses the distinction between immanence and transcendence, humanity and God, and time and eternity, his concept of history maintains a teleological orientation. In his essay "Was heißt und zu welchem Ende studiert man Universalgeschichte?" ("What Is, and to What End Do We Study, Universal History?") (1789), Schiller puts forth an idealist philosophy of history that follows in the footsteps of Kant's "Idee zu einer allgemeinen Geschichte in weltbürgerlicher Absicht" ("Idea for a Universal History with a Cosmopolitan Aim") (1784).[53] For Kant,

---

49. Friedrich Schiller, *Sämtliche Werke*, vol. 1, ed. Albert Meier (Munich: Hanser, 1962), 129–33, at 132.

50. Georg Wilhelm Friedrich Hegel, *Grundlinien der Philosophie des Rechts*, in *Gesammelte Werke*, vol. 14, no. 1, ed. Klaus Grotsch and Elisabeth Weisser-Lohmann (Hamburg: Meiner, 2009), 373. On Hegel's citation of Schiller, see Henning Ottmann, "Die Weltgeschichte (§§ 341–360)," in *G.W.F. Hegel, Grundlinien der Philosophie des Rechts*, ed. Ludwig Siep (Berlin: Akademie, 2014), 267–86, at 267.

51. See Eberhard Jüngel, *Ganz werden: Theologische Erörterungen V* (Tübingen: Mohr Siebeck, 2003), 327.

52. Compare Christian Hart-Nibbrig, "'Die Weltgeschichte ist das Weltgericht': Zur Aktualität von Schillers ästhetischer Geschichtsdeutung," *Jahrbuch der deutschen Schillergesellschaft* 20 (1976): 255–77, at 258–59.

53. On the influence of Kant's philosophy of history on Schiller, see Wolfgang Riedel, "'Weltgeschichte ein erhabenes Object': Zur Modernität von Schillers Geschichtsdenken," in *Prägnanter Moment: Studien zur deutschen Literatur der Aufklärung und Klassik*, ed. Peter-André Alt, Alexander Kosenina, and Hartmut Reinhardt (Würzburg: Königshausen & Neumann, 2002), 193–214, at 200; and John A. McCarthy, "Disciplining History: Schiller als Historiograph," *Goethe Yearbook* 12 (2004): 209–25, at 218. Kant famously claimed to be able to discern a "regular course" in the "play of freedom of the human will"—namely, a tendency toward the creation of cosmopolitan societies. See Immanuel Kant, "Idee zu einer allgemeinen

the teleological goal toward which history drives is a cosmopolitan state of freedom in which "humanity is able to develop completely its natural abilities."[54] Similarly, Schiller argues for a rational purpose as the teleological principle of world history: "The philosophical spirit draws this harmony out of itself and transplants it into the order of things, providing the course of the world with a rational purpose [*einen vernünftigen Zweck*] and *world history* with a teleological principle [*ein teleologisches Prinzip*]."[55]

Schiller can be considered, in the words of Daniel Fulda, "*the most important transformer* (*Transformator*) between the Enlightenment concept of history and historicism."[56] Combining attention to historical research—in works such as *Die Geschichte des Abfalls der vereinigten Niederlande* (*History of the Revolt of the Netherlands*) (1788) and *Geschichte des Dreißigjährigen Kriegs* (*The History of the Thirty Years' War*) (1792)—with the construction of world history as a universal and teleological system, Schiller notes that whereas the "course of the world" can be compared to an "uninterrupted flowing stream" (*ununterbrochen fortfließenden Strom*), the "course of world history" is full of gaps or "empty stretches" (*leere Strecken*).[57] These gaps disrupt the continuity of the stream of historical events, such that history appears as a series of isolated "waves."[58] Yet where historical research runs up against its limits, Schiller's teleological principle of a rational purpose fills in the gaps.

Schiller's thesis that "world history is the world's court of judgment" finds an heir in Hegel's philosophy of history. Like Schiller, Hegel participates in the secularization of eschatology, and he amplifies and extends Schiller's teleological model of history. In *Elements of the Philosophy of Right*, Hegel cites Schiller in arguing that the

---

Geschichte in weltbürgerlicher Absicht," in *Was ist Aufklärung? Ausgewählte kleine Schriften*, ed. Horst D. Brandt (Hamburg: Meiner, 1999), 3–19, at 3.

54. Kant, "Idee zu einer allgemeinen Geschichte," 14.

55. Friedrich Schiller, "Was heißt und zu welchem Ende studiert man Universalgeschichte?," *Schillers Werke: Nationalausgabe*, vol. 17, no. 1, ed. Karl-Heinz Hahn (Weimar: Böhlau, 1970), 359–76, at 374.

56. Cited in McCarthy, "Schiller als Historiograph," 209. McCarthy himself notes that "Schillers Leistung ist ein Markstein im beginnenden Historismus" (219).

57. Schiller, "Universalgeschichte," 372.

58. Schiller, "Universalgeschichte," 372.

"spirit of the world" exercises its right "in *world history* as *the world's court of judgment*."[59] In Hegel's view, the *Weltgericht* marks not the end of history but rather the "realization of a universal spirit" (*Verwirklichung des allgemeinen Geistes*).[60] Bringing the eschatological moment of judgment into the historical process, Hegel identifies the world's court of judgment with the progress of history from one epoch to the next.[61]

Accordingly, Hegel conceives of the teleological goal of history— its "ultimate purpose" (*Endzweck*)—as in equal measure rational and divine. In the *Enzyklopädie der philosophischen Wissenschaften im Grundrisse* (*Encyclopedia of the Philosophical Sciences in Basic Outline*) (1830), he describes the movement of history toward its ultimate purpose both as a "plan of providence" (*Plan der Vorsehung*) and as the manifestation of "reason in history" (*Vernunft in der Geschichte*).[62] Hegel's teleological concept of history is thus closely connected to his understanding of God as immanent to history. As Ernst Cassirer argues, "In Hegel's philosophy the Spinozistic formula *Deus sive natura* was converted into the formula *Deus sive historia*. But this apotheosis does not apply to particular historical events; it applies to the historical process taken as a whole. . . . [I]n the Hegelian system history is no mere appearance of God, but his reality: God not only 'has' history, he *is* history."[63] In the *Vor*-

---

59. Hegel, *Grundlinien*, 273.
60. Hegel, *Grundlinien*, 274.
61. See Glenn Magee, "Hegel's Philosophy of History and Kabbalist Eschatology," in *Hegel and History*, ed. Will Dudley (Albany: SUNY Press, 2009), 231–46, at 231. See also Jüngel, who claims that Hegel adapts an Old Testament idea of a *Völkergericht*, such that "das Weltgericht für ihn das sich innerhalb der fortschreitenden Weltgeschichte vollziehende Urteil über das [ist], was in dieser *Zukunft* haben soll" (*Ganz werden*, 338).
62. Georg Wilhelm Friedrich Hegel, *Enzyklopädie der philosophischen Wissenschaften im Grundrisse (1830)*, ed. Friedhelm Nicolin and Otto Pöggeler (Hamburg: Meiner, 1991), 426. Similarly, in his *Lectures on the Philosophy of History*, Hegel buttresses his claim that "reason rules the world" by arguing that "providence rules the world" (*eine Vorsehung die Welt regiere*) (Hegel, *Vorlesungen über die Philosophie der Geschichte*, in *Werke*, ed. Eva Moldenhauer and Karl Markus Michel, vol. 12 [Frankfurt am Main: Suhrkamp, 1970], 25).
63. Ernst Cassirer, *The Myth of the State* (New Haven, CT: Yale University Press, 1946), 262. On Hegel's formula "deus sive historia," see also Hans-Christian

lesungen über die Philosophie der Geschichte (*Lectures on the Philosophy of History*) (1837), for example, Hegel conceives of the "course of development" (*Entwicklungsgang*) of world history as "the true *theodicy*, the justification of God in history."[64] In other words, Hegel situates God squarely within the historical process as the foundation of its teleology.

Finally, Hegel's philosophy of history provided an impetus for the developmental model of history that prevailed in the nineteenth century. For Hegel, the "principle of development" (*Prinzip der Entwicklung*) provides the key formal determination of the way spirit realizes itself in world history.[65] While he recognizes that development is protracted, uneven, and conflicted,[66] Hegel perceives in the development of spirit as the "purpose and goal" (*Zweck und das Ziel*) of history a "progress as the progression [*Fortgang als ein Fortschreiten*] from the incomplete to the complete."[67] Indeed, invoking the key term "progress" as *Fortschritt*, Hegel emphatically states: "World history is the progress [*Fortschritt*] in the consciousness of freedom—a progress that we must recognize in its necessity."[68] Hegel's confidence in the progressive development of history provided a cornerstone of an emerging historicist approach to history.

In the course of the nineteenth century, the philosophies of history of Schiller and Hegel were succeeded by the approach to history known as historicism. The most prominent proponents of this movement were professional historians such as Droysen and Ranke, but historicism received a key impulse from Herder, whose *Auch eine Philosophie der Geschichte* (*This Too a Philosophy of History*) (1774) has been called "the grand foundational work of historicism."[69] Accounts of historicism often emphasize how it differs from the

---

Lucas, "Die Weltgeschichte als das Weltgericht: Zur Modifikation von Hegels Geschichtsbegriff in Heidelberg," *Hegel-Jahrbuch* (1981/82): 82–96, at 90.

64. Hegel, *Philosophie der Geschichte*, 540.
65. Hegel, *Philosophie der Geschichte*, 75.
66. See Hegel, *Philosophie der Geschichte*, 76: "Development is the tough, unwilling work against oneself."
67. Hegel, *Philosophie der Geschichte*, 78.
68. Hegel, *Philosophie der Geschichte*, 32.
69. See Frederick C. Beiser, *The German Historicist Tradition* (Oxford: Oxford University Press, 2011), 132.

philosophy of history—for example, by focusing on the uniqueness of epochs rather than on the larger, systematic structure of history[70]— but historicism shares with the philosophy of history an understanding of history that is rooted in ideas of progress and development. Taken together, historicism and the philosophy of history represent the two key pillars of the nineteenth-century concept of history.

In contrast to twentieth-century theorists of historiography who stress the construction of history from the standpoint of the present, historicists such as Ranke and Droysen were committed to an objective representation of history to be garnered through direct and unmediated access to the past.[71] According to Ranke's well-known dictum, the historian does not have the task of "judging the past," but merely of showing "how it actually was" (*wie es eigentlich gewesen*).[72] Similarly, Droysen claimed that the historian does not give structure to history but rather provides a mimetic representation of the course of history by reconstructing its genesis: "A *narrative account* represents the object under investigation as a course of events in the mimesis of its becoming; it reconstructs the object as a genetic picture."[73] A corollary to Droysen's genetic reconstruction of history can be found in Ranke's attention to the "causal nexus" (*Kausalnexus*) of events in which effects are deduced from causes.[74]

Ranke's and Droysen's focus on an objective and mimetic representation of the past is accompanied by a confidence in historical progress that has a strong affinity with the philosophy of history.

---

70. For example, Beiser comments: "Hegel wants the philosopher of world history to achieve a *conceptual* grasp of the whole, a *systematic* knowledge of how spirit realizes itself in the different epochs of world history. . . . For Ranke, the ultimate end of history is not to know the whole of world history, which is unattainable in any case, but to have an intuition of the particular, to feel the individual in all its infinite depth" (*The German Historicist Tradition*, 265).

71. As Ranke notes, "Unsere Aufgabe ist, sie [die Geschichte] bis auf den Grund ihrer Existenz zu durchdringen und mit völliger Objektivität darzustellen" ("Idee der Universalhistorie (1831/32)," in *Aus Werk und Nachlass*, vol. 4, ed. Volker Dotterweich and Walther Peter Fuchs [Munich: Oldenbourg, 1975], 72–89, at 81).

72. Leopold von Ranke, "Geschichten der romanischen und germanischen Völker," in *Sämtliche Werke*, vol. 33 (Leipzig: Duncker und Humblot, 1874), vii.

73. Johann Gustav Droysen, *Grundriss der Historik* (Leipzig: Veit & Comp., 1868), 23.

74. See Ranke, "Idee der Universalhistorie," 79.

Despite doubts about the uniformity of historical progress, Ranke was convinced that there is "an unremitting progress" (*ein[] unaufhörliche[r] Fortschritt*) in human ideas and believed that the course of history as a whole is characterized by cultural progress: "the progress of various nations and individuals toward the idea of humanity and culture," he claims, "is categorical."[75] With characteristic humility, Ranke notes that "only God knows world history," but he adds with confidence: "But nevertheless the existence of a unity, a progress [*Fortgang*], and a development [*Entwickelung*] is clear to us."[76] Even more emphatically, Herder describes history as "progress, progressive development" (*Fortgang, fortgehende Entwicklung*) in which epochs and cultures are connected by "a striving one after another in continuity" (*ein Streben auf einander in Kontinuität*).[77] Similarly, under the influence of Hegel's claim that freedom defines the purpose and goal of human history, Droysen argues for an "uninterrupted progress [*ununterbrochene[n] Fortschritt*] in the overall development of humanity."[78]

The historicist concept of progressive development is also connected to a teleology of history that bears a strong affinity to the philosophies of history of Schiller and Hegel. Droysen, for example, describes the "general conception of history as the development of humanity" as "a ceaseless pressing onward whose goal [*Ziel*] we can intuit from its direction."[79] Here the teleological orientation of history is induced by the direction that history takes. On a basic level, Droysen views the concept of a *Zweck* or purpose as being implicit in

---

75. Ranke, *Über die Epochen der neueren Geschichte* (Leipzig: Duncker und Humblot, 1899), 20–21.
76. Ranke, "Idee der Universalhistorie," 83.
77. Johann Gottfried Herder, "Auch eine Philosophie der Geschichte zur Bildung der Menschheit (1777)," in *Werke*, vol. 4, ed. Jürgen Brummack and Martin Bollacher (Frankfurt am Main: Deutscher Klassiker Verlag, 1994), 9–107, at 41–42.
78. Johann Gustav Droysen, "Theologie der Geschichte: Vorwort zur Geschichte des Hellenismus II (1843)," in *Historik: Vorlesungen über Enzyklopädie und Methodologie der Geschichte*, ed. Rudolf Hübner (Darmstadt: Wissenschaftliche Buchgesellschaft, 1960), 369–85, at 374. On the role of freedom in history, see Droysen, *Texte zur Geschichtstheorie*, ed. Günther Birtsch and Jörn Rüsen (Göttingen: Vandenhoeck & Ruprecht, 1972), 17.
79. Droysen, "Theologie der Geschichte," 385.

all movement: "All becoming and growth is movement toward a goal [*Bewegung zu einem Zweck*] that aims to fulfill and come to itself in this movement. In the moral world, purpose follows purpose in an infinite chain of rings."[80] This description of the teleology of the moral world as an "infinite chain of rings" (*unendliche[] Kette von Ringen*) provides a fitting image of continuity and causality as key historicist principles. The progress of history, for Droysen, consists in the movement in which these purposes are successively fulfilled.

In short, despite some notable differences, we can consider the work of Herder, Schiller, Hegel, Ranke, and Droysen under the collective umbrella of historicism. To be sure, there is a tension between the historicist view of the uniqueness of individual epochs and cultures and the commitment to universal history in philosophies of history. Nevertheless, an expansive concept of historicism, like the one proposed by Walter Benjamin, is capable of bridging these apparent oppositions.[81] In this expansive sense, historicism puts forth a concept of history in which the continuity, progress, and development of history are anchored by a teleological trajectory toward an end.

History and Eschatology in the Modernist Imagination

The modernist imagination of history as a complex of shaped times that cannot be mapped onto a schema of teleological development is a key legacy of the work of Barth, Rosenzweig, Kracauer, and Musil. On the surface, the members of this eclectic group of thinkers and writers could hardly be more different: Barth was a Swiss Reformed theologian and a central figure in Dialectical Theology; Rosenzweig was a German-Jewish philosopher who was steeped in the religious traditions of both Judaism and Christianity; Kracauer was a secular German Jew known for his work as a cultural critic and film theorist; Musil was an Austrian writer who was a key fig-

---

80. Droysen, *Grundriss der Historik*, 36.
81. As Matthias Fritsch notes, Benjamin uses the term "historicism" to refer both to the Historical School of Ranke and Droysen and to the philosophies of history of Kant and Hegel (Fritsch, *The Promise of Memory: History and Politics in Marx, Benjamin, and Derrida* [Albany: SUNY Press, 2005], 159).

ure in literary modernism. Yet despite these differences, all of these figures exemplify an effort in modernism to rethink history in new spatial terms, an effort that took place through a fusion of mathematics and religious thought. While numerous modernist writers—from Franz Kafka and Thomas Mann to Georg Simmel and Walter Benjamin—challenged the historicist paradigm and the modern regime of time,[82] the writers explored in this book form a unique constellation because they harnessed the nonintuitive possibilities of mathematics and religious traditions to achieve this end.

Collectively, Barth, Rosenzweig, Kracauer, and Musil push back against a teleology of history that conceives of God as immanent to history and articulate a liminal concept of eschatology that emphasizes the discontinuities and ruptures of historical experience. Drawing on the resources of modernist mathematics and spatial forms, they construct new shapes of time by turning to geometrical metaphors—from Barth's image of the circle defined by its liminal relation to a tangent to Rosenzweig's image of an arc spanning two eternities without a focal point. Similarly, Kracauer conceives of history as a set of time curves defined not by chronology or causality but by temporal exterritoriality, and he detemporalizes the relation of history and eschatology by spatializing the "last things before the last" as an anteroom area. In Musil's work, the antinomies and nonintuitive character of non-Euclidean spaces motivate a search for spatial images of the temporality of the other condition, as in, for example, his image of a bridge spanning an abyss whose passage from beginning to end cannot be imagined in terms of a narrative trajectory. In contrast to the historicist investment in narratives of chronological development guided by teleology—and the accompanying images of flowing time and time's arrow—modernism imagined history as a multidimensional space that is curved and folded back on itself.

The chapters that follow take up these themes in three parts. Chapter 1 demarcates the concept of space at stake in the book's argument, showing how non-Euclidean geometries, constructivist art, and new spatialized theories of historiography provided key contexts for the work of Barth, Rosenzweig, Kracauer, and Musil.

---

82. See Fuchs, *Precarious Times*, 19.

Chapters 2 and 3 explore how Barth's and Rosenzweig's theological texts use mathematical images and spatial forms to rethink history from an eschatological perspective. Finally, chapters 4 and 5 examine how Kracauer's and Musil's innovative concepts of history are informed by a pervasive undercurrent of eschatological thought, which they reconfigure with recourse to geometry, topology, and questions about the foundations of mathematics. Taken as a whole, these three parts show the manifold ways the emergence of new shapes of time in modernism was conditioned by a fertile interaction between religion and mathematics.

Chapter 1, "Constructivism, Mathematics, and the Space of History," delineates the groundbreaking understanding of space put forth by non-Euclidean geometries and shows how it motivated constructivist movements in the visual arts and the theory of historiography. Breaking with natural and empirical space—a single, uniform, three-dimensional space that serves as a container for objects—modernist mathematics recognized that space is constructed, multiple, and not bound to intuition. These insights enabled a re-envisioning of space in modernist art, as in, for example, the dynamic geometries of space in the constructivist art of El Lissitzky and Kazimir Malevich. Similarly, theorists of historiography such as Karl Mannheim, Karl Heussi, Ernst Troeltsch, and Oswald Spengler inaugurated new spatial approaches to history that were steeped in the rhetoric of constructivism. Arguing that historical objects are given shape by the historian in a process of figuration and reconfiguration, they repudiated a teleological concept of historical progress and conceived of history as a construction whose standpoint is the present moment. These contexts opened up new possibilities for thinking about the spatiality of time that Barth, Rosenzweig, Kracauer, and Musil pushed even further in the direction of nonintuitive timescapes.

Chapter 2, "The Eschatological Limit: Spatial Form in Karl Barth's Dialectical Theology," reads Barth's commentary in *The Epistle to the Romans* as a central document of modernism, showing how spatial figures and images played a key role in Barth's challenge to historicism. Treating Barth as a theoretician of the moment, who prior to Benjamin and Agamben grasped the significance of Paul's concept of "the time of the now," the chapter reconstructs the way Dialectical

Theology uses a geometrical language of intersecting lines, infinitesimal limits, and points without extension to represent the interruption and rupture of historical or narrative time. Drawing on the concept of the limit as it was developed in modern calculus, Barth redefines the end of history as a spatial boundary that stands in proximity to each moment in time. Above all, the nonintuitive character of modernist mathematics—and especially the paradox of intersecting parallel lines—serves as an inspiration for Barth's interrogation of the nonintuitive relationship of history and the eschatological now.

Chapter 3, "The Arc of History: Franz Rosenzweig's Figures of Time and Eternity," explores the theological and mathematical impulses behind Rosenzweig's dual-aspect theory of history, in which time is understood both as a chain of past, present, and future moments and as a springboard to eternity. In readings of *Der Stern der Erlösung* (*The Star of Redemption*) (1921) and Rosenzweig's letters, this chapter shows how Rosenzweig uses geometrical figures to represent Judaism and Christianity as embodiments of different spatializations of time. In the figure of the star, Rosenzweig draws on non-Euclidean geometry to show how the lines connecting God, human being, and world can be thought of as curves. Likewise, the projective geometry of Eugenio Beltrami, Felix Klein, and Henri Poincaré provides a model for mapping historical time onto the space of the end of history. Rosenzweig's unique figural thinking grasps eschatology in nonteleological terms as paradoxically equidistant to each moment in time, thereby complicating the historicist view of history as progressive development.

Taking as a point of departure the notion of temporal exterritoriality, in which the elements of history are located in a historical context and are part of time sequences to which they have no temporal proximity, chapter 4, "Temporal Exterritoriality: Siegfried Kracauer and the Shape of History," analyzes Kracauer's theory of shaped times. Finding an impetus for his approach in the mathematical discipline of topology, which explores the modal relations of objects that deform and change shape, Kracauer shows how the elements of urban space, photography, and history disintegrate and are reconfigured in new shapes and forms. For Kracauer, these manifold shapes of history represent the "last things before the last," but the end of history itself remains unrepresentable: the end can never take on an ultimate shape because shape itself is always subject to deformation. In the absence of a temporal end, Kracauer uncovers a space of eschatology in the rifts and fractures of historical time.

This idea of history as open-ended, provisional, and lacking in teleological development is pursued in its literary and aesthetic dimensions in chapter 5, "Images without End: Robert Musil's Narrative Ruptures." Detaching the concept of history from the assumption of a narrative thread, Musil uncovers a new shape of time in the epiphanic experience of the other condition, which emerges as a moment of narrative rupture in which time stands still. Giving shape to time as a space of images, Musil works with non-Euclidean geometries and the foundational crisis of mathematics to represent the narrative rupture of the other condition in images of bridges spanning an abyss, parallel lines that meet in infinity, interminable divisions, and endless knots. For Musil, the transgressive possibilities of eschatology are found not at the end of history but in the moment, not in the temporality of endings but in spaces of reflection.

When pressed to imagine the end of time, modernism responds by reframing the question. For Barth, Rosenzweig, Kracauer, and Musil, it was no longer a matter of representing the final moment in a temporal sequence, the climax of a teleological movement, or the last things. Instead, these writers conceive of the end of history in its proximity to the present moment by producing images of the suspension of time, the standstill of the moment, and the encounter with a time outside of chronology. Attention to the reimagination of history and eschatology in their work produces important insights into the status of history after historicism and the afterlives of religion in modernism. Indeed, the new shapes of time formulated by these figures show that religious thought had a wide resonance in modernist culture, and that the renewed interest in eschatology in the twentieth century—beyond its secularization in the philosophy of history—was saturated with the spatial forms of modernist mathematics.

1

Constructivism, Mathematics, and the Space of History

The exploration of shapes of time in modernist thought was conditioned by three key interrelated contexts: the development of new concepts of space in mathematics, the experimentation with spatial relations in constructivist art, and the emergence of new theories of historiography that parted ways with historicism and began to think about history as having a spatial structure and configuration. While constructivist art adopted insights from non-Euclidean geometry to probe the limits of three-dimensional space, twentieth-century historiography conceived of history as a constellation of mobile figures to be assembled and given form. As the concept of space was reconceived, new possibilities emerged for thinking about time itself as a kind of space.

Prior to modernism, space was organized by a central Renaissance perspective. In the visual arts, the depiction of space through a system of geometrical coordinates reached its apogee in the work of Filippo Brunelleschi and Albrecht Dürer and remained foundational through the nineteenth century.[1] Space was understood to be uniform, three-dimensional, and of unlimited extension. Empirical space was thought to obey the axioms and postulates laid out by Euclid, to be mappable using the Cartesian coordinates, and to provide the basis for the relation of forces in Newtonian physics. Kant subsequently gave space a metaphysical grounding, arguing that space is an a priori form in which we receive sense impressions and intuitions. In the *Kritik der reinen Vernunft* (*Critique of Pure Reason*), Kant conceives of space as the condition of possibility of outer experience and appearances, insisting on the unitary nature of space: "Space is essentially one.... [A]s far as space is concerned, an a priori intuition of it (i.e., one that is not empirical) underlies all concepts of space."[2] It follows from this definition of space that geometrical principles are derived not from concepts, but from a priori intuitions, guaranteeing their apodictic certainty. Kant provided a metaphysical foundation for Euclidean geometry by securing a concept of space grounded a priori in intuition.

In modernism, one can trace a decisive epochal shift in the representation of space, one that leaves behind the idea of a single space conceived as a given reality. As Mehrtens notes, modernist mathematics, music, art, and literature have a common cause in rejecting the representation of a given reality through linear perspective.[3] The uniform geometrical space of Brunelleschi and Dürer was dealt its death blow with the rise of Impressionism, Pointillism, and Cubism in late nineteenth- and early twentieth-century art. In contrast to Kant, modernist mathematics, beginning with Gauss, showed that it is possible to create multiple concepts of space that have no metaphysical foundation but are instead creations or constructions of mathematical language. Gauss dreamed of a "true home, where the chains of our sluggish bodies, the limitations of space [*die Schranken des Raums*], the scourge of earthly sufferings, and the taunts of

---

1. See Samuel Y. Edgerton, *The Mirror, the Window, and the Telescope: How Renaissance Linear Perspective Changed Our Vision of the Universe* (Ithaca, NY: Cornell University Press, 2009).
2. Immanuel Kant, *Critique of Pure Reason*, trans. Werner S. Pluhar (Indianapolis: Hackett, 1996), 78.
3. Mehrtens, *Moderne—Sprache—Mathematik*, 546.

our petty needs no longer weigh heavily on the awakened spirit."[4] The longing that Gauss expresses here is not simply a desire for transcendence; rather, it is a yearning for what Mehrtens calls a "immanent paradise of pure thought" (*diesseitiges Paradies reinen Denkens*)[5] in which mathematical constructions are no longer limited by an intuitive concept of space.

This insight that space is not given but constructed marked a fundamental break with the classic "problem of space," a leitmotif of nineteenth-century mathematics, physics, and philosophy that was concerned, as Mehrtens notes, with the question: "How is natural space to be described mathematically?" (*Wie ist der natürliche Raum mathematisch zu beschreiben?*)[6] In the wake of non-Euclidean geometries, which showed that new, previously unimaginable spaces could be conceived in mathematically exact terms, the concept of "natural space" became meaningless. Empirical space, the space of everyday experience, was shown to be just one of numerous possible mathematical spaces. As a result, mathematics could no longer claim a single definition of space; instead, it worked on the construction of "special" spaces such as Hilbert space or Hausdorff space.[7] Modernist mathematics is distinguished by the creative freedom through which it posits new worlds. In the context of geometry, this means that "the 'spaces' of modernist mathematics are not simply there; they too are created [*auch sie werden geschaffen*] and sometimes receive the name of their creator."[8] Accordingly, it was no longer possible, as it was for Kant, to speak of space as "essentially one." The multitude of spaces unleashed by non-Euclidean geometries was a decisive refutation of the Kantian idea of space as a priori.[9]

The concept of space as multiple and as an object of construction provides a framework though which modernist writers began to conceive of shapes of time. It is a space that is created, one that appears nonintuitive or out of sync with empirical space, a curved

---

4. Cited in Mehrtens, *Moderne—Sprache—Mathematik*, 26.
5. Mehrtens, *Moderne—Sprache—Mathematik*, 27.
6. Mehrtens, *Moderne—Sprache—Mathematik*, 44.
7. Mehrtens, *Moderne—Sprache—Mathematik*, 44–45.
8. Mehrtens, *Moderne—Sprache—Mathematik*, 45.
9. Mehrtens, *Moderne—Sprache—Mathematik*, 45.

space, one that engages us with paradoxes and pushes the limits of what we can visualize. In Barth, Rosenzweig, Kracauer, and Musil, these nonintuitive, curved spaces are transposed into discourses on time, history, and eschatology. This transposition takes place through spatial figures and metaphors that were inspired by the non-Euclidean revolution, by the constructivist turn in modernist aesthetics, and by spatial concepts of time and history in the theories of historiography of writers such as Karl Mannheim, Ernst Troeltsch, Karl Heussi, and Oswald Spengler. Taken together, modernist mathematics, constructivist art, and new directions in historiography are the key contexts for understanding the shapes of time developed by Barth, Rosenzweig, Kracauer, and Musil.

Spatial Constructions in Modernist Aesthetics and Mathematics

Constructivist movements in the visual arts—from Walter Gropius, Wassily Kandinsky, Paul Klee, László Moholy-Nagy, and others in the Bauhaus to Kazimir Malevich, El Lissitzky, and Russian Constructivism and Suprematism—were central to the re-envisioning of space in modernism. Malevich and El Lissitzky experimented with new constructions of space in which objects are neither tied to a horizon nor weighted to a center; likewise, Gropius saw the foundation of architecture in the construction of space (*den Raum gestalten*).[10] Malevich and El Lissitzky express constructivism's tango with nihilistic visions of modernity, where there is no orientation for the construct because the viewer must carry out the construction, whereas Gropius articulates a new spatial vision (*neue räumliche Vision*) in which architecture expresses the unity and totality of the finite experience of space.[11] The new visions and constructions of space in mod-

---

10. In his essay "Idee und Aufbau des staatlichen Bauhauses" (1923), Walter Gropius emphasizes that the artist can only construct (*gestalten*) space with finite means, aiming at a realization of the space of perception (*Raum der Schauung*) in the material world (*Die neue Architektur und das Bauhaus: Grundzüge und Entwicklung einer Konzeption*, ed. Hans M. Wingler [Mainz: Florian Kupferberg, 1965], 35).
11. See Gropius, *Die neue Architektur und das Bauhaus*, 10.

ernist aesthetics were closely connected, both in style and substance, to the non-Euclidean revolution in mathematics.

As T. J. Clark argues in *Farewell to an Idea: Episodes from a History of Modernism* (1999), Malevich's and El Lissitzky's paintings produce new dynamic geometries and experiment with spaces that stand in relations of tension, balance, and stress:

> The key to a Malevich painting is not geometry in general, but a geometry that "expresses the dynamic condition of forms." "We can only be aware of space," says Malevich at one point, "if we break away from the earth, if the fulcrum disappears." Now it seems to me that El Lissitzky was thoroughly in two minds about just this side of Suprematism.... [H]is conception of architecture was a metaphorical one. It was utopian. It meant a reunification of the arts, but it also meant their re-materialization. It meant the expression of specific weights and stresses, a bridging and balancing of parts.... The last thing that happens in an El Lissitzky painting is that the fulcrum disappears. No doubt the perceptual sums we are invited to do in order to discover where the fulcrum *is* are often mindboggling. Spaces are undecidable, solids and voids convert into one another at the drop of a hat. But the whole construction is tensed and stable. Architecture equals forces finally contained. And architecture in this sense is the ruling metaphor of El Lissitzky's art.[12]

For Malevich, breaking away from the earth entails a radically new concept of space, one that is not bound to stable geometrical coordinates but rather expresses a dynamism of forms in motion. The absence of a fulcrum indicates that space is not given or natural: there is no ground that provides a source of leverage for the fulcrum. Rather, space emerges in the act of painting itself. By contrast, El Lissitzky's constructivist approach discovers the elements of a new concept of space in the relation of forces to one another. Within a larger architecture of elements in tension, he discovers spatial relations that are undecidable, nonintuitive, and aporetic.

Despite the manifest differences in their approaches, Malevich and El Lissitzky both construct space using dynamic geometric forms, thereby incorporating key insights into the concept of space

---

12. T. J. Clark, *Farewell to an Idea: Episodes from a History of Modernism* (New Haven, CT: Yale University Press, 1999), 277.

that had been developed in non-Euclidean geometries. El Lissitzky comments on this exchange between art and mathematics in his essay "Kunst und Pangeometrie" ("Art and Pangeometry") (1925), which traces the relationship between various forms of "plastic configuration" (*plastische Gestaltung*) and their correlative mathematical concepts of space.[13] Plastic configuration begins, according to El Lissitzky, with the space of a two-dimensional surface and extends from there to a three-dimensional space through the discovery of perspective.[14] Renaissance perspective applies the laws of Euclidean geometry to produce a static three-dimensional space: "Perspective transformed space, in accordance with the view of Euclidean geometry, into a rigid 3-dimensionality, so that it appears . . . on a surface as a pyramid."[15] By fitting space into a system of coordinates, the law of perspective circumscribes and tames space, establishing its limits: "Perspective made space limited, finite, and closed" (*Die Perspektive hat den Raum begrenzt, endlich gemacht, abgeschlossen*).[16]

The emergence of forms of art that break with Renaissance perspective, in turn, was made possible by fundamental shifts in scientific knowledge, beginning with the Copernican revolution and culminating in the overturning of Euclidean space in the non-Euclidean geometries of Lobachevsky, Gauss, and Riemann:

> The geocentric world order of Ptolemy was replaced by the heliocentric world order of Copernicus. The rigid Euclidean space was destroyed by Lobachevsky, Gauss, and Riemann. The Impressionists were the first to begin to burst [*sprengen*] the inherited perspectival space. More significant was the method of the Cubists. They drew the horizon that concludes space into the foreground and identified it with the painted surface. They developed this fixed surface through psychic features (wallpaper glued to a wall, etc.) and with the elementary destruction of forms [*elementaren Formdestruktion*]. Starting from the pictorial surface [*Bildfläche*], they built forward into space. The latest outcomes are: the reliefs of Picasso and the counter-reliefs of Tatlin.[17]

---

13. El Lissitzky, "Kunst und Pangeometrie (1925)," in *Rußland: Architektur für eine Weltrevolution* (Braunschweig: Vieweg & Sohn, 1989), 122–29.
14. Lissitzky, "Kunst und Pangeometrie," 122–23.
15. Lissitzky, "Kunst und Pangeometrie," 123.
16. Lissitzky, "Kunst und Pangeometrie," 123.
17. Lissitzky, "Kunst und Pangeometrie," 124.

In El Lissitzky's view, the decisive innovation of modern art from Impressionism through Cubism and Constructivism consisted in a reimagining of space that challenges its rigid Euclidean form. The uniform three-dimensionality of Euclidean space was cast aside in favor of experimentation with the surface of the canvas; at the same time, the surface of the picture was taken as a point of departure for the development of new forms and new spaces.

Euclidean geometry, as developed in Euclid's *Elements*, describes figures on a plane and in three-dimensional space. Its central feature, as Jeremy Gray notes, is that the plane (and by extension, space) is "flat and featureless; it has no properties itself that could affect the objects in it."[18] Following fundamental definitions of objects such as points, lines, and planes, the *Elements* formulates a series of postulates: "Any two points can be joined by a line, and any line can be extended. . . . This extension can be of any size. . . . All right angels are equal to each other, and . . . circles can be constructed with any given center and radius."[19] It follows from these postulates that space will be the same at any point in Euclidean geometry. The most significant of Euclid's postulates, and from the perspective of the history of mathematics the most controversial, is the parallel postulate, which states: "That, if a straight line falling on two straight lines make the interior angles on the same side less than two right angles, the two straight lines, if produced indefinitely, meet on that side on which the angles are less than two right angles."[20] Accordingly, if a straight line falls perpendicularly on two straight lines, such that all interior angles are right angles, the two straight lines will be parallel to one another and will never meet. Although not explicitly postulated, Euclid's system also assumes the indeformability of figures in movement; that is, that a figure does not change shape when it moves in space.[21] Non-Euclidean geometry, by contrast, shows that there are possible spaces in which the parallel postulate does not hold, in which lines cut by a straight line with interior

---

18. Jeremy J. Gray, *János Bolyai, Non-Euclidean Geometry, and the Nature of Space* (Cambridge, MA: Burndy Library, 2004), 12.
19. Gray, *János Bolyai*, 13.
20. Cited in Gray, *János Bolyai*, 13.
21. See Henderson, *The Fourth Dimension*, 103.

angels equal to two right angles do in fact meet (or, alternatively, in which lines cut by a straight line with interior angles less than two right angles *never* meet). In other words, the parallel lines of Euclidean space can intersect in non-Euclidean space. Although such lines, strictly speaking, can no longer be defined as parallel, in the popular reception of non-Euclidean geometry it is common to refer to the meeting and intersection of parallel lines.

In his essay "Über den Ursprung und die Bedeutung der geometrischen Axiome" ("On the Origin and Meaning of Geometric Axioms") (1870), Hermann von Helmholtz gave an account of the non-Euclidean geometries of Gauss, Lobachevsky, Riemann, and Beltrami that made their ideas accessible to a nonmathematical audience. Helmholtz's essay served, as Linda D. Henderson notes, as "a primary vehicle for disseminating these first ideas about non-Euclidean geometry."[22] Helmholtz begins with the example of beings who live on a two-dimensional surface: "Let us imagine—it poses no logical impossibility—intelligent beings of just two dimensions that live and move on the surface of one of our solid objects."[23] Assuming that such beings do not possess the ability to perceive things beyond this surface, they would ascribe two dimensions to space in their geometry. Their concepts of points, lines, and surfaces would be exactly the same as in Euclidean geometry, except that the surface would be the most developed form of space that such beings could conceive. To demonstrate the possibility of non-Euclidean geometry, Helmholtz proceeds to vary the surfaces on which such beings live. If these beings lived on an infinite plane, their geometry would correspond to our planimetry, and they would claim that there is only one straight line between two points and formulate the parallel postulate as in Euclidean geometry. However, if such beings lived on the surface of a sphere, they would find that a straight line between two points would form a circle, and if the two points were endpoints of the same diameter as the circle, they would find that there are an

---

22. Henderson, *The Fourth Dimension*, 112.
23. Hermann von Helmholtz, "Über den Ursprung und die Bedeutung der geometrischen Axiome: Vortrag gehalten im Docentenverein zu Heidelberg 1870," in *Vorträge und Reden*, vol. 2 (Braunschweig: Vieweg, 1883), 1–31, here 8.

infinite number of straight lines between them. Moreover, the dwellers on the surface of a sphere would have no concept of parallel lines and would claim that intersecting lines meet at two points. Finally, such beings would find that sum of the angles of a triangle is greater than two right angles and that the angles of larger triangles have a greater sum than the angles of smaller triangles, such that the law of congruence does not hold.[24] On the basis of this thought experiment, Helmholtz concludes:

> It is clear that beings on the sphere, though equipped with the same logical abilities, would have to arrive at an entirely different system of geometrical axioms than beings on the plane and than we ourselves in our space of three dimensions. These examples show us that, depending on the kind of space in which they live, beings whose intellectual faculties correspond entirely to our own would have to arrive at different geometrical axioms.[25]

Helmholtz's thought experiment illustrates that it is possible to formulate geometrical axioms for other possible spaces that are entirely logically consistent but differ in decisive ways from our assumptions about Euclidean geometry. The space of Euclidean geometry would then be just one of any number of possible spaces.

Whereas Helmholtz mobilized the mathematics of non-Euclidean geometry to demonstrate the possibility of multiple forms of space, Ernst Mach showed that one need look no further than perceptual space—the space of vision, of touch, and of sensation in general— to find alternatives to the space of Euclidean geometry. "The space of Euclidean geometry," writes Mach, "is everywhere and in all directions constituted alike; it is unbounded and it is infinite in extent. On the other hand, the space of sight, or 'visual space,' as it has been termed by Johannes Müller and Hering, is found to be neither constituted everywhere and in all directions alike, nor infinite in extent, nor unbounded."[26] In contrast to the uniformity of

---

24. Helmholtz, "Bedeutung der geometrischen Axiome," 9.
25. Helmholtz, "Bedeutung der geometrischen Axiome," 10.
26. Ernst Mach, *Space and Geometry in the Light of Physiological, Psychological and Physical Inquiry*, trans. Thomas J. McCormack (La Salle, IL: Open Court, 1906), 5.

space in Euclidean geometry, in physiological space, or what Mach calls "the sensible space of our immediate perception," entirely different feelings are associated with objects that are above and below, near and far.[27] Indeed, Mach notes that the appearance of objects in the space of vision exhibits many of the same properties that are found in the spaces of non-Euclidean geometry. In particular, the deformability of figures in motion is an experience very familiar to visual perception:

> The apparent augmentation of the stones at the entrance to a tunnel as we rapidly approach it in a railway train, the shrinkage of the same objects on the train's emergence from the tunnel, are exceptionally distinct cases only of the fact of daily experience that objects in visual space cannot be moved about without suffering expansion and contraction,—so that the space of vision resembles in this respect more the space of the metageometricians than it does the space of Euclid.[28]

Similarly, Mach shows that the space of touch, or haptic space, differs significantly from geometric space. In particular, different parts of the skin show varying degrees of spatial sensibility. In Mach's example, a pair of dividers experienced as separate when placed on two adjacent fingers will be perceived as closed when carried down the forearm. Like the space of vision, haptic space is nonhomogeneous and has different properties in different directions.[29]

In the chapter "Space and Geometry" in *Science and Hypothesis* (1902), the French mathematician Henri Poincaré, who together with Helmholtz played an important role in popularizing non-Euclidean geometry, likewise distinguishes between geometrical space and what he calls "representative space."[30] In a manner comparable to Mach's discussion of physiological space, Poincaré defines representative space, in its visual, tactile, and motor forms, in contrast to geometrical space. Representative space, he writes, "is neither homogeneous nor

---

27. Mach, *Space and Geometry*, 5.
28. Mach, *Space and Geometry*, 6.
29. Mach, *Space and Geometry*, 8–10.
30. Henri Poincaré, "Space and Geometry," in *Science and Hypothesis* (London: Walter Scott, 1905), 51–71.

isotropic; we cannot even say that it is of three dimensions."[31] Criticizing the idea that we project the objects of our perception into geometrical space, Poincaré argues that "our representations are only the reproduction of our sensations; they can therefore only be arranged in the same framework—that is to say, in representative space. It is also just as impossible for us to represent to ourselves external objects in geometrical space, as it is impossible for a painter to paint on a flat surface objects with their three dimensions."[32] Representative space, according to Poincaré, is thus an "image of geometrical space" that has been "deformed by a kind of perspective."[33] This insight into the deformation of geometrical space in representative space leads Poincaré to posit the possibility of a "non-Euclidean world": because the space of Euclidean geometry is in no way imposed on our representations, "there is nothing, therefore, to prevent us from imagining a series of representations, similar in every way to our ordinary representations, but succeeding one another according to laws which differ from those to which we are accustomed."[34] The basic distinction between representative space and geometric space thus opens up the possibility of constructing non-Euclidean geometries.

To be sure, the principle of construction is fundamental to all systems of geometry, whether Euclidean or non-Euclidean. As Helmholtz comments: "It is well known that construction tasks [*Constructionsaufgaben*] play an essential role in the system of geometry."[35] The construction of figures in geometry both confirms their existence and demonstrates the conditions under which they are possible. As a result, geometrical operations demand that figures be constructed in a certain way. Helmholtz refers in this context to "points, straight

---

31. Poincaré, "Space and Geometry," 56.
32. Poincaré, "Space and Geometry," 56–57, translation corrected. As Gerhard Heinzmann notes, there is significant translation error in the standard English translation of Poincaré's text, which reads "they cannot therefore be arranged" ("Hypotheses and Conventions in Poincaré," in *The Significance of the Hypothetical in the Natural Sciences*, ed. Michael Heidelberger and Gregor Schiemann [Berlin: De Gruyter, 2009], 169–92, at 178n12).
33. Poincaré, "Space and Geometry," 57.
34. Poincaré, "Space and Geometry," 64.
35. Helmholtz, "Bedeutung der geometrischen Axiome," 6.

lines, or circles of the sort that must be constructed in the task."[36] But whereas in Euclidean geometry the constitution of space itself is taken as fixed and axiomatic, such that the geometer is concerned with the construction of points, straight lines, circles, and other figures, in non-Euclidean geometries of the sort envisioned by Riemann, space itself becomes an object of construction. In particular, Riemann showed that spaces can have a variable "degree of curvature" (*Krümmungsmaß*). Helmholtz notes that this insight allows for the construction of curved spaces:

> If this degree of curvature of space is now equal to zero everywhere, then such a space corresponds everywhere to the axioms of Euclid. In this case we can call it a *flat space* in contrast to other analytically constructible spaces that one could called curved, since their degree of curvature has a value other than zero. In the meantime the analytical geometry for spaces of the latter kind can be realized just as completely and consistently as the usual geometry of our actually existing flat space. If the degree of curvature is positive, we get a *spherical* space in which the straightest lines return to themselves and in which there are no parallel lines. Like the surface of a sphere, such a space would be unlimited but not infinitely large. By contrast, a negative constant degree of curvature results in a *pseudospherical* space in which the straightest lines run into infinity, and in which a bundle of straightest lines can be drawn through each point in every flattest surface that do not intersect other given straightest lines on that surface.[37]

The curved spaces of Riemannian geometry are, as Helmholtz puts it, spaces of analytical construction. Indeed, the construction of curved spaces can be carried out as completely and consistently as the construction of Euclidean space. That is, spherical and pseudospherical spaces have axioms and postulates that hold without contradiction within these spaces yet differ decisively from those of Euclidean space. In particular, the parallel postulate does not hold in these spaces. As a consequence of Riemann's non-Euclidean geometry, Euclidean space emerges as just one of many possible spaces, a space whose degree of curvature happens to be zero.

---

36. Helmholtz, "Bedeutung der geometrischen Axiome," 7.
37. Helmholtz, "Bedeutung der geometrischen Axiome," 18.

As Henderson notes in *The Fourth Dimension and Non-Euclidean Geometry in Modern Art* (2013), the possibility of curved space in non-Euclidean geometry, whether it emerged as an alternative to the parallel postulate or to the law of congruence, was of great interest to "artists in the early twentieth century, such as the Cubists and Marcel Duchamp."[38] In the wake of Einstein's application of non-Euclidean geometry to his concept of the space-time continuum, which too was defined by its variable curvature, this interest extended to Russia. By the early 1920s, as Henderson writes, "the presence of non-Euclidean geometry in the newly popularized Einsteinian General Theory of Relativity provided the major impetus for El Lissitzky and others to explore curved space in their art and theory."[39] In particular, El Lissitzky's *Proun* paintings exhibited "complex spaces with higher dimensional and non-Euclidean implications," spaces constituted by the interrelationships between dynamic forms.[40]

Indeed, the very principle of constructivism in El Lissitzky's *Proun* paintings is directed toward the construction of space. In his programmatic essay *Proun* (1922), published in the journal *De Stijl*, El Lissitzky expounds on his concept of art by way of the interrelated terms *Aufbau*, *Gestaltung*, and *Konstruktion*. He conceives of the artist as "a constructor [*Aufbauer*] of the new world of objects," while he defines *Proun* as "creative figuration [*Gestaltung*] (mastery of space) by means of the economical construction of the revalued material."[41] In a manner comparable to Cubism, El Lissitzky describes how the *Proun* paintings liberate themselves from the conventions of perspective by destroying "the sole axis of the painting that stands vertical to the horizon."[42] Instead, these paintings demand to be viewed from all sides, with the result that "we circle around and spiral into space" (*Umkreisend schrauben wir uns in den Raum hinein*).[43] The orbital motion of spiraling into space that El Lissitzky describes gives expression to the curvature of space discovered in

---

38. Henderson, *The Fourth Dimension*, 103.
39. Henderson, *The Fourth Dimension*, 374.
40. Henderson, *The Fourth Dimension*, 488.
41. El Lissitzky, "Proun," *De Stijl* 5, no. 6 (1922): 81–85, at 82.
42. Lissitzky, "Proun," 83.
43. Lissitzky, "Proun," 83.

non-Euclidean geometries. The essay concludes on an emphatic note with the claim that the *Proun* "proceeds to the construction of space [*schreitet zum Aufbau des Raumes*], structures it through the elements of all dimensions, and assembles a new multifaceted yet integrated figure [*neue vielseitige, aber einheitliche Gestalt*] of our nature."[44] In El Lissitzky's work, the process of configuration or *Gestaltung* mobilizes figures in relations of dynamic tension in order to give shape to new forms of space.

While non-Euclidean geometries provide an impetus for El Lissitzky's constructivist art, they also push the limits of what can be represented, not only in plastic form but also in terms of what can be intuitively imagined. In his essay "Art and Pangeometry," El Lissitzky reflects on how mathematical spaces and the concept of imaginary numbers broach this problem of representability. Beginning with a basic definition of "the imaginary thing i" as "the square root of negative one," he writes:

> We are entering a realm that is unimaginable, that is incapable of any intuitiveness, that follows from purely logical constructions, that is an elementary crystallization of human thought. What does it have to do with intuitiveness, with the sensuous knowability of art? In the vital desire for the enhancement of the figuration of art, some modern artists, some of my friends, believe that they can construct new, multidimensional, real spaces into which one can walk without an umbrella, where space and time have been made into a unity and are interchangeable with one another.[45]

Mathematical spaces and imaginary numbers, for El Lissitzky, are "purely logical constructions" and "elementary crystallizations of human thought" that are "not representable" (*nicht vorstellbar*) nor capable of "intuitive presentation" (*Anschaulichkeit*) or "sensuous knowability" (*sinnliche Erfaßbarkeit*). Beyond the epistemological questions raised by these mathematical paradoxes, the literary dimensions of which were explored by Robert Musil in *The Confusions of Young Törless* (see chapter 5), El Lissitzky raises the question

---

44. Lissitzky, "Proun," 85.
45. Lissitzky, "Kunst und Pangeometrie," 125–26.

of how multidimensional spaces and imaginary numbers can contribute to the figuration (*Gestaltung*) of space in the medium of art. His response to this question is nuanced: he refers to non-Euclidean geometries as "groundbreaking constructions of a new mathematical world" (*bahnbrechende Konstruktionen der neuen mathematischen Welt*), but he cannot withhold a certain skepticism about whether contemporary artists have any more than a surface-level understanding of these theories.[46]

El Lissitzky's deeper concern is whether the revolutionary aspects of non-Euclidean geometry, inasmuch as they exceed our powers of representation, can be given form at all in art. The nonrepresentability of non-Euclidean geometry does not diminish the mathematical import of the work of Lobachevsky, Gauss, and Riemann, but it does raise the question of to what extent their non-Euclidean spaces can be visualized. As El Lissitzky writes: "Lobachevsky and Gauss were the first to prove that Euclidean space is only one instance of an infinite series of spaces. Our senses are not capable of imagining this, but one of the characteristics of mathematics is that it is independent of our power of imagination. The result is that mathematically existing multidimensional spaces are not imaginable, not representable, not at all materializable."[47] The fact that we are unable to represent non-Euclidean spaces with our senses does not cast doubt on the legitimacy of such spaces, because mathematics is "independent of our power of imagination" (*von unserer Vorstellungsfähigkeit unabhängig*), but the nonintuitive character of non-Euclidean space does pose challenges to its use for constructivist art. With his claim that "mathematically existing multidimensional spaces" are "not imaginable, not representable, not at all materializable" (*nicht vorstellbar, nicht darstellbar, überhaupt nicht materialisierbar*), El Lissitzky appears to place such spaces beyond the purview of plastic configuration in modern art. At the very least, "Art and Pangeometry" suggests that constructivist art approaches the construction of space in such a way as to point to the nonintuitive and nonrepresentable character of multidimensional space.

---

46. Lissitzky, "Kunst und Pangeometrie," 126.
47. Lissitzky, "Kunst und Pangeometrie," 126.

In "On the Origin and Meaning of Geometric Axioms," Helmholtz gives a more nuanced definition of what it means to be able to represent something:

> Under the much-misused terms "to imagine" [*sich vorstellen*] or "to be able to conceive [*sich denken können*] how something takes place," I understand—and I do not see how one can understand something else without abandoning all sense of the expression—that one is able to envision the series of sensuous impressions that one would have if such a thing were to occur in a single case. If no sensuous impression is known that could be related to such a never observed event, as a movement in the fourth dimension would be for us, or a movement in our familiar third dimension of space would be for the being on the surface, then such an "imagination" [*Vorstellen*] is not possible, just as a person completely blind from youth will never be able to "imagine" colors, even if one could provide a conceptual description of them.[48]

To be able to represent something, for Helmholtz, entails being able "to envision the series of sensuous impressions that one would have" in a given case. The example of a person who has been blind from birth being unable to represent colors indicates that, for Helmholtz, the limits of representability are defined not by the constitution of the external world but by our "organs of perception" (*Sinneswerkzeuge*). However, whereas El Lissitzky appears to conflate non-Euclidean geometries and multidimensional spaces, Helmholtz makes a distinction between the two. The geometry of the sphere and the pseudosphere, he claims, make the spaces of non-Euclidean geometry "clear and intuitive" (*klar und anschaulich*) to us, while the movement of a figure from the third dimension into a fourth dimension exceeds the limits of our organs of perception.[49] Nevertheless, the crux of the intuitiveness and representability of non-Euclidean geometry, for

---

48. Helmholtz, "Bedeutung der geometrischen Axiome," 8.
49. Helmholtz, "Bedeutung der geometrischen Axiome," 15. Compare Henderson, who argues that "Helmholtz, on the contrary, held that the human mind could intuit, or represent to itself, non-Euclidean space. He defended his view by demonstrating such a process of intuition, using Beltrami's model for three-dimensional pseudospherical space.... [H]e took further precautions and distinguished his imaginable non-Euclidean space from a fourth dimension he believed was impossible to represent" (*The Fourth Dimension*, 115).

Helmholtz, rests on the analogy of two-dimensional beings that live on a surface. In order to represent the "intuition of space" (*Raumanschauung*) of such beings, we merely have to limit our own intuitions to a more restricted area. In other words, the analogy functions because it is subtractive rather than additive. As Helmholtz comments: "It is easy to imagine things without the intuitions that one possesses; but it is very difficult to imagine sensuously intuitions for which one has never had an analogue."[50] Helmholtz does not, however, claim that representing intuitions for which one has no analogue is impossible. Moreover, he leaves open the possibility that intuitions that appear to be unimaginable because of the constitution of our sense organs may in fact be representable if the proper analogy can be found.

In his 1884 novella *Flatland: A Romance of Many Dimensions*, which appeared fourteen years after Helmholtz's lecture, Edwin A. Abbott provides a fictional account of the problem of imagining dimensions that exceed the capabilities of our organs of perception. The novella is narrated by a Square who lives in Flatland, a two-dimensional space inhabited by geometrical figures whose social status is indicated by the number of their sides, ranging from lines to triangles, squares, and various polygons that approach the perfection of a circle.[51] In the second part of the novella, titled "Other Worlds," the Square visits the one-dimensional space of Lineland and tries to convince its monarch of the existence of a second dimension of width or breadth, telling him that "your Space is not the true Space. True Space is a Plane; but your Space is only a Line."[52] These efforts are in vain: the king is unable to understand what is meant by "left" and "right," since his field of vision is limited to a single line. Similarly, when the Square is visited by a Sphere from the three-dimensional space of Spaceland, the Sphere finds it nearly impossible to explain the

---

50. Helmholtz, "Bedeutung der geometrischen Axiome," 15.
51. Edwin A. Abbott, *Flatland: A Romance of Many Dimensions*, ed. Rosemary Jahn (Oxford: Oxford University Press, 2006). Another important precursor of Abbott's novella was Charles H. Hinton's 1884 essay "What Is the Fourth Dimension?," in *Speculations on the Fourth Dimension: Selected Writings by Charles H. Hinton*, ed. Rudolf v. B. Rucker (New York: Dover, 1980), 1–22.
52. Abbott, *Flatland*, 77.

third dimension to the Square. Unable to demonstrate the existence of a third dimension in terms that are intuitive to the Square, the Sphere attempts "the method of Analogy,"[53] using mathematics to show the properties of a cube via the progression from point to line to square.

In Abbott's narrative, the analogy functions in logical terms, as the Square is able to correctly derive the number of corners and sides of a cube, but it does not provide the Square with an intuition of three-dimensional space. Although the Square grasps the properties of three-dimensional space intellectually, he is unable to "see it all now."[54] At the climax of the novella, however, when the Square is plucked out of his plane and comes to Spaceland, he is finally able to behold the new world as "visibly incorporate," perceiving intuitively what before could only be "inferred, conjectured, dreamed."[55] In a brilliantly ironic twist, the Square, having overcome the limitations of two dimensions, uses the same method of analogy to posit a fourth and indeed an infinite number of dimensions, to which the Sphere responds: "The very idea of it is utterly inconceivable."[56] Having initiated the Square into the mysteries of the third dimension, the Sphere is unable to overcome the limitations of his own perceptual apparatus. This confirms Helmholtz's insight that it is much easier to restrict the scope of our perception than to expand our field of vision. Although the Square is unable to persuade the Sphere of the existence of a fourth dimension and becomes an outcast in Flatland for professing the existence of a third dimension, he has found an analogy that allows him to conceive of higher-dimensional spaces. The "other Space" of four and more dimensions is called "Thoughtland"—a concession, perhaps, that these spaces are mathematically consistent and thus conceivable in thought, yet from the perspective of sensuous perception completely nonintuitive.

While El Lissitzky and Helmholtz negotiate the problem of the nonrepresentability of non-Euclidean space, Oswald Spengler considers the nonintuitive character of such spaces to be a defining fea-

---

53. Abbott, *Flatland*, 89.
54. Abbott, *Flatland*, 91.
55. Abbott, *Flatland*, 95.
56. Abbott, *Flatland*, 103.

ture of modern mathematics. Spengler contrasts the optical, visual, and form-giving qualities of ancient mathematics to the abstract, disembodied musicality of modern mathematics:

> The Ancient mathematician only knows what he sees and grasps. His science comes to end where the limited and limiting visibility, the subject of his thought processes, breaks off. The Western mathematician, as soon as he is free of Ancient prejudices and belongs to himself, ventures into the completely abstract region of an infinite manifold of numbers of n—no longer of 3—dimensions, within which *his* so-called geometry is able to do without all intuitive help [*anschaulichen Hilfe*] and for the most part must do so.[57]

Intuitive representability, for Spengler, is no longer a precondition for modern geometry, which has left behind the limiting visibility of ancient geometry; and multidimensional geometry is not only able but compelled to dispense with such intuitive support. Nevertheless, Spengler considers the n-dimensional spaces of modern mathematics to be eminently real, with the caveat that reality is defined as more than just sensuous reality. Spengler's concept of morphology, which he goes on to apply to world history, has at its disposal a set of formations that are "entirely other than intuitive" (*ganz anders als anschaulich*).[58]

Spengler's claims about the nonintuitive character of modern mathematics were corroborated by the mathematician Hans Hahn in a 1933 lecture on "Die Krise der Anschauung" ("The Crisis of Intuition").[59] According to Hahn, modernist mathematics departs from the Kantian notion of space and time as pure forms of intuition. "The space of geometry" (*Der Raum der Geometrie*), he writes, "is not a form of pure intuition but rather a logical construction" (*nicht eine Form reiner Anschauung, sondern eine logische Konstruktion*).[60]

---

57. Oswald Spengler, *Der Untergang des Abendlandes: Umrisse einer Morphologie der Weltgeschichte* (Munich: C. H. Beck'sche Verlagsbuchhandlung, 1923), 112.
58. Spengler, *Der Untergang des Abendlandes*, 122.
59. Hans Hahn, "Die Krise der Anschauung," in *Krise und Neuaufbau in den exakten Wissenschaften: Fünf Wiener Vorträge* (Leipzig and Vienna: Franz Deuticke, 1933), 41–64.
60. Hahn, "Krise der Anschauung," 60.

To the objection that non-Euclidean and multidimensional geometries are of little use in describing our experiences because of their nonintuitive character, Hahn responds that all geometries—whether Euclidean or non-Euclidean—are logical constructions. What appears to be the intuitiveness of three-dimensional Euclidean geometries is in fact the result of a habituation to using a certain kind of geometry to order our experiences.[61] Whereas intuition was once considered to be fundamental to geometry, Hahn claims that modernist mathematics succeeded in shattering any trust in intuition and ultimately banished it completely from geometry.[62]

By decoupling the space of geometry from intuition, the non-Euclidean revolution opened up new possibilities for thinking about time and space. In place of the Kantian notion of time and space as forms of intuition with strictly separate sources, Einstein's theory of relativity (which applied key tenets of non-Euclidean geometry) not only showed the physics of curved spaces but also demonstrated the interdependency of space and time.[63] The insight into curved spaces in non-Euclidean geometry thus made possible concepts of the curvature of time and the deformation of the moment. This interweaving of curved spaces and curved times in mathematics and physics spilled over into modernist art. As El Lissitzky notes, artists in movements from Futurism to Suprematism and Constructivism worked with time as an element of plastic configuration: "Time now comes into question first and foremost as a new element of plastic configuration [*neuer Bestandteil der plastischen Gestaltung*]. In the studio of the modern artist, one squarely believes that one can configure [*gestalten*] a unity out of space and time, which thereby are able to be substituted for one another."[64] While El Lissitzky himself insists on a clear distinction between space and time, arguing that space is three-dimensional and time is one-dimensional and thereby criticiz-

---

61. Hahn, "Krise der Anschauung," 61.
62. Hahn, "Krise der Anschauung," 44. Hahn's account is consistent with the character of modernist mathematics that Mehrtens attributes to mathematicians such as Hilbert, Zermelo, and Hausdorff (*Moderne—Sprache—Mathematik*, 108–89).
63. On Einstein's revision of Kant's epistemology, see Hahn, "Krise der Anschauung," 42.
64. Lissitzky, "Kunst und Pangeometrie," 126.

ing the interchangeability of space and time, he also draws attention to the way modernist art represents time by registering motion and rhythm in space: "The achievements of the Futurists and Suprematists are static surfaces that indicate dynamics.... Tatlin and the Constructivists in Moscow symbolized this movement. The individual objects of the 'Monument to the Third International' revolve around their own axis with a velocity: a year, a month, a day."[65]

The spatial representation of time as dynamic motion in modernist aesthetics reflected a widespread awareness of Einstein's theory of a space-time continuum and the idea of curved space in non-Euclidean geometry. In Russia, the hyperspace philosophy of P. D. Ouspensky, who likewise conceived of a spatial fourth dimension in temporal terms, had already before Einstein informed artistic experimentation with the spatiality of time.[66] In *Tertium Organum: The Third Canon of Thought; a Key to the Enigmas of the World* (1911), Ouspensky developed a philosophically informed theory of time as the fourth dimension of space. "By time," he writes, "we mean *the distance* separating events in the order of their succession and binding them in different wholes. This distance lies in a direction not contained in *three-dimensional* space, therefore it will be *the new dimension of space. This new dimension satisfies all possible requirements of the fourth dimension.* ... It is perpendicular to all directions of three-dimensional space and is not parallel to any of them."[67]

Ouspensky's account of time as the fourth dimension does not simply posit a dimension of time that supplements three-dimensional space; rather, it conceives of time itself in spatial terms. Time is understood as distance that stands in a spatial and geometrical relation to the three dimensions of space: it is perpendicular to each dimension of space yet "not parallel to any of them." The idea of time as motion in space, which was key to its dynamic treatment as an element of

---

65. Lissitzky, "Kunst und Pangeometrie," 127.
66. For an introduction to Ouspensky's hyperspace philosophy and his influence on modernist artists, see Henderson, *The Fourth Dimension*, 377–86, 433.
67. P. D. Ouspensky, *Tertium Organum: The Third Canon of Thought; A Key to the Enigmas of the World*, trans. Nicholas Bessaraboff and Claude Bragdon (New York: Alfred A. Knopf, 1922), 45.

plastic configuration in modernist art, is for Ouspensky merely a proxy for our inability to see the spatiality of time:

> It is necessary to admit that by one term, *time*, we designate really two ideas—"a certain space" and "motion upon that space." This motion does not exist in reality, and it seems to us as existing only because *we do not see* the spatiality of time. That is, the sensation of motion in time (and motion out of time does not exist) arises in us because we are looking at the world as if through a narrow slit, and are seeing the *lines of intersection* of the time plane with our three-dimensional space only.[68]

Again adopting a geometrical language, Ouspensky notes that we perceive the perpendicular relation to time to the three dimensions of space only at their "*lines of intersection*," such that we are not aware of the spatial extension of the plane of time. The "narrow slit" through which we see the world—a notion comparable to Helmholtz's and El Lissitzky's consideration of the intuitive possibilities of our sense organs—gives rise to the sensation of motion in time, while the spatial extension of time eludes visual perception.

Ouspensky develops the spatial dimension of time from a theological perspective. Referring to extension in time as "extension into unknown space," he notes that we require the perspective of eternity in order to comprehend the spatiality of time: "From the standpoint of eternity, *time* does not differ in anything from the other lines and dimensions of space—length, breadth, and height."[69] Accordingly, a receptivity to the spatiality of time would entail a recognition of the simultaneity of past, present, and future: "There would ensue *the expansion of the moment,* i.e., all that we are apprehending *in time* would become something like a single moment, in which the past, the present, and the future would be seen at once."[70] Ouspensky's expansive concept of the *now*—encompassing "*before, now, after*"—provides a lens through which he can posit

---

68. Ouspensky, *Tertium Organum*, 46.
69. Ouspensky, *Tertium Organum*, 47.
70. Ouspensky, *Tertium Organum*, 49.

"time-sense" as "*the limit* or *the surface* of our 'space-sense.'"[71] A comparable treatment of the "time of the now" in geometrical terms guides Barth's reinterpretation of Paul's Epistle to the Romans.

Ouspensky explores the nexus between the visible world and the invisible world, between the space of three dimensions and the unknown space, seeking to define the point of contact between the space of three dimensions and the spatiality of time. The liminal relation of time and space suggested above, in which "time-sense" acts as a limit or surface of "space-sense," is supplemented in later chapters of *Tertium Organum* with a discussion of how these two senses are mediated. In one formulation of this relation, Ouspensky conceives of the sensation of motion as a projection of time-sense into the phenomenal world of space: "It is not difficult to realize," he writes, "that we are receiving as sensations, and projecting into the outside world as phenomena, *the immobile angles and curves of the fourth dimension*."[72] This projection of the immobile angles and curves of time, in Ouspensky's geometrical formulation, onto the phenomenal world, insofar as it expresses time as motion, does not allow the curvature of time as such to be perceived. In a second attempt to formulate the relation between time-sense and space-sense, Ouspensky raises the prospect of a translation of time-sense into space-sense: "The growth of the space-sense," he writes, "is proceeding at the expense of the time-sense. Or one may say that the time-sense is an imperfect space-sense (i.e., an imperfect power of representation which, being perfected, translates itself into the space-sense, i.e., into the power of representation in forms)."[73] In contrast to the projection of time-sense into the phenomenal world as motion, the translation of time-sense into space-sense allows time to represented as spatial form, thereby giving shape to the curvature of time. If time represented as motion can be considered an imperfect space-sense, then its perfection would entail the representation of time as spatial form itself.

---

71. Ouspensky, *Tertium Organum*, 49.
72. Ouspensky, *Tertium Organum*, 115.
73. Ouspensky, *Tertium Organum*, 119.

The translation of time into spatial form depends, Ouspensky notes, on the creation of a new language that is adequate for this transposition:

> Our language is absolutely inadequate to the *spatial expression of temporal relations*. We lack the necessary words for it, we have no verbal forms, strictly speaking, for the expression of these relations which are new to us, and some other quite new forms—*not verbal*—are indispensable. The language for the transmission of the new temporal relations must be a language without verbs. *New parts of speech* are necessary, an infinite number of new words.[74]

An approximation of such a "spatial expression of temporal relations" can be found in the language of geometrical and mathematical forms developed in different ways by Barth, Rosenzweig, Kracauer, and Musil. Their translations of time and history into the language of spatial form gave rise to new forms of historiography after historicism.

Constructivist Historiography and the Space of History

A second key context for the work of Barth, Rosenzweig, Kracauer, and Musil can be found in new approaches to historiography that emerged in the first decades of the twentieth century. Karl Mannheim, Karl Heussi, Ernst Troeltsch, and Oswald Spengler each made important contributions to the theory and practice of historiography. In contrast to the historicism of Droysen and Ranke, which sought to write history objectively, these writers conceived of history as an object of construction. Their "daring new constructions" (*verwegene Neukonstruktionen*)[75] of history insisted on the constructed nature of the historical object and formulated new historical architectures. In contrast to the philosophies of history of Kant and Hegel, which were dependent on the concept of a totality of

---

74. Ouspensky, *Tertium Organum*, 122.
75. Ernst Troeltsch, "Die Krisis des Historismus," *Die Neue Rundschau* 33, no. 6 (1922): 572–90, at 576.

history with an end or goal (*Ziel*) toward which humanity steadily progresses, figures like Mannheim and Troeltsch drew upon the self-reflective moment in the "philosophy of life" (*Lebensphilosophie*) of Wilhelm Dilthey to articulate approaches to history grounded in the material life of the present.[76] These new approaches to historiography sought to write history in the present and for the present.

While they did not push the concept of history as radically in the direction of nonintuitive and non-Euclidean spaces as the main figures considered in this book did, Mannheim, Troeltsch, Heussi, and Spengler inaugurated a spatial approach to history that, although not rigorously geometrical, was steeped in the rhetoric of constructivism. Indeed, the constructivist dimension of their historiographies is manifest in their prominent use of spatial forms and in their invocations of discourses of figures (*Gestalten*) and figuration (*Gestaltung*). Mannheim's theory of the perspectival standpoint of the historian, for example, transposes the condition of visibility of objects in space onto the temporal relationship of the historian to the historical material. The construction of history, for Mannheim, both imposes an organizing principle or structure on history and recognizes that the historical figures themselves are constantly changing shape. Similarly, Heussi argues that the historical object lacks any final and equivocal structure, such that the historian, engaging in *Gestaltung*, uses a vessel to give shape to a fluid and formless historical material. Introducing the concept of a construction (*Aufbau*) of history, Troeltsch views the present moment as the site of a historical architecture and argues for a stratification (*Schichtung*) of historical layers at work in the present. Finally, Spengler applies Goethe's theory of morphology to historical

---

76. On Troeltsch's and Mannheim's relationships to historicism, see Walter Bodenstein, *Neige des Historismus: Ernst Troeltschs Entwicklungsgang* (Gütersloh: Gütersloher Verlagshaus, 1959); Reinhard Laube, *Karl Mannheim und die Krise des Historismus: Historismus als wissenssoziologischer Perspektivismus* (Göttingen: Vandenhoeck & Ruprecht, 2004); and Laube, "Zwischen Budapester und Berliner Historismus: Eine Pathologie der 'Krise des Historismus' aus der Sicht eines ungarischen Emigranten," in *Krise des Historismus, Krise der Wirklichkeit: Wissenschaft, Kunst und Literatur, 1880–1932*, ed. Otto Gerhard Oexle (Göttingen: Vandenhoeck & Ruprecht, 2007), 207–46.

forms and figures, treating history as an ongoing process of figuration and reconfiguration.

For each of these writers, the spatiality and *Gestalt*-qualities of the construction of history have significant implications for the theory of historiography. In Mannheim they lead to a dynamic perspectivism that grasps the reciprocal interaction between given figures and constructed figures. In Heussi they imply that historical objects are elastic and change shape depending on the context in which they are read. Troeltsch's theory of the present moment as the site of a historical architecture makes the destruction of inherited historical models a precondition for new constructions of history. And Spengler's morphology of history provides an alternative to teleological concepts of history that envision a linear and continuous process of historical development. Taken together, these writers challenge the idea that there is a single space of history, just as modernist mathematics had called into question the Kantian premise of space as "essentially one." Rather than being something static and given, history is understood as mobile, multiple, and constructed.

Mannheim's essay "Historismus" ("Historicism") (1924) presents a self-reflexive historiography that reflects critically on the historicity of its own concepts, rather than attempting to describe the past "as it really was." Mannheim's approach represents a productive extension of historicism in which the historical material itself—and not an ideal vantage point—forms the basis for a dynamic philosophy of history. It is a historiography that seeks to give dynamic expression to the language of forms that history assumes at given moments. To derive the "organizing principle" that brings to light the "innermost structure" of historical becoming, Mannheim turns to a series of spatial images with geometrical import.[77] Differentiating between the "direction of the historical longitudinal section" (*Richtung des historischen Längsschnittes*) and the "direction of the historical cross section" (*Richtung des historischen Querschnittes*)—that is, between the diachronic development of historical figures and their synchronic relations—Mannheim treats historical motifs and formations as fig-

---

77. Karl Mannheim, "Historismus," *Archiv für Sozialwissenschaft und Sozialpolitik* 52, no. 1 (1924): 1–60, at 4.

ures (*Gestalten*) that undergo processes of evolution and transformation. In the historical longitudinal section, Mannheim's theory of historiography shows how "each later figure arises continuously and organically out of earlier figures" in a "metamorphosis of figures" (*Gestaltwandel*) along a horizontal axis of historical development. The historical cross section supplements this account of the metamorphosis of figures with an attention to the relations of apparently isolated figures to one another in "a synchronous stage" (*einem gleichzeitigen Stadium*) in what can be considered a vertical axis of synchronicity.[78]

The combination of longitudinal sections and cross sections provides insight into what Mannheim calls "the structure or the figure of this totality" (*die Struktur oder die Gestalt dieser Totalität*), which he develops on the basis of a manifold of historical forces in a process of change.[79] Mannheim uses the image of the *Gestalt* in two senses, referring both to the singular historical figure and to the totality of history whose form and structure are constituted out of a manifold of singular figures. This concept of history borrows the rhetoric of *Gestalten* from the Gestalt psychology of Max Wertheimer, Wolfgang Köhler, and Kurt Koffka, though with important differences. For Gestalt psychology there is an immediate perception of the whole of the figure, which is structured not by the Kantian understanding of space as an a priori form of intuition but by the objective qualities of the figure itself. The wholeness (*Ganzheit*) of the figure for Gestalt psychology is therefore not a summation of its discrete elements but itself gives form to its parts.[80] For Mannheim, by contrast, the *Gestalt* of a historical totality is one that must first be constructed out of a manifold of historical material. It is given only as the goal of a historicist perspective on history. Nevertheless,

---

78. Mannheim, "Historismus," 4.
79. Mannheim, "Historismus," 5.
80. See Mitchell G. Ash, *Gestalt Psychology in German Culture, 1890–1967: Holism and the Quest for Objectivity* (Cambridge: Cambridge University Press, 1995), 1. Gestalt psychologists such as Max Wertheimer, Wolfgang Köhler, and Kurt Koffka drew on Christian von Ehrenfels's notion of "Gestalt qualities" to assert that "dynamic structures in experience *determine* what will be wholes and parts, figure and background, in particular situations" (1).

in conceiving of historical totality as a changing figure or structure, Mannheim's concept of historicism shares the empiricist and anti-Kantian orientation of Gestalt psychology; and as a form of dynamic *Geschichtsphilosophie*, it is able to accommodate the shift or loss of figures that Gestalt psychology drew attention to.[81]

Mannheim's treatment of history in terms of its formal and figural qualities maintains a tension between the givenness of the historical material and the imposition of structure by the historiographer. The reciprocal interaction of given figures and constructed figures is a function of cognition itself, which, Mannheim claims, is "not pure contemplation, ... not a mere acceptance, but rather, like all sensoria, world-creative and world-receptive at once, streaming forward, creating in new forms, receiving in new forms at a single stroke."[82] Historical cognition involves both the receptive recognition of figures and the active production or construction of figures in the medium of new forms. Yet Mannheim is careful to reject the problematic distinction between form and content in which form designates a universal conceptual structure and content has a particular historical determination. The division of form and content, he argues, has as its basis a static "philosophy of reason" (*Vernunftsphilosophie*), according to which formal categories or values remain the same and are merely filled with new content. Mannheim compares this concept of static form to the image of "a vessel [*eines Gefäßes*] into which new liquid, or skins [*Schläuchen*] into which new wine can always be poured, while these vessels and skins can be taken as at all times invariable forms [*stets sich gleichbleibende Formen*]."[83] By contrast, Mannheim's image for a dynamic relation of form and content is that of "living, growing plants" in which "the 'form' and 'figure' of the plant grows and changes together with its ever renewing 'content.'"[84] The construction of historical *Gestalten*,

---

81. As Oliver Simons notes, the prominence of reversible figures and perceptual inversions in Gestalt psychology made visible an epistemological ambivalence typical of modernism: the shift or even loss of a figure becomes an experience that can be visualized (*Raumgeschichten*, 21).
82. Mannheim, "Historismus," 7.
83. Mannheim, "Historismus," 10.
84. Mannheim, "Historismus," 10.

in other words, does not entail the filling of preexistent forms with historical content; rather, historical content is itself the expression of dynamic forms. The distinction between form and content is problematic because history as a metamorphosis of *Gestalten* involves a dialectical relation between changing historical material and newly constructed systematic structures.

In Mannheim's theory of historiography, the historian occupies a unique spatial position in relation to the manifold of historical figures that are given structure while undergoing metamorphosis. Claiming that history can be constructed only from the "standpoint" of the present, Mannheim formulates a model of perspectivism that situates the historian within a visual field populated by historical objects. Yet in contrast to the law of Renaissance perspective, which establishes depth by placing objects within a system of coordinates, Mannheim's perspectivism emphasizes the limitations of being bound to a single viewpoint in space. The historian's "boundedness to a standpoint" (*Standortsgebundenheit*) entails a presentist mode of historiography that provides no other optic on history than that of the present. According to Mannheim, the "historico-philosophical standpoint of the observer" (*geschichtsphilosophische Standort des Betrachters*) extends into every element of historiography, from the categories with which objects are registered to the principles of their selection and the orientation of historical claims.[85] In making this claim, Mannheim departs radically from Ranke's ideal of reconstructing history "as it actually was": the consequence of the "the historico-philosophical (sociological) *boundedness to a standpoint* of all historical knowledge," he argues, is "that the historical image [*Geschichtsbild*] of the past changes with every epoch."[86]

Mannheim argues that the historical image of the past changes for each epoch not because the historical object itself changes but because the conditions under which it can be experienced (its *Erfahrbarkeit*) have changed.[87] In other words, the historical object can be known only from different historical standpoints and thus in different aspects.

---

85. Mannheim, "Historismus," 21–22.
86. Mannheim, "Historismus," 24.
87. Mannheim, "Historismus," 26.

Comparing the perspectivism of the historian to Husserl's theory of adumbrations (*Abschattungen*), Mannheim develops a model of perspectival space to describe the conditions of historical knowledge:

> Analogously to Husserl's observation—that it belongs to the essence of the thing in space that it is visible only in "adumbrations," that is, from certain localized standpoints [*von gewissen örtlichen Standorten aus*], and thus in each case is knowable only from certain sides and in certain perspectives—one could, in our view, put forward the thesis that it belongs to the essence of a historical-intellectual and psychological substance that it is penetrable [*durchdringbar*] only in "intellectual and psychological adumbrations," that is, in each case only in certain cross sections and dimensions of depth, which depend, in their characteristic features, on the intellectual-psychological standpoint [*geistigen-seelischen Standorte*] of the observing and interpreting subject.[88]

Mannheim transposes Husserl's insights into the conditions of the visibility of objects in space onto the temporal relationship of the historiographer to their historical material. In each case, the object is given to cognition only from a "localized standpoint" or "perspective." The visibility of the object, as the metaphorics of *Abschattungen* suggests, is given only in a faint and partial outline, while much of the object remains in shadows. Analogously, the sides of the historical object that present themselves to the historiographer offer specific "cross sections" (*Querschnitte*) and "dimensions of depth" (*Tiefendimensionen*) that depend on their standpoint and perspective. The object in space and the historical object appear in different aspects from different vantage points.

In making the leap from spatial to historical cognition, Mannheim makes a subtle shift from the optical register of visibility (*Sichtbarkeit*), with its connotations of surface appearances and outlines, to a geometrical or perhaps geological register of perviousness or permeability (*Durchdringbarkeit*), in which the object must be cut or penetrated with a cross section in order to expose its depths. Mannheim's transposition suggests that the visibility of the layers or strata (*Schichten*) of history, to borrow Reinhart Koselleck's im-

---

88. Mannheim, "Historismus," 26.

age, depends very much on the position (*Standort*) from which these layers are cut through.[89] Mannheim's comparison of the perspectival *Standortsgebundenheit* of the historiographer to the observer's perspective on the "visual knowability of the figure of a thing in space" (*visuellen Erfaßbarkeit der Gestalt eines Raumdinges*) allows him to emphasize the partiality and limited nature of historical knowledge.[90] Like the perspectival foreshortening of vision, the historiographer's perspective introduces a distortion that presents the object as closer than it is through a reduction of its distance and depth. At the same time, Mannheim's presentist historiography makes it clear that it is only through the limited perspective of the present that the visibility of historical objects is possible at all.

Similarly, Heussi's theory of historiography emphasizes both the limitations of historical knowledge and the imperative of giving shape to history. For Heussi, the historian is powerless before the terrifying magnitude of history and its sheer size. Indeed, he describes history as an awe-inspiring sublime object that exceeds the powers of human perception:

> The tremendous abundance of events [*ungeheure Fülle des Geschehens*], and even of particular temporally localized episodes like the Battle of Leipzig, cannot be grasped by any human being. What takes place in every moment on earth, and altogether what has been experienced up to this moment in the historically surveyable past, is of such a tremendous magnitude, exceeding all human imagination, that the human being cannot speak of it in any other way than to look for and emphasize particular moments in this magnitude, and even these moments can be only be conceptualized in thoughts that stand in for this abundance. The enigmatic "counterpart" [*Gegenüber*] is thus not actually graspable, it remains eternally "opposite" [*gegenüber*]; it is only immediately accessible to us and thus clear and distinct at a single point, namely within the limits of our own conscious life. In its totality it is, metaphorically speaking, the tremendous, eternally enigmatic current of reality [*Strom des Wirklichen*] that rolls through "time."[91]

---

89. See Reinhart Koselleck, *Zeitschichten: Studien zur Historik* (Frankfurt am Main: Suhrkamp, 2000).
90. Mannheim, "Historismus," 44.
91. Karl Heussi, *Die Krisis des Historismus* (Tübingen: Mohr Siebeck, 1932), 48–49.

Borrowing from the rhetoric of Kant's aesthetic of the sublime,[92] Heussi points to an overwhelming magnitude of history that makes it impossible for the imagination to grasp the historical object. The use of the term "counterpart" (*Gegenüber*) instead of the more usual "object" (*Gegenstand*) emphasizes this point: the historical counterpart cannot be fully constituted as an object by the history-writing subject; rather, the "Gegenüber" is a "limit concept" (*Grenzbegriff*) that emphasizes the challenge of defining objects in the face of the unfathomability of the historical process.[93]

For both Mannheim and Heussi, the dynamic movement of history makes a stable vantage point for its contemplation impossible. But for Heussi, this insight implies not only the perspectivism and subjectivism of the historian, but also the idea that the counterpart itself is not determined by any finality of form or significance:

> What is actually decisive for our consideration of the problem is not the fact of being bound to a standpoint [*Standortsgebundenheit*], but rather the difference between the view presented here and the approach to the problem of historical knowledge common around 1900. At that time one emphasized very strongly the "subjective" stake in all historical knowledge, but considered this stake merely as an unavoidable, more or less intense blurring of the view of a counterpart that is firmly given in itself and *clearly structured once and for all*. Accordingly, human ideas change, but the things remain the same. According to our view, the things are only structured *like this* in the thoughts of human beings.... The counterpart is not clearly and completely structured, it has no rigid magnitude [*starre Größe*]; rather, it is an inexhaustible stimulus to ever-new historical readings.[94]

For Heussi, both the subjective perspective of the historian and the constitution of the historical counterpart are in a state of flux. The historical counterpart not only appears different from different per-

---

92. For Kant, the sublime, in contrast to the beautiful, is "zweckwidrig für unsere Urtheilskraft, unangemessen unserm Darstellungsvermögen und gleichsam gewaltthätig für die Einbildungskraft" (*Kritik der Urteilskraft*, in *Gesammelte Schriften*, vol. 5, ed. Königlich Preußische Akademie der Wissenschaften [Berlin: Reimer, 1908], 245).

93. Heussi, *Krisis des Historismus*, 49n1.

94. Heussi, *Krisis des Historismus*, 56.

spectives, but also has no equivocal structure as a historical reality. As a result, Heussi's historiography can be considered constructivist: the historian must give structure and form to history by taking the historical counterpart as an "inexhaustible stimulus to ever-new historical readings" (*unerschöpflicher Anreiz zu immer neuen historischen Auffassungen*). The historical object thus comes into existence only in the particular constellation in which it is read; that is, always in relation to a concrete present that posits a historical relation.[95]

In treating the historical counterpart as a stimulus to ever-new historical readings, Heussi, like Mannheim, invokes the spatial rhetoric of figuration (*Gestaltung*). Recapitulating his central thesis, Heussi writes: "the rigid, clearly structured 'counterpart' is superseded by a fluid magnitude (*fließende Größe*) that prompts ever-new historical figurations."[96] The claim that the historical counterpart has a fluid rather than rigid magnitude can be compared to the disappearance of the concept of magnitude in modernist mathematics. As Mehrtens shows, in premodern mathematics, the concept of magnitude allowed for a connection between mathematical operations and a "representable something" (*vorstellbaren Etwas*) that is operated on.[97] Magnitude provided for the concrete measurement (*Konkretion*) that made premodern mathematics intuitive. In modernist mathematics and geometry, by contrast, the concept of magnitude disappears and is replaced by "elements without qualities" (*eigenschaftslose Elemente*).[98] Similarly, for Heussi, historiography works no longer with objects that have a stable and defined magnitude but rather with elements

---

95. Heussi's theory of historiography has a strong affinity with Benjamin's historical materialism in contrast to Ranke's historicism. In Benjamin's work, the historian engages with the past in terms of its relation to the present moment: "Er [der Historiker] erfaßt die Konstellation, in die seine eigene Epoche mit einer ganz bestimmten früheren getreten ist" ("Über den Begriff der Geschichte," 704). Benjamin's presentist historiography thus situates itself in opposition to Ranke's claim to write history "as it actually was": "Vergangenes historisch artikulieren," writes Benjamin, "heißt nicht, es erkennen 'wie es denn eigentlich gewesen ist.' Es heißt, sich einer Erinnerung bemächtigen, wie sie im Augenblick einer Gefahr aufblitzt" ("Über den Begriff der Geschichte," 695).
96. Heussi, *Krisis des Historismus*, 64.
97. Mehrtens, *Moderne—Sprache—Mathematik*, 43.
98. Mehrtens, *Moderne—Sprache—Mathematik*, 43.

whose qualities are in flux. As a fluid magnitude, the historical counterpart gives rise to ever-new historical figurations or *Gestaltungen*.

While rejecting any finality of form or structure for the historical object, Heussi claims that the historian's fundamental act is to give form or structure to history through a process of *Gestaltung*. Referring to a lecture by Karl Brandi titled "Geschichte als Gestaltung" ("History as Figuration") (1921), Heussi writes: "The expression '*figuration*' [Gestaltung] . . . can be applied to the view of 'historical thinking' presented in the text: the historian 'configures' [*gestaltet*] history ('history in the subjective sense')."[99] Heussi's version of such a figuration of history borrows the Kantian concept of a categorical schema to describe how the historian's approach to history imprints a certain structure on the historical material: "If we want to gain historical thoughts, we have to approach the counterpart with a certain *question*, category, or 'schema.' It is at this point, in its approach [*im Ansatz*], that historical thinking includes a 'systematic' element. We have to have a vessel [*Gefäß*], so to speak, in order to scoop something out of the tremendous current of life [*Lebensstrom*], out of the 'counterpart.'"[100] To the pairing of "fluid magnitude" and "figuration" that we saw above, Heussi adds here the image of a vessel that scoops figures out of the "tremendous current of life." The historian's approach, understood as a vessel, gives shape and form to a historical material that in itself is fluid and formless.

Consequently, the shape of history, for Heussi, is constantly undergoing displacements in accordance with the relative positions of the historian and his or her counterparts. Indeed, there is a double displacement of observer and counterpart: "It is the same with the large historical complexes that we call 'epochs.' The great periods of art history are absolutely nothing rigid [*nichts Starres*], nothing that lies finished and structured once and for all in the counterpart; rather, they move with the shifting standpoint of the observer."[101] The changing shape of history in Heussi's theory of historiography resembles the deformability of figures as they move in non-Euclidean

---

99. Heussi, *Krisis des Historismus*, 49n2.
100. Heussi, *Krisis des Historismus*, 52.
101. Heussi, *Krisis des Historismus*, 59.

spaces. The resulting multiplicity of possible historical relations leads Heussi to consider an eschatological perspective from which these relations could be grasped in their entirety: "Only at the end of days is the ideal possibility available to survey all relations of the magnitudes that now long belong to the past. Past magnitudes that we perhaps do not consider particularly important can still potentially have significant repercussions in a time that for us is still in the future. Seen in this perspective, the past is nothing rigid, but rather something living, something constantly changing and growing."[102] This gesture toward an eschatological account of history is marked by a tension: while Heussi argues for the liveliness of history in taking on changing forms in the future, the ideal vantage point at the "end of days" (*am Ende aller Tage*) that might permit an overview of all historical relations would represent a finality of the historical perspective, such that history would no longer be defined by change and growth. Yet the very impossibility of this vantage point is precisely what prevents the past from becoming rigid and petrified.[103] Paradoxically, the figure of an eschatological end suggests that the final word on the past and the form it will take has yet to be spoken. For Heussi, the imperative of constructing history means that the shape it is given will remain provisional.

Mannheim's emphasis on the perspectival *Standortsgebundenheit* of the historiographer and Heussi's insights into the changing shapes of the historical counterpart inaugurated a new presentist and constructivist approach to historiography. Once the historical object was understood as a counterpart with no fixed shape, the structure of history was opened up for new formations. For Troeltsch, the construction of history took the form of an *Aufbau* (structure, formation, or construction), a term with architectural resonances that was borrowed from Wilhelm Dilthey's *Der Aufbau der geschichtlichen Welt in den Geisteswissenschaften* (*The Formation of the Historical World in the Human Sciences*) (1910).[104] The site of a historical

---

102. Heussi, *Krisis des Historismus*, 69.
103. Heussi's thought experiment has a strong similarity to Kracauer's treatment of the figure of Ahasuerus (see chapter 4).
104. Dilthey argued for the construction of the historical world via the hermeneutic relationship between the "experience" (*Erlebnis*) of the historical subject

architecture, for Troeltsch, can only be the present moment. It is from the vantage point of the present that the historiographer perceives the historical layers that are at work in contemporary life. Here Troeltsch supplements the architectural metaphor of an *Aufbau* with the geological motif of a stratification (*Schichtung*) of layers, giving the fourth chapter of his book the title "The Stratification of the Formation" (*Die Schichtung des Aufbaus*). Troeltsch's construction of history entails the isolation and selection of the historical impulses that exercise a decisive force in the present:

> The idea of construction [*Idee des Aufbaus*], as represented from the standpoint of the present, requires that we extract and accentuate those major periods in which decisive elements of contemporary life arose. The decisive and militant forces of the present are not singular scattered historical heirlooms or the totality thereof, but rather the great spirits of entire periods. It is necessary to retrieve the latter as decisive for life and make clear how they became fused and intertwined in the course of time. Everything else is merely a matter for scholarly historical research and elucidation, for the investigation of past connections, transitions, and preparations. It belongs to the exact historical sciences, not to the humanly meaningful and captivating image of history through which we understand ourselves.[105]

Troeltsch's theory of historiography parts ways with the historicist paradigm of historical research into the past and its causal relations. Instead, insight into the past is oriented toward self-understanding in the present. That is, the past is connected to the present no longer through the mediation of time but rather in terms of past forces that can be retrieved and extracted from their integration in the present. Troeltsch articulates a nontemporal constructivism that builds up the past from the ground of the present rather than seeking to reconstruct historical causalities. His idea of *Aufbau* can be represented only "from the standpoint of the present" (*vom Standort der*

---

and the objectified forms of historical knowledge in its reflection by the subject. See Wilhelm Dilthey, *Der Aufbau der geschichtlichen Welt in den Geisteswissenschaften* (Berlin: Verlag der Königlichen Akademie der Wissenschaften, 1910), 33.

105. Ernst Troeltsch, "Der Historismus und seine Probleme (1922)," in *Gesammelte Schriften*, vol. 3 (Aalen: Scientia, 1977), 757.

Gegenwart) and in relation to the life of the present. Historiographical construction in the early twentieth century was thus conceived as a construction of the present moment.

While Troeltsch adopts Dilthey's concept of *Aufbau*, he rejects Dilthey's understanding of history as *Entwicklungsgeschichte* (a history of development).[106] Although he acknowledges continuities between the modern world and its history, Troeltsch seeks to make historiography relevant to a historical situation defined by a loss of faith in a narrative of universal history. His approach isolates the "basic forces" (*Grundgewalten*) of the past that are active in constituting the modern world:

> This world has an extraordinarily rich and colorful intellectual content that has its origin not only in itself, but also largely in our entire history since the Greeks. Therefore, the horizon inevitably extends to encompass the entire universal historical development leading up to it. But the idea of construction requires only that we retrieve the great elemental *basic forces* that are immediately meaningful, operative, and intuitive—and not merely so for scholarly historical knowledge and its use in school lessons. To render intelligible these basic forces in their originary sense, in their emergence out of the historical movement, to thus give our historical memory the decisive emphasis and arrange it in relation to the present, and finally to grasp the relationship of these basic forces, as they take shape in the modern world, to one another and to modern life: this is the idea of a construction of European cultural history.[107]

For Troeltsch, the retrieval (*herausholen*) of the basic elemental forces of the past that play an immediate role in the constitution of the present does not require the mediation of historical knowledge. His construction of history is therefore decoupled from nineteenth-century historicism. The originary potential of these basic forces, for Troeltsch, is not tied to their historical moment but exceeds their

---

106. Dilthey, *Aufbau der geschichtlichen Welt*, 32. Dilthey's concept of *Entwicklungsgeschichte* is related to its Hegelian predecessor but works with a more rigorous historicism and severs the connection, still decisive for Hegel, to the metaphysical absolute. See Fulvio Tessitore, *Kritischer Historismus: Gesammelte Aufsätze* (Cologne: Böhlau, 2005), 26ff.

107. Troeltsch, "Der Historismus und seine Probleme," 765.

historical specificity in a manner comparable to Aby Warburg's concept of the afterlife (*Nachleben*) of pictorial representations.[108] Historical memory would then consist in arranging (*gliedern*) the elemental historical forces in the relation they take to one another in the present. For such a constructivist historiography, the question of causality and development falls by the wayside; the aftershocks of historical forces may arrest the present across the greatest historical distance.

Troeltsch conceives of his theory of historiography as a union of "critique, constructing representation [*konstruierende Darstellung*], and historico-philosophical synthesis."[109] By constructing history in relation to the present, he aims to renew and enliven the past by dispelling its petrified images. The construction of new shapes of history is thus closely connected to a critique and destruction of its accumulated and inherited forms: "In order to liberate ourselves from history and acquire sovereign rule over it, we plunge into an ocean of historical critique and reconstruction [*Ozean historischer Kritik und Rekonstruktion*]. But through this work, which in reality distances the past from us, we give the past new life [*beleben*] and revive [*wecken*] its spirits more freshly, originally, vitally, and actively than

---

108. Warburg explored what he called an ongoing "process of dialectical engagement with . . . surviving figural representations [*nachlebenden bildlichen Vorstellungen*]" of earlier cultures ("Italienische Kunst und internationale Astrologie im Palazzo Schifanoja zu Ferrara (1912)," in *Die Erneuerung der heidnischen Antike: Kulturwissenschaftliche Beiträge zur Geschichte der europäischen Renaissance*, ed. Horst Bredekamp and Michael Diers [Berlin: Akademie, 1998], 459–81, at 478; Warburg, "Italian Art and International Astrology in the Palazzo Schifanoia Ferrara," in *The Renewal of Pagan Antiquity: Contributions to the Cultural History of the European Renaissance*, trans. David Britt [Los Angeles: Getty Research Institute for the History of Art and the Humanities, 1999], 563–91, at 586, translation modified). Warburg was interested in the ways images, figures, symbols, and entire strategies of representation reappear—in modified form, in different costumes and guises—at discontinuous historical moments. As Sigrid Weigel has noted, the term *Nachleben* is notoriously difficult to translate: "The term signifies neither afterlife nor survival, since it expresses something like 'live' impact, in modified form, of a figure or symbol following its historical life—often, its reappearance after a long period of absence or hidden, marginalized existence" ("Warburg's 'Goddess in Exile': The 'Nymph' Fragment between Letter and Taxonomy, Read with Heinrich Heine," *Critical Horizons* 14, no. 3 [2013]: 271–95, at 272–73n4).

109. Troeltsch, "Der Historismus und seine Probleme," 723.

they were in the guise of previous historical conventions."[110] The critique and reconstruction of history estranges and distances the past, yet also allows its originary force to become active again. For Troeltsch, the historical image must be destroyed so that a new image may arise in its place: "This destruction is at once the revitalization and renewal [*Wiederbelebung und Erneuerung*] of a deeper and re-envisioned image of past things understood from the outlook of the present and its expansion of the field of research."[111] Invoking a language of resurrection with terms such as *beleben, wecken, Wiederbelebung*, and *Erneuerung*, Troeltsch argues that the historical image must first be brought to the threshold of death by estranging its conventional representation before it can be resurrected in a new historical construction.

By emphasizing the destructive side of historical critique, which allows the past to enter new relationships with the present, Troeltsch stands squarely within an antihistoricist strand of historiography that bears a strong affinity to Benjamin's theory of history. Like Troeltsch, Benjamin seeks to destabilize the past as the historical past so that its energies can be won for the present. In his notes on historiography and the theory of allegory in *Das Passagen-Werk* (*The Arcades Project*) (1940), Benjamin argues that both the historian and the allegorical reader engage in a practice of citation in which a historical object or figure is excerpted from its historical context, drained of life, and reinserted, as an isolated fragment, in a new context of meaning.[112] The citation of the historical object, for Benjamin, depends upon a destructive moment that shatters any sense of historical continuity: "The destructive or critical momentum of materialist historiography is registered in that blasting of historical

---

110. Troeltsch, "Der Historismus und seine Probleme," 723.
111. Troeltsch, "Der Historismus und seine Probleme," 723–24.
112. In his work on Baudelaire, Benjamin described the theory of allegory as follows: "That which the allegorical intention has fixed upon is sundered from the customary contexts of life: it is at once shattered [*zerschlagen*] and preserved" (Walter Benjamin, *Das Passagen-Werk*, in *Gesammelte Schriften*, ed. Rolf Tiedemann and Hermann Schweppenhäuser, vol. 5 [Frankfurt am Main: Suhrkamp, 1980], 414–15; Benjamin, *The Arcades Project*, trans. Howard Eiland and Kevin McLaughlin [Cambridge, MA: Belknap Press of Harvard University Press, 1999], 329).

continuity [*Aufsprengung der historischen Kontinuität*] with which the historical object first constitutes itself.... Materialist historiography does not choose its objects arbitrarily. It does not fasten on them but rather springs them loose from the order of succession [*sprengt sie aus dem Verlauf heraus*]."[113] Benjamin's historiography treats the treats the historical object as a fragment, which, through the act of citation, is violently separated from the totality of the epoch to which it belonged. The act of bursting the historical object out of its historical context breaks decisively with the historicist presupposition of historical development.[114]

Likewise, for Troeltsch the historiographer is charged with the task of perpetually shattering the composition of history in order to recover its autochthonous and originary potential:

> [Our culture] must shatter mythical and conventional history again and again, whenever it arises, in order to maintain contact, like Antaeus, with the maternal ground of its becoming.... Out of this process historical concepts and dogmas again emerge that correspond to new needs, only to be shattered and replaced [*zertrümmert und ersetzt*] again for the same reason. This continues as long as the sovereign power of remelting, streamlining, deepening, and reinvigorating historical possessions, the retrieval of ever-new original forces, and the shedding of ever-new accumulated ballast is present.[115]

Troeltsch's theory of historiography emphasizes the process of constructing history rather than its result, such that the liveliness of history is concentrated in the moment of historical synthesis. Similarly, in Benjamin's theory of allegory each allegorical image is subject to being taken from its context and mobilized for a new present. Both the reference to the myth of the giant Antaeus and the formulation "in order to liberate ourselves from history and acquire sovereign rule over it" suggest that the historiographer can never gain the upper hand over history. Just as the historiographical Antaeus regains contact with the nourishing strength of its historical ground, he is

---

113. Benjamin, *Das Passagen-Werk*, 594; Benjamin, *Arcades Project*, 475.
114. For more on Benjamin's antihistoricism, along with that of Erich Auerbach and Aby Warburg, see McGillen, "Auerbach and the Seriality of the Figure."
115. Troeltsch, "Der Historismus und seine Probleme," 724.

again torn away by the building of concepts, and each instance of sovereign mastery over history produces yet another historical image from which the historiographer must free themself. In this sense, the potential of Troeltsch's historiography is concentrated in the creative moment of giving shape to history while ceaselessly discarding constructions that have become ossified.

Spengler's theory of historiography, by contrast, relinquishes the notion of a larger historical structure (*Aufbau*) in favor of the construction of morphological relations between historical figures (*Gestalten*). The concept of a morphology of history suggests that history is an accumulation of shapes, forms, and figures whose contours define the space of history. This space, however, is dynamic and constantly in motion; unlike a morphology of nature, whose forms can be surveyed in their totality, the morphology of history is always in a state of becoming:

> [The] idea of a *morphology of world history, of the world as history* [*Idee einer* Morphologie der Weltgeschichte, der Welt als Geschichte], ... in contrast to the morphology of nature, ... combines once again all figures and movements of the world in their deepest and ultimate meaning, but in an entirely different order, not as the total picture of everything known, but as the image of life, not of what has become, but of becoming.[116]

Spengler's theory of historiography contains numerous echoes of Nietzsche's work, both in his use of organic and biological metaphors of life and "becoming" and in his characterization of his approach as a "philosophy of the future."[117] Spengler draws a sharp contrast between his understanding of the morphology of history and key tenets of historicism and the philosophy of history. In particular, he considers the construction of historical trajectory from antiquity to the Middle Ages to modernity to be problematic because it posits a "tendency of movement" in world history that is "spun in the figure of a *line* [*in Gestalt einer* Linie]."[118] In criticizing the

---

116. Spengler, *Der Untergang des Abendlandes*, 6.
117. Spengler, *Der Untergang des Abendlandes*, 6.
118. Spengler, *Der Untergang des Abendlandes*, 25.

linearity and continuity of such a concept of history, evident in the implicit image of a thread (*Faden*) that is spun (*fortgesponnen*), Spengler notes that it makes little difference whether a historian interprets this line as "leading upward or downward."[119] Regardless of whether the trajectory of world history is understood as one of progress or decline, such a concept of history conceives of the shape of world history in terms of a linear development.

By contrast, Spengler argues that history is characterized by a discontinuous collection of figures and forms that, as living entities, are determined by a necessary mortality. Cautioning against constructing "the *shape* of world history" (*die* Gestalt *der Weltgeschichte*) as if it were an "intellectual possession,"[120] Spengler points to the plurality of historical forms and their insistent modulation and decline:

> But with respect to the history of higher mankind there is an unrestrained optimism with regard to the course of the future, one that disdains all historical and organic *experience*, so that everyone discerns in what is coincidentally present the "rudiments" of an especially prominent linear "progression" [*Weiterentwicklung*], not because it has been scientifically demonstrated, but because it is desired. One counts on limitless possibilities—never with a natural end—and devises from the situation of each moment a completely naive construction of its continuation.... Instead of that tedious image of a linear world history that one can only maintain by closing one's eyes to an overwhelming number of facts, I see the spectacle of a plurality of powerful cultures that blossom with primordial force from the womb of a maternal landscape to which each is strictly bound in the entire course of its existence. Each of these cultures imprints on its material, on mankind, its *own* form; each has its *own* idea, its *own* passions, its *own* life, volition, feeling, its *own* death.... I see in world history the image of an eternal figuration and reconfiguration [*Gestaltung und Umgestaltung*], of a delightful becoming and passing away of organic forms. But the historian by trade sees history in the figure of a tapeworm that ceaselessly "adds" [*ansetzt*] epochs.[121]

Whereas for Troeltsch the construction (*Aufbau*) of history has a synthetic function that explicates the concretization of the history

---

119. Spengler, *Der Untergang des Abendlandes*, 25.
120. Spengler, *Der Untergang des Abendlandes*, 21.
121. Spengler, *Der Untergang des Abendlandes*, 28–29.

of the past in the life of the present, articulating the structure in which the past continues to inform the constitution of the present, Spengler's criticism is directed against the ideology of progress that extends this historical construction into the future. The underlying assumption of such an understanding of history is not only a continuity of historical development but also a notion of a unified subject of history—that is, a unified historical construction to which epoch upon epoch is added. Spengler, by contrast, sees a plurality of historical cultures, each of which has its own unique forms and a finite life span. World history, then, is the story of perpetual figuration and transfiguration: the historical figures do not permit a synthetic construction but rather represent singular historical forms.

Spengler's criticism of the optimistic belief in continued development (*Weiterentwicklung*) goes hand in hand with a critique of a teleological concept of history directed toward the present as its necessary result. A teleology of history, Spengler notes, is foundational to modernity's stylization of itself as the end and goal of world history: "Evidently it is a requirement of the western European sense of self to define its own appearance as a kind of conclusion [*eine Art Abschluß*]."[122] Here Spengler has in mind the philosophies of history of Herder, Kant, and Hegel: "Herder called history the education of mankind, Kant the development of the concept of freedom, Hegel the self-unfolding of the world spirit."[123] But he also reconstructs the genealogy of such concepts of history as teleological development in Lessing's "'Education of Mankind' (with the stages of the child, the youth, and the man)" and notes that Lessing adopted his triadic schema from Joachim von Floris's division of history into "the ages of the Father, the Son, and the Holy Spirit."[124] Spengler's genealogy corresponds closely to Taubes's argument for the emergence of a "philosophical eschatology" in German Idealism.[125] But Spengler's own theory of historiography marks a sharp break with these eighteenth- and nineteenth-century philosophies

---

122. Spengler, *Der Untergang des Abendlandes*, 26.
123. Spengler, *Der Untergang des Abendlandes*, 26.
124. Spengler, *Der Untergang des Abendlandes*, 26.
125. See Taubes, *Abendländische Eschatologie*, 125–63.

of history. "Humanity," he writes, "has no goal, no idea, no plan [*kein Ziel, keine Idee, keinen Plan*], no more than the species of butterflies or orchids has a goal."[126] Spengler's rejection of the teleological concept of historical progress could hardly be stronger, and it provides the basis for his conviction that a "decline of the West" is as necessary as the "decline of antiquity" was in its time.

Yet despite his repudiation of teleology, Spengler's theory of historiography remains a philosophy of the future, one that ventures, in a form of fatalism, to "determine history in advance," as he puts it in the opening lines of *Der Untergang des Abendlandes* (*The Decline of the West*) (1923).[127] Spengler therefore diverges from Mannheim's presentist claim that history can be written only from the standpoint of the present. Indeed, in the second chapter of *The Decline of the West*, titled "Das Problem der Weltgeschichte" ("The Problem of World History"), Spengler argues that the historiographer requires "*distance* from the object" in order to grasp the morphological relations of historical figures.[128] Despite the modern historian's claim to objectivity, Spengler argues, historiography has yet to achieve the distance necessary to "contemplate the present, in the overall picture of world history, . . . as something infinitely distant and foreign [*etwas unendlich Fernes und Fremdes*]."[129] Without such distance, history is written from the "contingent standpoint [*zufälligen Standort*] of the observer in some—his—'present'" and thereby gives the historical material "an arbitrary, surface-level

---

126. Spengler, *Der Untergang des Abendlandes*, 28.
127. In *Meaning in History*, Karl Löwith argues that Spengler's key concept of *Schicksal* (fate) implies a "noncyclical, 'historical' time, directed toward the future," but notes that the futural orientation of Faustian culture, in Spengler's work, stands in tension with "the acceptance of a definite outcome" (*Meaning in History*, 12). Interpreting Spengler's citation of Schiller's line "Die Weltgeschichte ist das Weltgericht" at the conclusion of the second volume of *The Decline of the West*, Löwith writes: "The Faustian soul cannot help interpreting fate in the perspective of an *eschaton*" (12). Löwith argues that Spengler is divided between an ancient, cyclical model of history and a Judeo-Christian model of history oriented toward a future messianic or eschatological end. Yet even though Spengler no doubt extends the cyclical model of history to encompass the future, his morphology of world history is articulated in decidedly non-teleological terms.
128. Spengler, *Der Untergang des Abendlandes*, 127.
129. Spengler, *Der Untergang des Abendlandes*, 127.

form" that is imposed by the "temporally and spatially limited perspective" (*zeitlich und räumlich beschränkten Perspektive*) of the historian.[130] Presentist historiography, for Spengler, is problematic because it entails an implicit teleology: fragments of the past are selected by the presentist historiographer whose goal is the contingent present moment.[131]

Spengler aims to differentiate his morphology of history from an arbitrary imposition of structure on history by emphasizing the dynamism and mobility of historical figures. Accordingly, he distinguishes between the principles of the figure (*Gestalt*) and of the law (*Gesetz*) and, analogously, between mathematical and chronological numbers. Whereas figures and mathematical numbers are connected to organic metaphors of life and becoming, Spengler associates the law and chronological numbers with inorganic metaphors of lifelessness and rigidity (*Starrheit*). Spengler's organic treatment of morphology is derived, as he notes, from Goethe's morphology of natural forms: "The more purely a world is beheld as eternally becoming, the more unquantifiable [*zahlenfremder*] is the impalpable abundance of its figuration [*ungreifbare Fülle ihrer Gestaltung*]. 'The figure is mobile, it becomes and passes away. The theory of figures is a theory of metamorphosis. The theory of metamorphosis is the key to all signs of nature,' is how Goethe puts it in a note from his unpublished works."[132] Goethe's insight that the theory of figures is a theory of metamorphosis emphasizes the mobility and becoming of figures, such that the figuration of forms resists quantification (in Spengler's terms, it is "alien to number" or *zahlenfremd*). Spengler's key intervention consists of applying Goethe's morphology to the realm of history rather than to the realm of nature.[133] In this context, Spengler

---

130. Spengler, *Der Untergang des Abendlandes*, 127.
131. See Spengler, *Der Untergang des Abendlandes*, 128.
132. Spengler, *Der Untergang des Abendlandes*, 132. The organic metaphors of Spengler's historiography provide an example of what Mannheim calls an "anschaulich-organischen Darstellung der Gestalten" in contrast to a "logisch-dialektischen Aufbau der Geschichtsphilosophie" ("Historismus," 31).
133. As Gregor Streim notes, a similar application of Goethe's morphology to the domain of history can be found in Theodor Lessing's *Geschichte als Sinngebung des Sinnlosen* (1919): "Lessing selbst spricht von einer 'Gestaltenkunde des Lebens,' wobei er das Verfahren von Goethes vergleichender Morphologie auf die Geschichte

conceives of the "theory of figures" (*Gestaltenlehre*) as an antidote to historical research grounded in chronology, which he describes as "that net of numbers that is inwardly entirely foreign to becoming and yet never perceived here as strange, which ensnares and penetrates the historical world of figures as a scaffolding of years or as statistics."[134] History conceived in chronological terms, according to Spengler, erects an external structure or scaffolding (*Gerüst*) around the historical world of figures that is foreign to the morphological qualities of the figures themselves. Such a "net of numbers" (*Zahlennetz*), in Spengler's second metaphor, captures historical figures as in a spider's web, treating them as dead forms rather than as mobile figures in a process of becoming.

Spengler's concept of the mathematical number, by contrast, provides a means of outlining the shapes of historical figures in terms of their possibilities: "The chronological number designates the unique real [*das einmalig Wirkliche*], the mathematical number the constantly possible [*das beständig Mögliche*]. The latter circumscribes figures and works out the contours of epochs and facts for the perceptive eye; it *serves* history. The former *is itself the law* that it is supposed to determine, the end and goal of research."[135] Like Mannheim and Heussi, Spengler is committed to an open structure of history in which the significance of historical figures is not limited to a single reality but remains full of enduring possibilities. In place of the historicist concepts of development, linear continuity, and chronological relations of cause and effect, Spengler treats history as a morphological space of mobile shapes and figures.

Mannheim, Heussi, Troeltsch, and Spengler conceive of historiography as a construction of history. Whereas historicism spun history along the thread of a narrative, connected by causalities, evolutions, and developments and unfolding along a temporal axis, Mannheim, Heussi, Troeltsch, and Spengler think of history as an assembly of forces and a constellation of figures in a manifold space of history. In

---

überträgt" ("'Krisis des Historismus' und geschichtliche Gestalt: Zu einem ästhetischen Geschichtskonzept der Zwischenkriegszeit," in *Literatur und Geschichte*, ed. Daniel Fulda and Silvia Serena Tschopp [Berlin: De Gruyter, 2002], 463–88, at 469).

134. Spengler, *Der Untergang des Abendlandes*, 133.
135. Spengler, *Der Untergang des Abendlandes*, 133.

this respect, constructivism in historiography has a strong affinity with Malevich's and El Lissitzky's approach to art. By revealing the absence of a fixed structure of the historical object, twentieth-century historiography shows that the shape of time does not follow the law of Renaissance perspective and cannot be mapped as a temporal geometry whose focal point is the vanishing point of a historical teleology. In the work of Mannheim, Heussi, Troeltsch, and Spengler, history can be better compared to the undecidable, dynamic, and tensed spaces that we find in constructivist art. Troeltsch pulls the weightless and suddenly undetermined basic forces (*Grundgewalten*) of history into the orbit of a constructivist foundation, making the present the ground of a new historical architecture. Like El Lissitzky's tensed containment of forces in space, Troeltsch's *Aufbau* of history realigns historical forces in terms of an orienting historical architecture. A similar attention to the structure of history as a mobile totality, organized from the perspective of the present, can be found in Mannheim's theory of historiography.

The theories of historiography of Heussi and Spengler come closer to the geometry of the dynamic conditions of form that Clark points to in Malevich. Malevich's interest in a kind of space not bound by a pivotal orienting force, one that breaks away from the earth, has its counterpart in Heussi's concept of history as a sublime current whose magnitude exceeds human perception and whose objects lack a fixed structure. Similarly, Spengler, in his criticism of a teleological historical trajectory in favor of a morphology of historical figures, conceives of history as a space without a pivotal orienting force. The mobile and shifting figures of Heussi's and Spengler's historiographies assert a dynamic condition of temporal forms that lacks an organizing fulcrum, just as Malevich's space of dynamic geometric forms breaks away from any support or anchor. Their historiographies can be considered as temporal geometries that function without defined historical horizons.[136]

---

136. Compare to Clark's verdict about Malevich's painting: "My judgement, you will gather, is that the best Maleviches really do *not* turn on the relation of their parts to a built-in horizon, drawn or implied, or even to the picture's overall shape conceived as a finite, generative entity, having a top and bottom, and ultimately dictating the behavior and gravity of the forms within it" (*Farewell to an Idea*, 285).

In the work of Barth, Rosenzweig, Kracauer, and Musil, the constructivist approach to historiography of Mannheim, Heussi, Troeltsch, and Spengler is combined with a non-Euclidean concept of space. Adapting insights from the curved spaces of non-Euclidean geometry, they present images of the curvature of time. Just as non-Euclidean and multidimensional spaces stretched the limits of intuitive visibility, or *Anschaulichkeit*, so too did Barth, Rosenweig, Kracauer, and Musil produce nonintuitive and liminal relations of history and eschatology. Their accounts of the deformation of the moment, bent and inflected by the end of history, would be unthinkable without the constructivist turn in mathematics, art, and historiography.

2

THE ESCHATOLOGICAL LIMIT

Spatial Form in Karl Barth's Dialectical Theology

In his essay "Spatial Form in Modern Literature" (1945), Joseph Frank argues that "modern literature, exemplified by such writers as T. S. Eliot, Ezra Pound, Marcel Proust, and James Joyce, is moving in the direction of spatial form." "This means," he clarifies, "that the reader is intended to apprehend their work spatially, in a moment of time, rather than as a sequence."[1] Frank makes use of Gotthold Ephraim Lessing's distinction between pictorial space and narrative time, but he shows that these analytic categories do not map exclusively onto the opposition between painting and literature. In modernist literature, he argues, "the time flow of narrative is halted" and

---

An earlier version of parts of this chapter was published as "Theology's Weimar Moment: History before the Eschatological Limit," in *The Weimar Moment: Liberalism, Political Theology, and Law*, ed. Leonard Kaplan and Rudy Koshar (Lanham, MD: Lexington Books, 2012), 267–87.

1. Joseph Frank, "Spatial Form in Modern Literature: An Essay in Two Parts," *Sewanee Review* 53, no. 2 (1945): 221–40, at 225.

the reader is concerned with "the unification of disparate ideas and emotions into a complex presented spatially in an instant of time."[2]

This definition of spatial form as the condensation of an image in a moment outside of narrative sequence is symptomatic not only of a specific mode of modernist literature but also of the more general situation of cultural and intellectual history in the early twentieth century. What Frank describes in terms of aesthetic perception had equally significant manifestations in the conceptual languages of philosophy and theology, and indeed, these aesthetic and conceptual languages were deeply intertwined. Modernism's interest in a moment that stands outside of narrative sequence can be accounted for in part by the collapse of historicist modes of historiography, which had maintained the continuity of the present with the past by upholding an idea of historical progress and development. Without the anchor of narrative sequence, the complex image of the moment as spatial form was cut adrift and opened up to new figurations.

A prime example of modernism's interest in time as spatial form can be found in the Dialectical Theology of Karl Barth. A Swiss Reformed theologian who came of age in the first decades of the twentieth century, Barth turned radically against the tradition of German Protestant liberal theology, which, following the impetus of Friedrich Schleiermacher, stressed the immanence of religious experience in cultural life.[3] Influenced by the German philosophical tradition from Kant to Hegel and Nietzsche, and more immediately by the writings of Søren Kierkegaard and Franz Overbeck, Barth is best known for his claim that God is "wholly other" (*der ganz Andere*), a claim that emphasizes the radical distance between humanity and God. In *Der Römerbrief (The Epistle to the Romans)* (1922), his commentary on Paul's Epistle to the Romans, Barth presents a groundbreaking theory of the moment that offers a critique of historicism.

---

2. Frank, "Spatial Form," 231, 226.

3. As Emil Brunner argues, Schleiermacher's emphasis on inner religious experience as the intensification (*Steigerung*) of life through an immediate relation to the divine expresses, in theological terms, an ideology of progress in which individuals ascend, exceeding themselves, to higher and higher forms of life (*Erlebnis, Erkenntnis und Glaube* [Tübingen: Mohr Siebeck, 1923], 7–9).

The moment for Barth is always a present moment, and yet it is equally radically absent and hence a moment of nonpresence. Barth's thought maintains a tremendous tension between the nearness and distance, the presence and nonpresence, of the moment, which are submitted to dialectical oscillation. A centerpiece of Barth's theory of the moment can be found in his interpretation of what Paul calls the "time of the now," a concept whose eschatological energies imply both a radical negation of historical time and its paradoxical redemption. Barth's work can rightly be called Dialectical Theology insofar as it conceives of the prospect of messianic fulfillment as bound up with its dialectical opposite: the crisis and rupture of historical time. As a result of this ambivalence, the otherness of the time of the now is both utterly alien to the historical present and contained within each historical moment.

In his articulation of these paradoxes, Barth's work bears the stamp of modernity.[4] In particular, he conceptualizes the temporal dislocation of the moment by drawing on the resources of what Frank called spatial form. Barth negotiates the temporal aporias of the moment through prominent metaphors of spatial proximity and distance, such as limits, thresholds, and hollow spaces. This complex array of spatial metaphors expresses his unique sense of what it means to be out of the moment.[5] In this respect Barth plays a key role in the critique of historicism, drawing on not only the thought of Kierkegaard and Nietzsche but also the language of modernist aesthetics and mathematics. Barth's relation to modernist aesthetics is not "expressionistic,"

---

4. The claim that Barth was embedded in modernist artistic and cultural production and produced a modernist concept of history runs counter to scholarship that argues for Barth as a critic of modernity. See Graham Ward, "Barth, Modernity, and Postmodernity," in *The Cambridge Companion to Karl Barth*, ed. John Webster (Cambridge: Cambridge University Press, 2000), 274–95; and Trutz Rendtorff, "Karl Barth und die Neuzeit: Fragen zur Barth-Forschung," *Evangelische Theologie* 46 (1986): 298–314.

5. As Hent de Vries notes in his discussion of Barth, Taubes, Adorno, and Benjamin, an analysis of the post-secular dimension of their theologies requires help from "what might seem an unlikely ally: the language and formalism of mathematical, geometric precision (speaking of tangents and circles, points without spatiotemporal extension, number and set theory, and the like)" ("Inverse versus Dialectical Theology: The Two Faces of Negativity and the Miracle of Faith," in *Paul and the Philosophers*, ed. Ward Blanton and Hent de Vries [New York: Fordham University Press, 2013], 466–511, at 496).

as some critics have argued with reference to the explosive language and rhetoric of crisis in *The Epistle to the Romans*.[6] Rather, Barth appropriates the spatial forms of constructivism and modernist geometry to think about history from an eschatological perspective.

Indeed, the language of mathematics is ubiquitous in Barth's *Epistle to the Romans*, and it serves as a vehicle for expressing a nonintuitive relationship between humanity and God, between the historical moment and the eschatological now. With reference to mathematical limits and the intersection of parallel lines, Barth reconfigures the "end" of history in nonteleological terms: not as the culmination of a narrative trajectory, but as coextensive with the present and, hence, as the limit of each moment of time. Just as apparently parallel lines meet at the poles of the sphere in non-Euclidean spherical geometries, so too do primal history (*Urgeschichte*) and the history of the end (*Endgeschichte*) converge in Barth's theology. In Barth's work, the end of history is folded back on itself, such that the end stands in the closest proximity to each present while remaining at a distance.

Dialectical Theology's Perception of Time at Its Limit

Barth's use of spatial form to bolster his antihistoricist concept of the moment can be observed in his commentary on Paul's Epistle

---

6. Prominent critics who have argued that Barth's theology is "expressionistic" include Jacob Taubes and Hans Urs von Balthasar. Taubes claimed that "Barth's commentary on Paul's Epistle to the Romans was 'expressionistic' because it was an attempt to express a new, even unheard-of, situation in theological language: to interpret theologically the eclipse of God that became manifest post Hegel" ("Theodicy and Theology: A Philosophical Analysis of Karl Barth's Dialectical Theology," in *From Cult to Culture: Fragments toward a Critique of Historical Reason*, ed. Elisheva Fonrobert and Amir Engel [Stanford, CA: Stanford University Press, 2010], 177–94, at 193–94); whereas Balthasar described the second edition of Barth's *The Epistle to the Romans* as "theological expressionism, especially in its methodology" (*The Theology of Karl Barth*, trans. Edward T. Oakes [San Francisco: Ignatius Press, 1992], 83, quoted in Ian R. Boyd, *Dogmatics among the Ruins: German Expressionism and the Enlightenment as Contexts for Karl Barth's Theological Development* [Oxford: Peter Lang, 2004], 176). Boyd argues for an expressionistic dimension of Barth's theology in *The Epistle to the Romans* in its language, vocabulary, and rhetoric, but also in its atmosphere and sense of crisis (*Dogmatics among the Ruins*, esp. 174–79).

to the Romans, where he makes widespread use of mathematical and geometrical metaphors to reflect upon the displacement of the historical present. Barth first introduces his theory of the moment in a reading of Christ's two natures. The problems that arise from this duality are familiar to readers of Ernst Kantorowicz's *The King's Two Bodies* (1957), which showed how the logic of incarnation and the dual natures of Jesus Christ—both human and divine—were mapped onto the legal theories of monarchical sovereignty in Tudor jurisprudence.[7] In Kantorowicz's account of this early modern political theology, the natural body of the king forms a union with the immortal body of sovereign authority as long as the king lives. When the natural body perishes, the immortal body survives and the office of kingship passes to another natural body. The union of the king's two bodies is modeled on the incarnation of Christ, who is at once human and divine. Unlike Kantorowicz, Barth is interested in the theory of two natures because of its implications for the philosophy of history and the problem of temporality. For Barth, the figure of Jesus names a historical determination, whereas the figure of Christ stands for the suspension of historical time:

> *"Jesus Christ our Lord"* .... In this name two worlds meet and part from one another; two planes intersect [*sich schneiden*] here, a known and an unknown one. The known plane is the world created by God, which has fallen out of its original unity with God and is therefore in need of redemption: the world of the "flesh," the world of humanity, of time, and of things, our world. This known plane is cut [*wird geschnitten*] by another unknown plane, by the world of the Father, the world of original creation and final redemption.... —The point on this line of intersection [*Schnittlinie*] where [the relation between humanity and God] can be seen and is seen is *Jesus*, Jesus of Nazareth, the "historical" Jesus, *"born of the seed of David according to the flesh."* "Jesus" as a historical determination stands for the fracture [*Bruchstelle*] between the world known to us and an unknown one, . . . [for] that point . . . that makes visible [*sichtbar werden lässt*] the concealed line of intersection [*Schnittlinie*] of time and eternity, thing and origin, humanity and God. (R 5)

---

7. Ernst H. Kantorowicz, *The King's Two Bodies: A Study in Mediaeval Political Theology* (Princeton, NJ: Princeton University Press, 1957).

The gulf between Barth's understanding of Christ and the view articulated in *The King's Two Bodies* is immediately apparent. Whereas the latter allows for the saturation of the earthly body with divine authority—if only provisionally, contingent upon the health of the earthly body—Barth views the figure of Jesus Christ as fundamentally paradoxical. He represents for Barth the convergence of two worlds that are radically other, so much so that the point of contact between them can only be conceived as a wound or fracture. The violence of the intersection of the two worlds is especially apparent in Barth's German, whose mathematical precision depends upon a series of metaphors of cutting: the "line of intersection" (*Schnittlinie*) between the world of the flesh and the world of the Father is literally the "line of a cut." Similarly, the verb "to intersect"—in German: *schneiden*—keeps the literal referent of cutting at the forefront of the reader's mind.

The Barth scholar Bruce McCormack has noted that the 1922 edition of *The Epistle to the Romans* allows for very little "present realization," "to the point of making the incarnation itself problematic."[8] Barth's statement that "in the name Jesus Christ two worlds meet and part from one another" bears this claim out. The meeting of the human and the divine is immediately clarified as a parting of ways. The spatial metaphors of Barth's work, which operate with a pared-down geometric language, raise this separation to the level of the infinitesimal: the line of intersection between humanity and the divine, Barth argues, is visible only at a single point, that of the historical Jesus.[9] This point represents a break or "fracture" (*Bruchstelle*): a

---

8. Bruce L. McCormack, *Karl Barth's Critically Realistic Dialectical Theology* (Oxford: Oxford University Press, 1995), 21. Barth's skepticism, in the second edition of *The Epistle to the Romans*, toward the idea of eschatological completion reflects what McCormack calls a shift from a "process eschatology" in Barth's work from 1915 to 1920 (culminating in the first edition of *The Epistle to the Romans*) to a "consistent eschatology" in the 1920s (and especially in the second edition of *The Epistle to the Romans*). Whereas "process eschatology" "preserved the tension between present realization and future fulfillment," the "consistent eschatology" of 1920s pictured at best a liminal point of contact between history and eschatology.

9. On Barth's use of the "mathematics of the infinitesimal" to describe the "indescribable event" of God's "non-identical relationality to man," see Smith, "The Infinitesimal as Theological Principle," 586–87.

rupture of the historical continuum. At this stage in Barth's reading, the redemptive possibility of messianic time seems remote, and instead we are meant to understand the brokenness and fragmentation of creaturely life that the figure of Christ reveals.

While for Barth this rupture of the historical continuum can be localized to the figure of the historical Jesus, it is equally a rupture that is possible at any moment in time: "The time of revelation and discovery are thus the years 1–30. This is the time, as a glance at David shows, in which the new, different, and divine determination of *all* time can be *seen*, and which itself suspends [*aufhebt*] its singularity among other times insofar as it introduces the possibility that every time could be a time of revelation and discovery" (R 5). The point at which the concealed line of intersection of time and eternity becomes visible, even though it is historically determined and corresponds to the years 1–30, simultaneously suspends its historical specificity insofar as the plane of eternity that becomes visible in this historical moment reveals itself as having cut through all time. At stake, then, is a revelation that opens up the possibility that every moment in time is a potential time of revelation, a breaching of history that gains historical force not through the temporal movement of history, not by being sustained, but by being enacted again and again. This phenomenon might be thought of as repetition were this term not itself bound to a temporal structure. Instead, we might think of it as an incessant, persistent negation of history in history, or, to borrow a phrase from Kierkegaard, a movement whose effect is to "negate the historical historically."[10]

In his subsequent reflections on the infinitesimal point of contact between historical time and eternity, Barth is keen to establish the radical negativity of this point:

> This point on the line of intersection [*Schnittlinie*] itself, however, just like the entire unknown plane whose existence it announces, has no extension [*Ausdehnung*] whatsoever onto the plane known to us. The outward radiating forces [*Ausstrahlungen*] or rather the astonishing craters

---

10. In Kierkegaard's formulation: "geschichtlich das Geschichtliche zunichtemachen lassen." Søren Kierkegaard, "Philosophische Brocken" (1844), trans. Emanuel Hirsch, in *Gesammelte Werke*, vol. 10 (Düsseldorf: Eugen Diederichs, 1952), 56.

of impact [*Einschlagstrichter*] and hollow spaces [*Hohlräume*] through which this point makes itself felt within historical visibility [*Anschaulichkeit*], even when they are called "the life of Jesus," are not that other world that touches upon our world in Jesus. And insofar as our world is touched in Jesus by the other world, it ceases to be historical, temporal, material, and directly visible [*anschaulich*]. (R 5)

Barth's use of the mathematical concept of the point is precise. A point, in contrast to a line or a plane, has no extension into a second or third dimension. The historical visibility of this point is therefore conceived entirely in negative terms. The divine becomes present in time and history not as an "outward radiating force" (*Ausstrahlung*) but through lacunae or "hollow spaces" (*Hohlräume*): at each moment in which the divine makes itself felt within time, time continues to exist only as the negation of itself.[11] The severity of this dialectical movement is striking. Either there is no contact with the other world—and historical time is marked by the absence of the divine—or else there is a contact and our world ceases to be historical and temporal. In either case, the moment of presence can be known only as a moment of nonpresence.

Barth's use of the spatial metaphors of *Einschlagstrichter* and *Hohlräume* to describe the historical impact of Jesus marks a departure from the theology of Schleiermacher, who conceived of Christ as a historically charged "impulse" that extends outward, through a sort of charismatic radiance, from the historical moment of his revelation. Schleiermacher's notion of Christ as a historical force is summed up nicely by Emil Brunner: "With Christ a new force entered into the historical world, a thrust, so to speak, with a particular intensity and direction, and thus a historical movement came into being that can now seize the individual, a certain force field into

---

11. In keeping with this sense of the negativity of historical time, expressed as a negative space, Friedrich Gogarten, in his canonical essay "Zwischen den Zeiten" (1920), writes: "Das ist das Schicksal unserer Generation, daß wir zwischen den Zeiten stehen. Wir gehörten nie zu der Zeit, die heute zu Ende geht. Ob wir je zu der Zeit gehören werden, die kommen wird? . . . So stehen wir mitten dazwischen. In einem leeren Raum" (in *Anfänge der dialektischen Theologie; Teil II: Rudolf Bultmann, Friedrich Gogarten, Eduard Thurneysen*, ed. Jürgen Moltmann [Munich: Kaiser, 1967], 95–101, at 95).

whose range he can be drawn."[12] Schleiermacher's historicization of the figure of Christ had a counterpart in the historical research on the life of Jesus (*Leben-Jesu-Forschung*) of Ferdinand Christian Baur, David Friedrich Strauß, and Ernest Renan. This research, to which Barth obliquely refers in the passage above, applied historical-critical methods of scholarship in order to reconstruct the personality of the historical Jesus as a counterweight to a dogmatic Christology. Albert Schweitzer provided a critical account of this research in *Von Reimarus zu Wrede: Eine Geschichte der Leben Jesu-Forschung* (*The Quest of the Historical Jesus: A Critical Study of Its Progress from Reimarus to Wrede*) (1906).[13] For Barth, by contrast, the strictly historical impact of Jesus can be conceived only in negative terms, as an index of the absence of the "other world" in this one.

In his consideration of the historical Jesus, Barth mobilized geometrical metaphors to emphasize the interruption and rupture of the historical continuum. When he turns to the figure of Christ as the Messiah, though, he invokes the eschatological possibility of a new world as a limit phenomenon, which again oscillates between nearness and distance, presence and nonpresence:

> Jesus as *the Christ*, the messiah, is the end of time, he can only be understood as paradox (Kierkegaard), as the victor (Blumhardt), as primal history [*Ur-Geschichte*] (Overbeck). Jesus as the Christ is the plane unknown to us that cuts through [*durchschneidet*] the known plane vertically from above. . . . In the resurrection the new world of the Holy Spirit touches the old world of the flesh. But it does so as the tangent touches the circle, without touching it, and precisely insofar as it does *not* touch, the new world touches the old as its limit [*Begrenzung*], as a *new* world. (R 6)

We can see here the rigor with which Barth expresses Kierkegaard's concept of Christ as paradox. Like the figure of the point discussed above, the tangent to the circle allows for a moment of contact at the infinitesimal limit, but it is a moment of contact that touches only by

---

12. Emil Brunner, *Die Mystik und das Wort: Der Gegensatz zwischen moderner Religionsauffassung und christlichem Glauben dargestellt an der Theologie Schleiermachers*, 2nd ed. (Tübingen: Mohr Siebeck, 1928), 199–200.
13. See Albert Schweitzer, *Von Reimarus zu Wrede: Eine Geschichte der Leben-Jesu-Forschung* (Tübingen: Mohr Siebeck, 1906).

not touching. Barth's metaphor of a circle touched by a tangent at a single liminal point emphasizes both the proximity and the distance of eschatological time: it at once limits history and is in this way constitutive of it, and yet it does not permit any actualization or realization in history and thus remains a moment of nonpresence.[14]

In mathematics, the concept of the limit was fundamental to the development of calculus by Newton and Leibniz in the late seventeenth century, though the concept itself was not formalized until the early nineteenth century. In his *Principia Mathematica* (1687), Newton conceives of the limit in terms of the convergence of ratios in what he calls "evanescent quantities." For example, he proposes that for a chord AB on the arc AB of a curve, and for the corresponding tangent AD that passes through A, the arc, the chord, and the tangent will converge as the points A and B approach one another (figure 1). For Newton, the approach to the limit takes place as quantities are diminished to an infinitesimal difference:

> By the ultimate ratio of evanescent quantities (i.e., ones that are approaching zero) is to be understood the ratio of the quantities not before they vanish, nor afterwards, but with which they vanish.... Those ultimate ratios with which quantities vanish are not truly the ratios of ultimate quantities, but limits towards which the ratios of quantities decreasing without limit do always converge; and to which they approach nearer than by any given difference, but never go beyond, nor in effect attain to, till the quantities are diminished *in infinitum*.[15]

As C. H. Edwards notes in *The Historical Development of the Calculus*, while Greek mathematicians such as Archimedes provided key impulses for calculus, they lacked an explicit limit concept, avoided the concept of the infinite, and did not make the connection

---

14. In this sense, Hent de Vries describes Barth's position as a "dual aspect theory of reality": "The unredeemed and redeemed world are co-extensive in every respect and yet touch upon one another merely *as a tangent upon a circle*; in other words, the point at which they touch has no extension in historical time and empirical space" ("Inverse versus Dialectical Theology," 498–99).

15. Scholium to Section I of the *Principia Mathematica*. Cited in C. H. Edwards, *The Historical Development of the Calculus* (New York: Springer, 1979), 225.

between tangent lines and "rates of change."¹⁶ That is, Greek mathematicians viewed tangent lines as merely touching lines and did not understand that the tangent can be used to define the slope or rate of change of a function at a given moment in time.¹⁷ In Book III of *The Elements*, Euclid defines the tangent as follows: "A straight line is said to touch a circle, which meeting the circle and being produced, does not cut the circle."¹⁸ In Proposition 16, Euclid goes on to show that the tangent touches the circle at a single point and that no other line through the point does so: "The straight line drawn at right angles to the diameter of a circle from its extremity will fall outside the circle, and into the space between the straight line and the circumference another straight line cannot be interposed."¹⁹ In contrast to "the static Greek view of a tangent line as a line touching the curve in only one point,"²⁰ modern calculus used the tangent to calculate infinitesimal differences or differentials.

The implicit concept of the mathematical limit in Newton's work was given a formal definition by the French mathematician Jean le Rond d'Alembert in a contribution to Diderot's *Encyclopédie* (1765): "One says that a magnitude is the *limit* of another magnitude when the second may approach the first more closely than by a given quantity, as small as one wishes, moreover without the magnitude which approaches being allowed ever to surpass the magnitude that it approaches; so that the difference between such a quantity and its *limit* is absolutely unassignable."²¹ This definition of the limit considers the approach of a variable to a fixed magnitude in terms of proximity or nearness, such that the difference between the variable and the limit becomes infinitely small, or indeed "absolutely unassignable." As

---

16. Edwards, *Development of the Calculus*, 75.
17. See Edwards, *Development of the Calculus*, 75.
18. Euclid, *The Elements*, book III, definition 2, cited in J. L. Coolidge, "The Story of Tangents," *American Mathematical Monthly* 58, no. 7 (1951): 449–62, at 450.
19. Cited in Coolidge, "The Story of Tangents," 450.
20. Edwards, *Development of the Calculus*, 122.
21. Cited in David M. Bressoud, *Calculus Reordered: A History of the Big Ideas* (Princeton, NJ: Princeton University Press, 2019), 143.

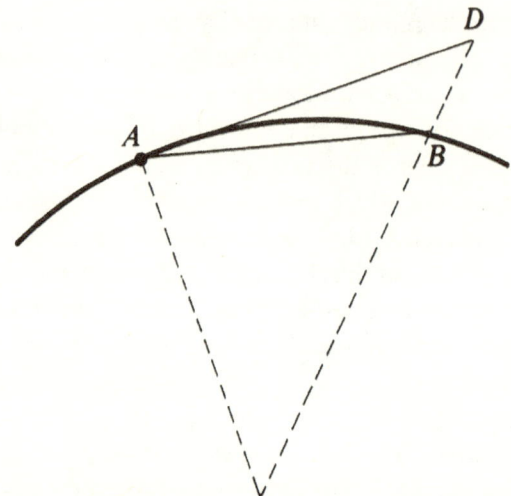

Figure 1. The convergence of ratios in Newton's concept of the limit. C. H. Edwards, *The Historical Development of the Calculus* (New York: Springer, 1979), 225. Reprinted with permission from Springer Verlag.

John L. Bell puts it, the infinitesimal difference is "greater than nothing but less than anything."[22] This concept of the limit was formalized by Augustin-Louis Cauchy in *Cours d'analyse* (1821): "When the successive values attributed to a variable approach indefinitely a fixed value so as to end by differing from it by as little as one wishes, this last [fixed value] is called the *limit* of all the others. Thus, for example, an irrational number is the limit of diverse fractions which furnish more and more approximate values of it."[23] Cauchy's concept of the limit, which became foundational for the subsequent development of calculus, allowed for a new definition of the infinitesimal, no longer as an "infinitely small fixed number," but as a "variable with zero as its limit."[24]

---

22. John L. Bell, *The Continuous, the Discrete and the Infinitesimal in Philosophy and Mathematics* (Cham, Switzerland: Springer, 2019), xi.
23. Cited in Edwards, *Development of the Calculus*, 310.
24. Edwards, *Development of the Calculus*, 310.

Barth's image of the tangent that touches the circle without touching it departs from the Euclidean understanding of the tangent as a line that touches a curve at a single point. His account of the tangent can be compared instead to a mathematical limit: just as a variable approaches a fixed value yet differs from it by an infinitely small quantity—an infinitesimal difference—so too does the old world of the flesh approach the new world as its limit, coming infinitely close yet never making contact. From the point of view of Greek mathematics, the tangent to a circle defines a single point of contact rather than a liminal relationship. Yet Barth projects the concept of the mathematical limit onto the geometry of the circle, introducing a gap that is at once infinitely proximate and unbridgeable. By applying the theory of infinitesimal limits from modern calculus, Barth turns the geometry of the circle in the direction of the nonintuitive. The mathematical *Grenzwert* (limit) becomes a spatial metaphor for the resurrection as the limit (*Begrenzung*) of the world of flesh.

Barth's geometrical imagination plays a central role in his critique of historicism. Indeed, in his image of the tangent that touches the circle as its limit, without touching it, the thresholds of geometrical space are mapped onto an unstable boundary between historical time and a time of redemption. In the move from considering Jesus as a historical figure to considering Jesus as Christ, Barth's spatial metaphors undergo a subtle shift: as Jesus, he is the point of fracture that reveals the plane of eternity; as Christ, he *is* this unknown plane and as such cannot be grasped in historical terms. Nevertheless, as the unknown plane, Christ cuts through the plane of the known world from above; he is that transcendent point of reference that bears a vertical relationship to the world at all points in history.[25] The resurrection, however, regardless of its reference to each moment

---

25. Barth's imagery of planes and vertical points of intersection has a striking affinity with the work of P. D. Ouspensky, who was a key figure in the philosophical reception of non-Euclidean geometries and four-dimensional spaces (see chapter 1). Like Barth, Ouspensky uses the figure of the vertical perpendicular to describe the relationship of time to eternity: "But in reality *eternity* is not the infinite dimension of time, but the one *perpendicular to time*" (*Tertium Organum*, 42, 47).

in history, must be understood, according to Barth, as the limit that both conditions and suspends history; through the resurrection, Christ is both the end of time and "primal history" (*Urgeschichte*).

This limit at which the new world touches the old world without touching it can be thought of as an eschatological limit.[26] Indeed, Barth engages in a critical contestation of the value of history in light of the eschatological thought of the last things.[27] Dialectical Theology calls for an engagement with history from the perspective of a "now" in which the contingency of the present historical moment becomes apparent. From the vantage point of such a now, the present takes on meaning in relation to its determination by the eschatological limit. What is at stake, then, is no simple negation of the value of history but the thought of history in other terms, terms that are no longer historical in the sense of historicism.[28] As Barth argues in *Die Auferstehung der Toten* (*The Resurrection of the Dead*) (1924), an interpretation of Paul's Epistle to the Corinthians, it is important to work with a strong sense of the "end" when considering the notions of the "end of history" or the "end of time"; to speak of the "finitude of history [*Endlichkeit der Geschichte*], of the finitude of time" is thus at once to speak of that which "*grounds* [begründet] all time and everything that takes place in time. The history of the end [*Endgeschichte*] would have to be ... synonymous with *primal* history [*Urgeschichte*]; the limit of time [*Grenze der Zeit*] ... would have to

---

26. On the concept of the "eschatological limit" in Barth's work, see McGillen, "Theology's Weimar Moment"; and Michael McGillen, "Eschatology and the Reinvention of History: Theological Interventions in German Modernism, 1920–1938" (PhD diss., Princeton University, 2012). On the importance of eschatology for Barth's thinking, see also Smith, "The Infinitesimal as Theological Principle," 584–86.

27. Barth's attention to his historical moment makes him a key figure in the intellectual thought of the 1920s. As Douglas Cremer argues, Dialectical Theology needs to be understood within the broader context of Weimar intellectual thought ("Protestant Theology in Early Weimar Germany: Barth, Tillich, and Bultmann," *Journal of the History of Ideas* 56, no. 2 [1995], 289–307).

28. Compare to Folkart Wittekind, who argues that eschatology was central to Dialectical Theology's overcoming of nineteenth-century "weltimmanente Geschichtsphilosophien" ("Zwischen Deutung und Wirklichkeit: Überlegungen zum Bildcharakter eschatologischer Aussagen," in *Die Gegenwart der Zukunft: Geschichte und Eschatologie*, ed. Ulrich H. J. Körtner [Neukirchen-Vluyn: Neukirchener Verlag, 2008], 55–84, at 35).

be the limit of each and every time, and thus necessarily the *origin* of time."[29] For Barth, the eschatological end is not a temporal end of history but rather the ground of history as its limit, and thus its condition of possibility.[30]

With the concept of the limit (*Grenze, Begrenzung*) and the image of the tangent that touches a circle without touching it, Barth draws on a mathematical and geometrical imaginary to represent in spatial terms the eschatological problematic of the end of time. In defining this liminal relation between history and eschatology, Barth makes it clear that the visibility or "intuitiveness" (*Anschaulichkeit*) of the historical world is at stake.[31] The other world that becomes manifest in the figure of Christ has no extension in our world, Barth argues, and hence the historical visibility that does pertain to Christ is not the other world as such. Yet the liminal point of contact deprives our world of its historicity, its temporality, and its intuitive visibility. The space of eschatological time, in Barth's interpretation, thus has the same nonintuitive character as the non-Euclidean geometries described by Helmholtz, Poincaré, Spengler, El Lissitzky, and Ouspensky.

Mathematics and the Visualization of the Nonintuitive

Barth's use of mathematical images to explore key theological paradoxes is a recurrent feature of his commentary in *The Epistle to*

---

29. Karl Barth, *Die Auferstehung der Toten: Eine akademische Vorlesung über 1. Kor. 15* (Munich: Kaiser, 1924), 58.

30. The concept of eschatology underwent a semantic shift in Barth's work: it no longer referred to the doctrine of the "last things" with apocalyptic connotations but came to stand for an end in the sense of a limit (*Grenze*). On the new concept of eschatology in Barth, see Keith Innes, "Towards an Ecological Eschatology: Continuity and Discontinuity," *Evangelical Quarterly* 81, no. 2 (2009), 126–44, at 126: "Eschatology has traditionally been the study of the 'last things' and the doctrine that issues from such study. However, the great Swiss theologian Karl Barth saw eschatology as dealing not with the future but with the breaking of eternity into time."

31. Smith notes, "In so far as mathematics is employed in order to represent mysteries of religion, its paradoxical nature plays the role of allowing us to visualize precisely the non-visualizable ('dieses unanschauliche positive X' for Barth)" ("The Infinitesimal as Theological Principle," 565).

the Romans.[32] In particular, Barth draws on mathematics to illustrate the coexistence of apparently mutually exclusive possibilities and to articulate liminal situations. Mathematics allows for the visualization of the nonintuitive, and in doing so, it stretches the limits of what can be visualized. Consider, for example, Barth's discussion of "the new human being" in Romans 5:1–11, in which he takes pains to describe the repredication of humanity through faith. Those who have faith simultaneously belong to the world of the flesh and to that of the spirit, or, as Barth puts it, "we are not only what we are; we are also through faith what we are *not*" (R 136). The paradox that in faith "we" (*wir*) are "not we" (*wir nicht*) is expressed with the help of the geometrical image of a hyperbola:

> What extends into the everyday human life in "infinite passion" (Kierkegaard), nonintuitively, intuitively only as a vacuum [*unanschaulich, nur als Vakuum anschaulich*], what in terms of all human comprehension can only be negated always and everywhere and yet for this very reason is attested to always and everywhere, what can only appear, as seen by us, as the zero point between two branches of a hyperbola that end in infinity [*Nullpunkt zwischen zwei im Unendlichen auslaufenden Hyperbelarmen*], and as such is in an unprecedented manner end and beginning—this is the new human, the subject of the predicate "faith." (R 136)

The persistent motif of negation, expressed above in the image of "hollow spaces" (*Hohlräume*), is likewise at the center of this image. Because the new human being, the subject of the predicate "faith," is for Barth fundamentally "nonintuitive" (*unanschaulich*), all attempts to visualize this new human being must take place in the mode of negation. The new human being is intuitable only as a vacuum, as that which we are not. Tellingly, in describing the image of the two arms of a hyperbola that extend into infinity, Barth focuses not on the liminal approach of the arms to the asymptotes but on the "zero point" (*Nullpunkt*) that lies between them at their center. From the perspective of human understanding, this zero point is the end and

---

32. On Barth's use of mathematical images in *The Epistle to the Romans*, see Hans-Joachim Vollrath, "Mathematische Bilder in Karl Barths Römerbrief," *Mathematica Didactica* 11 (1988): 3–10; and Smith, "The Infinitesimal as Theological Principle," 583–87.

negation of the human being, the point that is farthest away from the hyperbola arms. Yet from the perspective of the new human being, the zero point is a beginning that marks the intersection of the two asymptotes, hence the point of convergence of the infinite limit. With this image, Barth seeks to make visible the negation of intuitive visibility.

Similarly, in his commentary on Romans 8:5–9, Barth compares the suspension of humanity between the worlds of flesh and spirit to the two focal points of an ellipse, an image that reflects his view that we cannot choose between spirit and flesh. Barth sets up the opposition between these realms in accordance with a series of additional oppositions, between time and eternity, condemnation and election, death and life. Yet his ultimate aim is to collapse these mutually exclusive oppositions:

> In time it *is* decided that we all are in the flesh; in eternity it *is* decided that we all are in spirit. In the flesh we *are* abandoned; in spirit we *are* chosen. In the world of time, things, and human beings we *are* condemned; in the Kingdom of God we *are* justified. Both decisions, rejection and election, judgment and justice, death and life, as focal points of an ellipse [*Brennpunkte einer Ellipse*] that move ever closer together until they become one as the center of a circle [*Mittelpunkt eines Kreises*], thus the unity of both decisions, but (and this cannot be represented mathematically) not as the unity of an *equi*librium [Gleich*gewichts*], but as the unity of an infinite *pre*ponderance [Über*gewichts*], as the unity of an eternity that devours time, as the unity of the infinite victory of spirit over the flesh, as the unity of the path traveled from here to there, the absolute moment, the flash of insight, the flash of resurrection, the flash of God [*Blitz Gottes*]. (R 292)

In place of a choice between condemnation or election, judgment or justice, death or life, Barth imagines the unification of these opposites in the image of the focal points of an ellipse converging as the center of a circle. Yet he immediately clarifies that this unification of opposites implies not equilibrium or harmony but rather the predominance of spirit over flesh, of eternity over time. The geometrical image of the ellipse transforming into a circle provides a means of visualizing the both / and structure of these dual determinations. Nevertheless, mathematics is unable to represent the dialectical implications of this metaphor of reconciliation. This impossibility underscores Barth's tendency to use mathematics to stage the failure to represent the nonintuitive character of the divine.

Finally, in his commentary on Romans 8:14–17, Barth addresses the problem of theodicy with reference to the intersection or nonintersection of parallel lines, showing his familiarity with non-Euclidean geometries. Here Barth confronts the "unintuitiveness, impossibility, and paradox" (*Unanschaulichkeit, Unmöglichkeit und Paradoxie*) of Paul's statement that we are children of God. The impossibility of this prospect, for Barth, is bound up with the facts of "human suffering, human guilt, human fate"—the manifold ways in which humanity stands in horror before itself, cannot be justified, and suffers, and thus can never be crowned with the concept of God (R 308–9). And yet the suffering of humanity, its unanswered questions, and its powerless rebellion make it an *Other* to itself and hence capable of being accepted by God as a child. This acceptance of humanity by God is for Barth the fulfillment of the problem of theodicy. Barth describes this "human suffering, human guilt, human fate"—that is, those elements of human existence that separate humanity from God—not only in existential and historical terms but also in mathematical terms. They are revealed, he writes, "in the banal violence of our most primitive necessities of life and in the ideological unworldliness of our knowledge and conscience, in the terror of birth and death, in the enigma of nature that cries out of every stone and every tree bark and in the inconclusiveness of the cycles of world history, in the squaring of the circle and in the two parallel lines that never intersect in finitude [*in der Quadratur des Zirkels und in den beiden Parallelen, die sich im Endlichen nie schneiden*]" (R 308). These examples emphasize the finitude and limitations of human existence and the failure of human understanding to surpass these limitations. The classical geometrical problem of "squaring the circle" was proven to be impossible in 1882 by the German mathematician Ferdinand von Lindemann. Meanwhile, the image of parallel lines that never meet in finite space, but that (though Barth leaves this unsaid) might meet in infinity, points to the challenge of representing non-Euclidean geometrical forms in three-dimensional space.

Barth was familiar with the problem of parallel lines meeting in infinity from his reading of Dostoevsky, which he pursued together with Eduard Thurneysen.[33] In particular, Barth's image of parallel lines in his commentary on Romans 8:14–17 alludes to Ivan Karamazov's discussion of parallel lines that intersect in infinity in his rebellion against God in *Die Brüder Karamasoff* (*The Brothers Karamazov*) (1880).[34] In the Piper edition of *Die Brüder Karamasoff*, which appeared in German in 1914 as part of Dostoevsky's *Sämtliche Werke*, and which both Barth and Thurneysen read, Dostoevsky writes:

> If God exists and if He really did create the world, then, as we all know, He created it according to the geometry of Euclid and the human mind with the conception of only three dimensions in space. Yet there have been and still are geometricians and philosophers, and even some of the most distinguished, who doubt whether the whole universe, or to speak more widely the whole of being, was only created in Euclid's geometry; they even dare to dream that two parallel lines, which according to Euclid can never meet on earth, may meet somewhere in infinity [*daß zwei parallele Linien, die doch nach Euklid nie und nimmer und unter keiner Bedingung auf Erden zusammenlaufen können, vielleicht doch irgendwo in der Unendlichkeit zusammenlaufen*]. I have come to the conclusion that, since I can't understand even that, I can't expect to understand about God. I acknowledge humbly that I have no faculty for settling such questions, I have a Euclidian earthly mind [*einen euklidischen, einen irdischen Verstand*], and how could I solve problems that are not of this world?[35]

Although the reference is only implicit, there is no doubt that Barth had this passage in mind when he invoked the image of parallel lines, for just a few lines later in his commentary, he refers to Ivan

---

33. As Katya Tolstaya notes, "Barth and Thurneysen read Dostoevsky in the *Sämtliche Werke*, but occasionally quote from the 25-volume *Insel-Ausgabe*, edited by Stefan Zweig, as well. Both felt attracted to the world of the novels and recognized kindred themes in them" ("Literary Mystification: Hermeneutical Questions of the Early Dialectical Theology," *Neue Zeitschrift für Systematische Theologie und Religionsphilosophie* 54, no. 3 [2012]: 312–31, at 318).

34. See Tolstaya, "Literary Mystification," 317n13.

35. Fyodor Dostoevsky, *The Brothers Karamazov*, trans. Constance Garnett (New York: Macmillan, 1922), 247; Fjodor Michailowitsch Dostojewski, *Die Brüder Karamasoff*, trans. E. K. Rahsin (Munich: Piper, 1914), 467.

Karamazov as a figure who has perceived the existence of human suffering and guilt without any presupposition of providence or harmony.[36] Indeed, both Barth and Dostoevsky explore the limitations of Euclidean geometry as a means of reflecting on the relationship between humanity and God in general and on the problem of theodicy in particular. At stake in Karamazov's comments on non-Euclidean geometry is the capacity of human beings to fathom God. In keeping with the Kantian definition of space as a form of human intuition, Karamazov claims that human understanding is capable of grasping only the three dimensions of space. But he also maps the distinction between Euclidean and non-Euclidean geometries onto the gulf that separates humanity from God: Karamazov's inability to imagine the convergence of parallel lines in infinity corresponds to his inability to comprehend God. As a result, his Euclidean understanding of three-dimensional space is an "earthly understanding" (*irdische[r] Verstand*). Likewise, Barth mobilizes the image of parallel lines that never meet in finite space in order to emphasize the radical distance of human suffering, guilt, and fate from the concept of God: in the failure of the parallel lines to meet, as in the impossibility of squaring the circle, the inability of human understanding to surpass its limitations is manifest.

Although Barth follows in Dostoevsky's footsteps by using the problem of non-Euclidean geometry as a metaphor for the gulf between humanity and God, he ultimately reaches a very different conclusion. Whereas Karamazov refuses to accept the justification of human suffering in a final moment of eternal harmony, thus rejecting the possibility of a theodicy, Barth takes the reality of human suffering, the absence of harmony, and the protest and rebellion against suffering as the starting point for the acceptance of human beings as the children of God. Barth's theodicy, in other words, does not justify

---

36. See *R* 309. Hong Liang, by contrast, argues that Barth's image of parallel lines cannot be considered a reference to Dostoevsky because this image could refer to "den von Barth postulierten unendlichen Gegensatz zwischen dem Göttlichen und dem Menschlichen" (*Leben vor den letzten Dingen: Die Dostojewski-Rezeption im frühen Werk von Karl Barth und Eduard Thurneysen (1915–1923)* [Neukirchen-Vluyn: Neukirchener Theologie, 2016], 127n239). Yet Barth's explicit reference to Ivan Karamazov in his commentary makes this reading difficult to support.

human suffering but rather discovers in the self-alienation of suffering the possibility of a new humanity as something other. In announcing his refusal to accept the justification of human suffering, Karamazov returns to the idea of parallel lines that meet in infinity: "Even if parallel lines do meet [*Mögen sich sogar die Parallellinien treffen*] and I see it myself, I shall see it and say that they've met, but still I won't accept it."[37] Even the visual evidence of parallel lines converging, taken here as a metaphor for the intuition of that which exceeds the faculties of human understanding, would not be enough to shake Karamazov's determination to rebel against God. Barth, by contrast, inverts Karamazov's position and maintains that the failure of parallel lines to converge, as an index of the absence of God in the world and a figure of the impossibility of justifying human suffering, is precisely the form in which God accepts the human being as his child. To be sure, both Dostoevsky and Barth had an imperfect understanding of non-Euclidean geometry, which, as Helmholtz noted, can be demonstrated on the surface of a sphere on which parallel lines meet in finite space.[38] Dostoevsky likely derived his idea of parallel lines meeting in infinity from the Russian mathematician Nikolai Lobachevsky, who attempted to demonstrate non-Euclidean space by analyzing the parallax of stars.[39] Nevertheless, the non-Euclidean formulation of the parallel problem provides Barth with a spatial representation of the gulf between humanity and God and between time and eternity, and simultaneously represents the liminal possibility of their convergence.

Indeed, Barth turns to mathematics first and foremost as a means of working through paradoxes, incommensurabilities, apparent contradictions, and nonintuitive possibilities. Consider, for example, his gloss on grace and sin as "incommensurable magnitudes" (*inkommensurable Größen*): "They are, mathematically speaking,

---

37. Dostoevsky, *The Brothers Karamazov*, 248; Dostojewski, *Die Brüder Karamasoff*, 469.

38. See Helmholtz, "Bedeutung der geometrischen Axiome," 18. Helmholtz's popularization of non-Euclidean geometry is discussed in chapter 1.

39. On Lobachevsky's influence on Dostoevsky, see Tolstaya, "Literary Mystification," 317n13. On Lobachevsky's contributions to the development of non-Euclidean geometry, see Gray, *János Bolyai*, 77–78.

not only points on different planes but points in different spaces [*Punkte in verschiedenen Räumen*], of which the second excludes the first. Even the question of the 'relation' of the two, of the possibility of getting from here to there, is excluded" (R 186). An analogue to these "points in different spaces" can be found in Barth's discussion of the eschatological now in Romans 3:21–26 and Romans 8:18–25. In these passages, a nonintuitive image of space provides a framework for thinking about the incommensurability of eschatological time with the time of history. In the third chapter of *The Epistle to the Romans*, drawing attention to the significance of Paul's "But now" (*Jetzt aber*), Barth compares the "nontemporal time" (*unzeitliche Zeit*) of eschatological crisis to a "nonspatial place" (*unräumlichen Ort*):

> "But now." We stand before a comprehensive and irresistible suspension [*Aufhebung*] of the world of time, things, and human beings, before a penetrating and ultimate crisis [*aufs Letzte gehende Krisis*], before a rolling up of all being by his sovereign nonbeing.... The "But now" designates the nontemporal time, the nonspatial place, the impossible possibility, the light of uncreated light upon which are founded the tidings of the turning point, of the Kingdom of God come near, of affirmation in negation, of redemption in the world, of absolution in condemnation, of eternity in time, of life in death. "I saw a new heaven and a new earth; for the first heaven and the first earth passed away." (R 72–73)

The other temporality of the now (*Jetzt*) presupposes a crisis driven by eschatological thought, a "crisis of the last things" (*aufs Letzte gehende Krisis*). The Pauline "But now" (*Jetzt aber*) is rhetorically powerful because it suggests that the rupture in time, in which eternity breaks in for a moment, comes suddenly and can happen at any moment. And yet this now is by no means commensurable with any present in time. It is instead a time without time, a turning point in which the extremes of eternity and time, life and death, are seen to be constitutive of one another, not as oppositions but as non-mutually-exclusive possibilities. And just as the now is a nontemporal time, it is also a nonspatial space, or, as we observed in Barth's interest in non-Euclidean geometry, a space that exceeds intuitive representation.

While avoiding an identification of the eschatological now with the present historical moment, Barth does explore, in his reading of Romans 8:18–25, the time in which the now can become manifest.[40] His translation of Paul, which anticipates his interpretation of the time of the now, runs as follows: "For I reckon that the sufferings of the time of the now [*Zeit des Jetzt*] will carry no weight compared to the glory that will be revealed to us. . . . For we know that everything created sighs in unison and labors together up to the now [*bis auf das Jetzt hin*]" (R 311). Barth's interpretation of the passage hinges upon his reading of Paul's expression *tou nun kairou*, which he correctly translates as "the time of the now" (*die Zeit des Jetzt*) as opposed to the more familiar but grammatically implausible "the present time" (*die jetzige Zeit*).[41] What at first glance appears to be a slight change in emphasis turns out to be significant for understanding the relation of historical time to the moment of the now.

Drawing an explicit connection between the time of the now and the "But now" considered above in the context of Romans 3:21, Barth suggests that this moment of rupture is paradoxically a moment of nonpresence:

> We thus see the *time* in which we live characterized as "the time of the now" [*die Zeit des* Jetzt]. . . . For truth is the now (3:21), the moment outside of all time in which the human being stands naked before God, the point from we come that is not a point beside other points, Jesus Christ the crucified and resurrected. What exists before and after this moment of all moments [*Augenblick aller Augenblicke*], what surrounds as

---

40. Barth's reading of the "now" runs counter to the common understanding that Christianity, in contrast to Judaism, already lives in a fulfilled present. For example, Gustav Stählin suggests: "Dennoch gilt für das AT ganz allgemein: Jedes Jetzt schaut wieder voraus auf ein neues Einst; das *nun* des AT steht schließlich doch immer *vor* der ersehnten Zeitenwende, das *nun* des NT fällt mit ihr zusammen" ("Nun," in *Theologisches Wörterbuch zum Neuen Testament*, vol. 4, ed. Gerhard Kittel [Stuttgart: Kohlhammer, 1942], 1099–1117, at 1107). By contrast, Barth argues that the time of the now can have but a liminal relation to the time of history.

41. Were *nun* to have an adjectival function, it would be declined. The novelty of Barth's translation is apparent in contrast to Martin Luther's translation: "Denn ich halte es dafur / das dieser Zeit leiden der Herrligkeit nicht werd sey / die an vns offenbaret werden" (*Die gantze heilige Schrift*, ed. Hans Volz, vol. 2 [Bonn: Lempertz, 2004], 2282).

a surface this point that itself has no extension—this is time. Time arises in relation to this now, to eternity, as its negation, as an always already former past and a future always still to come. We call it the "time of the now" according to that which it conceals, points to, is measured by, and without which it would not be. (R 313–14)

In place of a temporal continuum of past, present, and future in the historicist mold, Barth defines the now as a moment without extension in time that stands in relation to time as its negation. He uses spatial forms and geometrical figures to express this moment of nonpresence: the now is a "point" (*Punkt*) that is surrounded by the "surface" (*Fläche*) of time yet has no "extension" (*Ausdehnung*) on this surface.[42] The eschatological now, in Barth's understanding, does not exist on the same plane as time, for it is not a point beside other points in time. Barth further develops these spatial relations with the image of the now as a "submarine island" (*submarine Insel*) that has been flooded by time yet remains completely intact "under its visible surface" (*unter dessen anschaulicher Oberfläche*) (R 313).

In theorizing a moment that both constitutes and negates time, and accordingly a time that refers to the eschatological now and conceals it, Barth reconfigures the temporal proximity of an imminent return of the messiah in terms of spatial proximity. The eschatological now, for Barth, is not futural but present in the mode of nonpresence: "In the shadow of the day of Jesus Christ that has *not* dawned but is infinitely *near*, we see our day of life unfold; in the shadow of the now [*im Schatten des Jetzt*], we see time unroll" (R 314). In figuring the passing of time in the shadow of the now, Barth discovers a spatial image for the proximity of an end whose temporal deferral makes any sense of nearness paradoxical. The now is the limit that stands not at the extremes of past and future but always in the closest proximity.[43]

---

42. Barth's topology of the point is reminiscent of Spengler's characterization of the status of the *Punkt* in modern mathematics: "*Wir* kennen im Grunde nur das abstrakte Raumelement des Punktes, das ohne Anschaulichkeit, ohne die Möglichkeit einer Messung und Benennung, lediglich ein Beziehungszentrum darstellt" (*Der Untergang des Abendlandes*, 112).

43. Michael Beintker provides an eloquent summary of Barth's claim that each moment stands in potential relation to the eschatological now: "Thus, Barth

Barth's concept of a time of the now as a moment cut loose from the historical continuum embodies an antihistoricist gesture that borrows resources from what Frank called "spatial form," a form that dispenses with narrative trajectory. Barth's representation of the paradoxical confrontation of historical and messianic time, like Frank's vision of modern literature, halts "the time flow of narrative" and presents the reader with a moment of time rather than with a sequence. Barth uses the language of spatial forms to represent the interruption and rupture of historical—or narrative—possibilities. The figures of liminal points, lines, and thresholds allow him to focus on a moment of time rather than on historical trajectory. Barth's scrutiny of the moment, however, seizes on the alienation of the moment from itself—and this too is an index of his embeddedness in modernist culture.

Modernism's sense of being out of the moment reflects an acute destabilization of the modes of historical representation. The aesthetic manifestations of this situation—from Woolf to Joyce to Kafka and Musil—are familiar to us: the thread of narrative continuity is interrupted and undercut by series of images whose temporal location is not easy to define. Yet the philosophical and theological indexes of these problems are equally noteworthy. Barth's Dialectical Theology points to an underlying sense of absence that makes historical and narrative continuity impossible to sustain. "There *are* high points [*Höhepunkte*] in history," Barth ventures: "They are to be found where history points beyond itself, where *in* history an alienation [*ein Befremden*] and horror [*ein Entsetzen*] *about* history takes place" (R 71). In this way, the concept of a moment cut loose from the historical continuum was crucially conditioned by a substrate of theological thought.

---

developed in *Romans* II distinct features of an 'eschatology of the *hic et nunc*,' in which all moments of our time and history can be thought of as being in the *same* nearness to the eschaton" (*Die Dialektik in der "dialektischen Theologie" Karl Barths: Studien zur Entwicklung der Barthschen Theologie und zur Vorgeschichte der "Kirchlichen Dogmatik"* [Munich: Kaiser, 1987], 54). For more on Barth's eschatology of the "here and now," see Walter Kreck, *Die Zukunft des Gekommenen: Grundprobleme der Eschatologie* (Munich: Kaiser, 1961), 40–50. Beintker and Kreck grasp the potential of time to become a parable of the eternal moment in Barth, but they do not recognize Barth's contribution to a modernist concept of history.

Contemporaneity and the Critique of Historicism

Barth's theorization of a now as a "moment of all moments" (*Augenblick aller Augenblicke*) does not take place in an abstract, ahistorical mode. On the contrary, his work shows an uncommon attention to its historical moment and cannot be understood without considering the historical context of the 1920s in Europe.[44] Far from being a mere symptom of a cultural crisis in the aftermath of the First World War, Barth's work played the critical role of producing a crisis of culture from the point of view of theology. The dialectical theologian, in his dual role as pastor and theologian, can thus be understood as an intellectual whose task was to generate a certain urgency about his time. In a number of Barth's texts from the early 1920s, we can observe an exemplary self-diagnosis of the "situation" (*Lage*) of humanity in the historical present and a critical reflection on how this moment might be grasped in relation to the charged moment of the eschatological now.[45] As Jörg Kreienbrock has noted, the concept of *Lage* or *Situation* had widespread currency among German intellectuals in the 1920s and 1930s.[46] While the semantics of *Lage* refer to concrete existence in space, the term is frequently used in discourses on temporality and in philosophies of history—for example, in Karl Jaspers's *Die geistige Situation der Zeit*

---

44. Similarly, John C. McDowell suggests that in Barth's work "eschatological discourse specifically serves as an interrogation of life in the present" (*Hope in Barth's Eschatology: Interrogations and Transformations beyond Tragedy* [Aldershot: Ashgate, 2000], 8). McDowell is right to note the importance of the present moment for Barth, but it is important to understand how his "historiography of the present" depends upon a critique of historicism.

45. See Friedrich-Wilhelm Marquardt, *Theologie und Sozialismus: Das Beispiel Karl Barths* (Munich: Kaiser, 1972), 1–25. Marquardt was one of the first Barth scholars to address the impact of the social, political, and historical context in which Barth moved for his theology. As McCormack notes, "The great merit of Marquardt's book . . . was its insistence that Barth's theology was always *zeitgemäß*; that is, it was always directed to a particular situation and really had no intention of being 'timeless'" (*Karl Barth's Critically Realistic Dialectical Theology*, 26–27).

46. Jörg Kreienbrock, "Erkenne die Lage! Medien des Konkreten im Nachkrieg," presentation at the Institut für Medienwissenschaft, Ruhr-Universität Bochum, April 19, 2016, https://ifmlog.blogs.ruhr-uni-bochum.de/forschung/mediendenken/kreienbrock-erkenne-die-lage/.

(*The Spiritual Situation of the Age*) (1931), in Erich Rothacker's *Geschichtsphilosophie* (*Philosophy of History*) (1934), in Martin Heidegger's *Sein und Zeit* (*Being and Time*) (1927), and in Gottfried Benn's letters (1936).⁴⁷ As Karl Heussi writes in *Die Krisis des Historismus* (*The Crisis of Historicism*) (1932): "There is no such thing as historical knowledge or a critique of historical knowledge as such, so to speak in a vacuum [*im luftleeren Raum*]; rather, all historical knowledge and the critique thereof is necessarily and indissolubly bound to a historical situation [*an eine geschichtliche Situation gebunden*]."⁴⁸ Barth's reflections on the concrete situation of his historical moment are thus part of a larger discourse of thinking about history and time in modernity in spatial terms.

In *The Epistle to the Romans*, Barth's interrogation of the situation (*Lage*) of his moment is pitched as a kind of *Positionsbestimmung* (determination of the location) of humanity in its relation to God. In stark terms, Barth underscores the alienation from God that follows from humanity's assertion of its autonomy and sovereignty: "The human being is his own master. His unity with God has been torn [*zerrissen*] in a manner that makes its restoration entirely inconceivable to us. His creatureliness is his fetter. His sin is his guilt. His death is his fate. His world is a formlessly surging and undulating chaos of natural, psychological, and various other forces. His life is an illusion. That is our situation" (R 13–14). Barth's laconic statement "that is our situation" (*Das ist unsre Lage*) sums up a picture of humanity in all its fallenness, perdition, and creatureliness. This account, however, is by no means a description of the secularity of a modern culture that has turned its back on religion. Rather, for Barth "our situation" (*unsre Lage*) is fundamentally defined by "the general

---

47. See Karl Jaspers, *Die geistige Situation der Zeit* (Berlin: De Gruyter, 1931); Erich Rothacker, *Geschichtsphilosophie* (Berlin: Oldenbourg, 1934), esp. 44–51; Martin Heidegger, *Sein und Zeit*, 11th ed. (Tübingen: Niemeyer, 1967), 299–300; and Gottfried Benn, "Brief an Friedrich Wilhelm Oelze, 2. Oktober 1936," in *Die Zeit*, July 1, 1977: https://www.zeit.de/1977/27/erkenne-die-lage-rechne-mit-deinen-defekten. On Rothacker's philosophy of history, see Michael Grossheim, "Erkennen oder Entscheiden: Der Begriff der 'Situation' zwischen theoretischer und praktischer Philosophie," in *Internationales Jahrbuch für Hermeneutik*, ed. Günter Figal, vol. 1 (Tübingen: Mohr Siebeck, 2002), 279–300, at 291–92.

48. Heussi, *Krisis des Historismus*, iv.

situation between God and human being" (*die allgemeine Lage zwischen Gott und Mensch*), a situation that consists in the fact "that we cannot grasp, nor hunt down God, that he is and remains for us the absolutely other, the foreign, the unknown, the unapproachable [*der schlechthin Andere, Fremde, Unbekannte, Unnahbare*]" (R 383).

The situation of humanity, according to Barth, is thus defined by its precarity and offers no stable position or perspective. It is a *Lage* that arises out of an affliction, as we can see in Barth's rhetoric of "affliction" (*Not*), "sighing" (*Seufzen*), "questioning" (*Fragen*), "searching" (*Suchen*), and "crying out" (*Schreien*) (R 401). In his essay "Not und Verheißung der christlichen Verkündigung" ("The Affliction and Promise of the Christian Proclamation") (1923), which opens the first volume of the journal *Zwischen den Zeiten*, Barth notes that the germ cell of his commentary on Paul's Epistle to the Romans can be found in the "familiar situation [*bekannte Situation*] of the pastor at his desk on Saturday," who is charged with the task of speaking both to the "tremendous contradictions of life" and to the "incredible message of the Bible."[49] Indeed, he suggests that theology in general is "the expression of this aporetic situation [*dieser ausweglosen Lage*] and question of the pastor, the truest possible description of the difficulties the human being gets into when he dares to take on this task, a call out of great affliction and great hope for redemption."[50] In his commentary on Paul's Epistle to the Romans, Barth writes, he did not find a way out of this "aporetic situation" (*ausweglosen Lage*) or "critical situation" (*kritischen Situation*); rather, the aporetic or critical situation itself became the basis of his exposition of theology.[51] But this aporetic situation, he clarifies, is foundational only as a point on which one cannot stand: "what I can call at best 'my theology,' if I look closely, consists ultimately in a single point and this is not ... a *stand*point, but rather a *mathematical* point on which one cannot stand [*ein mathematischer Punkt, auf dem man also nicht stehen kann*], merely a point of *view*."[52] In contrast to the spatial representa-

---

49. Karl Barth, "Not und Verheißung der christlichen Verkündigung," *Zwischen den Zeiten* 1 (1923): 3–25, at 5.
50. Barth, "Not und Verheißung," 5.
51. Barth, "Not und Verheißung," 5.
52. Barth, "Not und Verheißung," 3.

tion of perspective in Mannheim's and Heussi's historiographies, which emphasized the *Standortsgebundenheit* of the historian, Barth claims to occupy not a "standpoint" (*Standpunkt*) but a mere "point of view" (*Gesichtspunkt*) that has no solid foundation. As in *The Epistle to the Romans*, Barth's reduction of the point to a mathematical point underscores its liminal qualities: the point occupied by theology is a point without extension, one that cannot claim to fathom God in merely human terms.

Barth's effort to isolate the situation (*Lage*) of humanity, in keeping with the wider application of the term to discourses on temporality, has a historical index: for Dialectical Theology, humanity stands neither at the end of time nor on the brink of an impending end, but rather "between the times" (*zwischen den Zeiten*).[53] For Barth, this phrase implies not a hiatus between epochs but rather a liminal state between historical consciousness and the otherness of the time of the now. Humanity is between the times because it has been graced with the revelation of another now and yet is still burdened by its entanglement in temporality. As we saw above, Barth reads the figure of Christ as a point of intersection between time and eternity, one that can be located at a specific moment in history yet inheres in each moment in time. For Barth, this incision in the fabric of the historical continuum was not merely a historical anomaly but is also a possibility that arises in the present, and specifically in his own present. In claiming that every present is a potential time of revelation, Barth seeks to generate urgency by producing a crisis of historical consciousness that is not a crisis *in* history (for the events of the recent past provided sufficient forms of such crisis) but rather a crisis *of* temporal and historical understanding.

The point of departure for Dialectical Theology is therefore a strong sense of the present, a *Gegenwart* so rigorously problematized that it reveals a presence that does not submit to temporal and historical reckoning, the presence of the now. Whereas for Schleiermacher this present takes the form of a "lived experience" (*Erlebnis*)

---

53. Hence the name of the journal that Barth founded together with Eduard Thurneysen and Friedrich Gogarten: *Zwischen den Zeiten*. The journal was published from 1923 to 1933 by the Christian Kaiser Verlag.

that is subject to an infinite intensification within time and history, for Barth the present is a juncture where a problem, a crisis, or an aporetic situation arises, and only from the "adversity" (*Not*) and "affliction" (*Bedrängnis*) of this situation can something like the revelation of a fulfilled now emerge. The fullness of the now inheres in the poverty and emptiness of the present, but one perceives this "yes" only within a "no," never on its own terms. Whenever Barth turns his gaze toward "the nonintuitive focal point" (*jenem unanschaulichen Blickpunkt*) of divine absolution and grace, the coordinates of this visual space of spectatorship can be found in "the contemporary situation [*die gegenwärtige Lage*] (in Rome in the first century and in all places at all times) in its full concreteness" (R 450). The spatial dimensions of our concrete situation, condensed in the semantics of the concepts of *Lage*, are for Barth inseparably tied to the *Aktualität* of our temporal location in the present moment; that is, to a historically specific "situation in this moment" (*Lage in diesem Augenblick*).[54]

In *The Epistle to the Romans*, Barth notes that this possibility that "our time," in its historical specificity, could be a moment of moments is a question "at all times" (*zu jeder Zeit*). To pose this question, he claims, is to initiate a "crisis of the situation, of the moment": "Whether our time is qualified time, now time [*Jetztzeit*] (8:18; 13:11), time full of *meaning* by which one can orient oneself, that is the *question* at all times. Therefore: *serve* time! *Into* the crisis of the situation, of the moment! [Hinein *in die Krisis der Lage, des Augenblicks!*]" (R 481). In this passage, Barth returns to his earlier discussion of the time of the now in Romans 8:18 and coins the term "now time" (*Jetztzeit*)—which Walter Benjamin picks up in his *Arcades Project* (1927–40) and in his essay "On the Concept of History" (1940)—as a figure for the moment of all moments. Here Barth is concerned specifically with the question of whether our time can be now time, a time "qualified by the moment of the great divine disruption" (*qualifiziert durch den Augenblick der großen göttlichen Störung*) (R 481). The possibility that our time could be now time marks a moment of

---

54. Karl Barth, "Das Problem der Ethik in der Gegenwart," *Zwischen den Zeiten* 2 (1923): 30–57, at 33.

rupture in the historical continuum, and Barth notes that to pose this question is "certainly not 'timely' [*zeitgemäß*]" (R 481). Barth's intervention in the philosophy of history, and his critique of historicism, is thus predicated on a shock to the historical continuum, which he describes as a moment of crisis. In naming this crisis "the crisis of the situation, of the moment" (*die Krisis der Lage, des Augenblicks*), Barth again emphasizes the entanglement of spatial and temporal concepts in his account of the liminal situation of the historical moment.

In Barth's concept of history, each moment stands potentially in relation to the time of the now. This potential for relation not only interrupts the continuum of history, but also establishes a contemporaneity of each present with the eschatological now. Barth defines this contemporaneity as follows: "We suspend all identity between the moment of the last trumpet and everything that comes before and after, and thereby announce the contemporaneity of all times [*Gleichzeitigkeit aller Zeiten*], of everything that comes before and after, because we are no longer able to discern a before and after that does not stand, in its complete otherness, in the light of this moment and participate in its dignity and meaning" (R 99). Here Barth imagines a shape of time in which the historical relations of before and after recede before the contemporaneity of all times before the last things. As much as eschatology circumscribes a limit to history, this view does not imply the annulment of what Barth, drawing on Kierkegaard, called "the 'infinite qualitative difference' between time and eternity" (*den 'unendlichen qualitativen Unterschied' von Zeit und Ewigkeit*) (R xx). Yet insofar as each present borders on the eschatological now, the movement of history is suspended and the moment enters into a relationship to eternity characterized by contemporaneity. In this respect, Barth gives Kierkegaard's concept of *Gleichzeitigkeit* a strong eschatological reading.[55] Whereas Ernst Bloch notes a "noncontemporaneity of the contemporaneous" (*Ungleichzeitigkeit der Gleichzeitigen*) in which various peoples or social classes can belong to differing historical stages of development despite their temporal contemporaneity,[56] Barth insists

---

55. Compare Kierkegaard, "Philosophische Brocken," 63–68.
56. Ernst Bloch, *Erbschaft dieser Zeit* (Zurich: Oprecht & Helbling, 1935).

on a contemporaneity of the noncontemporaneous (*Gleichzeitigkeit der Ungleichzeitigen*): the most historically charged moment becomes contemporaneous with the remotest of historical times in light of the eschatological now—that is, in view of the limit imposed on time by eternity.[57]

Barth's concept of a *Gleichzeitigkeit* that runs through history and connects disparate moments in history via their relation to the "unhistorical" is a key aspect of his critique of historicism in nineteenth-century theology, and it defines his opposition to research in the early twentieth century on the life of the historical Jesus.[58] Barth's response to the rigorous historicization of the biblical text in relation to a model of philological-historical criticism is grounded in his reading of Paul's Epistle to the Romans. By considering Abraham's justice not in terms of his inheritance of the law but purely in terms of his faith, Paul contested the theological significance of historical circumstances.[59] Similarly, Barth reads the biblical text not as history but in its contemporaneity with the present moment.[60] Drawing on Nietzsche, he argues that the past has a value only insofar as it speaks to the question of how we ought to live in the present. In this respect, Barth's concept of history can be compared to the presentist historiographies of Mannheim and Troeltsch.

---

57. Spengler develops a comparable concept of the simultaneity of historical distant figures, albeit via a homology of historical forms rather than from an eschatological standpoint: "Aus der Homologie historischer Erscheinungen folgt sogleich ein völlig neuer Begriff. Ich nenne '*gleichzeitig*' zwei geschichtliche Tatsachen, die, jede in ihrer Kultur, in genau derselben—relativen—Lage auftreten und also eine genau entsprechende Bedeutung haben. . . . *Gleichzeitig* vollzieht sich die Entstehung der Ionik und des Barock. Polygnot und Rembrandt, Polyklet und Bach sind *Zeitgenossen*. Gleichzeitig erscheinen in allen Kulturen die Reformation, der Puritanismus, vor allem die Wende zur Zivilisation" (*Der Untergang des Abendlandes*, 152).

58. For an account of efforts to reconstruct the historical life of Jesus, see Schweitzer, *Von Reimarus zu Wrede*. For a general introduction to Barth's critique of historicism, see Rudy Koshar, "Where Is Karl Barth in Modern European History?," *Modern Intellectual History* 5, no. 2 (2008): 333–62.

59. See Barth's reading of Romans 3:31 (*R* 98ff).

60. For de Vries, Barth's mode of reading is thus "historicist and antihistoricist at once" and defines a stance of "actualizing interpretation" ("Inverse versus Dialectical Theology," 475–76).

The contemporaneity of the noncontemporaneous provides the cornerstone of a theory of history defined by a discontinuous relation of historical figures. Barth understands contemporaneity as a communicative structure in which the past can speak to the present and the present can listen to the voice of the past. This structure implies a relationship of the past to the present that exceeds historical causality. Barth's theory of contemporaneity thus has a strong affinity with modernist accounts of "historical afterlives," "allegorical citation," and "figural interpretation," as developed by Aby Warburg, Walter Benjamin, and Erich Auerbach, respectively.[61] Warburg's, Benjamin's, and Auerbach's historiographies each sought to overcome a linear history of development by positing the potential of the historical moment to enter into a diverse set of discontinuous historical configurations.[62] Barth's theory of contemporaneity shares with these modernist historiographies an antihistoricist impulse insofar as it places the distant past and the present into a direct relationship with one another.

Yet Barth's antihistoricism is unique because it establishes the discontinuity of history in relation to an eschatological now that is unhistorical and nonintuitive. Whereas the theory of allegory entails the capacity of a historical moment to "speak otherwise"[63] in a different historical context, Barth's theory of contemporaneity involves a "monologue" (*Selbstgespräch*) that suspends time and history. The contemporaneous element in the past and the present, for

---

61. For an analysis and comparison of Warburg's, Benjamin's, and Auerbach's modernist historiographies, see McGillen, "Erich Auerbach and the Seriality of the Figure."

62. Warburg, Benjamin, and Auerbach each developed a historical optics through which the past enters into a figural or allegorical relation in the present. See *Kulturwissenschaftliche Bibliographie zum Nachleben der Antike*, eds. Bibliothek Warburg, Hans Meier, Richard Newald, and Edgar Wind (Leipzig: B. G. Teubner, 1934); Erich Auerbach, "Figura (1939)," in *Gesammelte Aufsätze zur romanischen Philologie*, ed. Fritz Schalk (Bern: Francke Verlag, 1967), 55–92; Benjamin, "Über den Begriff der Geschichte"; and Benjamin, *Das Passagen-Werk*.

63. Heinz Drügh notes that the alterity of the allegorical figure is implicit in the etymology of the term: the literal meaning of the Greek *allos agoreuein* is "anders als auf dem Marktplatz reden" (*Anders-Rede: Zur Struktur und historischen Systematik des Allegorischen* [Freiburg im Breisgau: Rombach, 2000], 8).

Barth, points to what exceeds human intuition and comprehension; namely, the end and beginning of history:

> History can have a use. The past can speak to the present. For in the past and in the present there is something contemporaneous [*ein Gleichzeitiges*] that can heal the muteness of the past and the deafness of the present, prompting the former to speak and the latter to listen. This contemporaneity, in its monologue that suspends and fulfills time, announces and perceives the unhistorical, the nonintuitive, and the incomprehensible that is the end and beginning of all history [*das Unhistorische, Unanschauliche, Unbegreifliche, das aller Geschichte Ende und Anfang ist*]. (R 132)

The alterity and discontinuity of past and present, for Barth, are a function of the otherness of each moment in time in relation to the contemporaneity of the eschatological now. Barth participates in a historiography of the present, yet his concept of contemporaneity also calls into question the self-identity of the present. Recognizing that the past has a claim on us, that Abraham's situation is not only his situation but also our situation—a matter for our time in its historical specificity—Barth uses the figure of contemporaneity to show that the present moment is contemporaneous not with itself but rather with a discontinuous, nonintuitive, and noncontemporaneous moment at the limit of history.

The concept of contemporaneity thus plays a key role in Barth's critique of historicism from the point of view of eschatology. Eschatology is the threshold at which the singular historical moment is related, precisely as a particular historical moment, to a moment of eternity in which history is released from its particularity and becomes contemporaneous. Whereas Hegel envisions the becoming history of the absolute—that is, the realization of spirit in history—Barth conceptualizes a becoming absolute of history in which history is given a grounding and yet simultaneously released from temporality.[64] In this sense Barth's historiography is concerned with uncovering in history those traces of the unhistorical that enact a

---

64. In his *Lectures on the Philosophy of History*, Hegel argues that world history is the "course of development and the real becoming of spirit" (*Philosophie der Geschichte*, 540). As Löwith notes, "in Hegel, the historical process is understood on the pattern of the realization of the Kingdom of God" (*Meaning in History*, 54).

crisis of historical consciousness. The contemporaneity of history in light of the unhistorical is at once the point at which history is driven to its breaking point and the moment in which history serves as a witness of what exceeds the historical specificity of the moment: "Is it not apparent," Barth writes, "that the frame of history is burst [*der Rahmen der Geschichte gesprengt wird*] in the moment in which history discloses its secret? We have no reason to shun the light of history; it can do nothing but *bear witness*" (R 126). In the moment in which history reveals the traces of primal history, Barth ventures, the present is no longer constituted by the historical continuum but exceeds the frame of history. The recognition of the unhistorical in history therefore has an explosive power; it displaces the historical frame and lets history reveal and bear witness to that which lies at its limit.

In Barth's understanding of history, which can be compared to the "bursting of historical continuity" (*Aufsprengung der historischen Kontinuität*) in Benjamin's theory of historiography, temporality is not structured by a principle of continuity and causality. Instead, past and present are brought together in a relation to that which bursts (*sprengen*) the frame of history.[65] Barth challenges, not the perspective from which history is written, as did Mannheim, Heussi, Troeltsch, and Spengler,[66] but rather the shape and contours of history itself, as defined by its liminal relation to the eschatological now. History, for Barth, does not proceed horizontally from the past through moments of fulfillment culminating in a fulfilled future at the end of time. Yet neither is the relation of noncontemporaneous moments to the contemporaneity of the unhistorical simply a vertical relationship without a historical index. Instead, Barth argues for

---

65. See Benjamin, *Das Passagen-Werk*, 594.
66. For a detailed discussion of the historiographies of Mannheim, Heussi, Troeltsch, and Spengler, see chapter 1. The broader point made here is that Barth's questioning of the limit of history goes beyond his critique of liberalism and the dominance of the "history of religion" in theology. Critics like Rudy Koshar are right to note Barth's rejection of the liberalism of his teacher Adolf Harnack as an important element of his antihistoricism ("Where Is Karl Barth in Modern European History?" 343), but Barth's challenge to historicism goes deeper and depends upon his liminal concept of eschatology.

a tension between the particularity of the concrete historical moment and the way its determination by the eschatological limit explodes the frame of history.[67] As an alternative to the poles of horizontal development and vertical abstraction, Barth suggests a diagonal cut of unhistorical presence through historical reality: "the resurrection takes place *diagonally through* [quer hindurch] the life and death of human beings; it is the history of salvation that goes its own way through the other history [*die Heilsgeschichte, die ihren eigenen Weg geht durch die andere Geschichte*]."[68] For Barth there is no realization of a history of salvation in human history, contrary to the Hegelian supposition that world history marks the processual unfolding of the absolute.[69] Rather, Barth gives shape to history at the liminal point of contact between our concrete situation and the nonintuitive space of eschatology.

Representing the Unrepresentable: Constructing the Eschatological Moment

Barth uses the paradoxes of spatial form, finally, in new constructions of history and the eschatological moment. As outlined in chapter 1, the early twentieth century saw the emergence of constructivism in historiography by figures such as Mannheim, Heussi, Troeltsch, and Spengler. Like these writers, Barth was a strong critic of historicism and formulated a new antihistoricist approach to historiography. Yet whereas Mannheim, Heussi, Troeltsch, and Spengler produced new

---

67. The problem with which Barth wrestles is how a God that is radically other can be revealed and known in history. According to McCormack, Barth seeks "to speak of a presence of God (revelation, the Kingdom of God, the new humanity, etc.) *in* history in such a way as to make it clear that these realities are not *of* history" (*Karl Barth's Critically Realistic Dialectical Theology*, 209). The eschatological limit, in other words, reinvents the understanding of history even as it is utterly incompatible with history.

68. Barth, *Die Auferstehung der Toten*, 122, emphasis added.

69. Compare de Vries, who, citing Agamben's remark that "Hegel, however, thinks the pleroma not as each instant's relation to the Messiah, but as the final result of the global process," notes that "both Adorno and Barth, then, are resolute anti-Hegelians in this respect" ("Inverse versus Dialectical Theology," 509).

constructions of history, Barth takes as the object of his construction an eschatological moment that stands at the limit of history, a moment defined by its nonintuitive and nonrepresentable character. In his paradoxical effort to represent the unrepresentable, Barth gives shape to the eschatological moment in constructivist terms, grappling with the nonintuitive character of eschatology in a manner comparable to the representation of nonintuitive spaces in mathematics.

In keeping with this theological form of constructivism, Barth's contemporary and fellow theologian Paul Tillich argued that Troeltsch's project of a construction or *Aufbau* of history was unable to overcome historicism because it sought to do so with the materials of history itself: in Troeltsch's terms, "To overcome history *through history*."[70] "There can be no doubt," Tillich claims, that "Troeltsch and the time for which he stood did not overcome historicism, did not name the place from which historicism is overcome."[71] Troeltsch took "our moment of life" (*unser Lebensmoment*) as the standpoint for his construction of history, seeking a perspective on history that would be relevant to contemporary culture. By contrast, in Tillich's perspective, which has an important kinship to Barth's, the "place" from which history can be overcome is a standpoint that cannot be occupied, a standpoint that shakes and unsettles all historical constructions:

> For if a place is to be found that lies above the highest place on which a herald of the present can stand, this may not be a place on which it would be possible to stand [*so darf dieses kein Ort sein, auf dem es möglich wäre zu stehen*]. As such it would be only a standpoint that could be opposed to another standpoint. That would not accomplish anything. It can only be such a standpoint that unsettles and suspends every standpoint and calls it to account. But that would be, metaphorically speaking, a point, a direction, whence and whither, a summit above every possible summit, thus the absolute, incomparable summit. Only the completely inaccessible, the incomparable, the unconditional [*das schlechthin Unzugängliche, Unvergleichliche, Unbedingte*] frees us from historicism.[72]

---

70. Troeltsch, "Der Historismus und seine Probleme," 772, emphasis added.
71. Paul Tillich, *Kairos: Zur Geisteslage und Geisteswendung* (Darmstadt: Reichl, 1926), 2.
72. Tillich, *Kairos*, 2–3. For a similar rejection of the possibility of a theological "standpoint," compare Barth, "Not und Verheißung," 3.

According to Tillich, Troeltsch's model of the present moment as the place from which history can be overcome is flawed because the present is merely one historical standpoint among others. The overcoming of history, Tillich suggests, requires an eschatological standpoint that defines the limits of historical construction. Yet this standpoint, paradoxically, is one that destabilizes and negates each and every standpoint. In Barth's terms, it is a point without extension, an inaccessible standpoint whose nonintuitive character both demands metaphorical exposition and defies comparison. Barth and Tillich thus critique Troeltsch's historiographical construct in favor of the construction of eschatological time, providing figurative and geometric representations of eternity as a vector that touches upon historical time without being identified with it.

Barth's and Tillich's constructions of eschatology take their cues from the new directions in historiography that emerged in the twentieth century. Heussi's claim that the historical counterpart does not have a fixed historical structure provides a background for Dialectical Theology's treatment of history in relation to that which is outside history. In contrast to Heussi, Barth sees the historical relation, not as a fruitful source of ever-new historical readings, but instead as a conduit for perceiving "the unhistorical, nonintuitive, and inconceivable that is the end and beginning of all history" (R 132). Nevertheless, the perspectivism of historiography in the early twentieth century, with its attention to the inexhaustible alterity of the present moment, is at the heart of Barth's treatment of the moment as at once historical and contemporaneous with an eschatological time of the now. The open structure of history therefore serves as a springboard to Barth's theology of history and its eschatological constructivism. The thrust of Barth's constructivism is to figure an event that cannot be subsumed in history and hence to broach an impossible historiography. In contrast to Troeltsch and Spengler, Barth and Tillich appropriate history for the construction of an eschatological moment that defines the limit of history.

In constructing the eschatological moment, Barth revels in the paradoxes of the nonintuitive and nonrepresentable, which evoke non-Euclidean geometry. The ahistorical end or limit of history, he claims, is also the origin of history. Rather than resolving tension in a moment of transcendent completion, Barth argues, the eschatological moment is "primal history" (*Urgeschichte*). The suspension of history and the shattering of the frame of history do not mark the terminal end of history but rather generate history. For Barth, *Urgeschichte* is an originary force within history, yet it ceaselessly recedes into the darkness of historical consciousness, lit up only in moments of contemporaneity. Drawing on Nietzsche, Barth understands *Urgeschichte* as the unhistorical condition of history that is apparent only at the critical threshold separating the knowable from the unknowable: "But we would like to discern, precisely on the critical 'line that separates the visible and bright from the unilluminable and dark' (Nietzsche), the unhistorical [*ungeschichtliche*], that is *primal historical* [ur-geschichtliche] conditionality of all history, the light of the *logos* of all history and all life" (*R* 127). This paradox guides Dialectical Theology's concept of eschatology as *Urgeschichte*: the unhistorical negates and suspends history in the moment of contemporaneity, yet it is also the ground, origin, and condition of history. *Urgeschichte*, according to Barth, is the light that gives history meaning, but it is a light that appears only as twilight, on the verge of nonrecognizability. It is a matter, not of crossing the critical line of *Urgeschichte*, but of perceiving the threshold and limit that it traces.

Barth's concept of *Urgeschichte* both deflects the pervasive historicization of thinking that was endemic to his time and affirms the historical significance of a nonhistorical origin, suggesting both continuities and discontinuities between historicism and the modernist concept of history. The liminal concept of *Urgeschichte*, in which history is determined by the tension between origin and end, is indebted to the work of Franz Overbeck, whose notebooks were published posthumously in 1919 by Carl Albrecht Bernoulli under the title *Christentum und Kultur* (*Christianity and Culture*).[73] Barth read Overbeck's notebooks during the fertile years between the publication of the first edition of *The Epistle to the Romans* (1919) and the completely revised second edition (1922). In 1920 Barth and

---

73. See Franz Overbeck, *Christentum und Kultur: Gedanken und Anmerkungen zur modernen Theologie von Franz Overbeck*, ed. Carl Albrecht Bernoulli (Basel: Benno Schwabe, 1919).

his colleague Eduard Thurneysen published a review of Overbeck's notebooks in the form of a series of reflections "on the inner situation of Christianity" (*zur inneren Lage des Christentums*).[74] In his review, Barth articulates the significance of Overbeck's notion of *Urgeschichte* from the point of view of eschatology, to which *Urgeschichte* stands in a relation of mutual tension.

The condition of history in *Urgeschichte*, Barth argues, becomes manifest at the eschatological limit: the limit of the origin coincides with the limit of the end, and what we call the world is maintained in a state of tension between these poles. Characteristic of Barth's interpretation is the interchangeability of beginning and end: in light of the eschatological moment, beginning is end and end is beginning:

> According to Overbeck, two points, each of which are both points of departure and end points [*Ausgangs- und Endpunkte*], determine and characterize the existence of human beings and of humanity. With the concept of *"primal history"* or genesis he denotes the first, with the concept of *death* the second. We come from a supratemporal, unfathomable, incomparable [*überzeitlichen, unerforschlichen, unvergleichlichen*] primal history that consists of nothing but beginnings, in which the boundaries that separate the individual from the whole are still fluid. We are heading toward the singular, unimaginably meaningful moment of death, in which our life enters into the same sphere of the unknown in which everything that lies beyond the world known to us already stands during our lifetime. We have perhaps gazed too deeply into the foundation of things, *we know too much* about all things, even the most veiled and impenetrable, about the things of which we actually cannot know anything, *about the last things.* "*We are beyond help with this knowledge and we have to live with it.*" What stands *between* these "last things," that is the *world*, our world, the comprehensible world that is given to us.[75]

Drawing on Overbeck, Barth understands the eschatological limit as a primal historical condition of history. The end posed to humanity by death has the same threshold character on the border of something incomparable and inscrutable as does humanity's origin in *Urgeschichte*. We know something about the last things, the *es-*

---

74. Karl Barth and Eduard Thurneysen, *Zur inneren Lage des Christentums: Eine Buchanzeige und eine Predigt* (Munich: Kaiser, 1920).
75. Barth and Thurneysen, *Zur inneren Lage des Christentums*, 7–9.

chaton, despite the fact that we cannot know them, because they are the ground—and the abyss—of all things in the world. If *Urgeschichte* is simultaneously *Endgeschichte*, then life and history are suspended in tension between creation and redemption. Thus, Barth argues that Overbeck's critical understanding of *Urgeschichte* and death contains a "profound recognition of the dialectic of creation and redemption."[76]

Barth's analysis of the imbrication of *Urgeschichte* and eschatology constructs a mathematical imaginary with which to frame the nonintuitive character of these concepts. The world is conceived as suspended between primal history and death, yet each of these points is at the same time a "point of departure" (*Ausgangspunkt*) and a "terminus" (*Endpunkt*). The reversal of beginning and end suggests that *Urgeschichte* and eschatology are not temporal but liminal concepts. In particular, they define an epistemological limit. What lies beyond the limit is "unfathomable" (*unerforschlich*), "incomparable" (*unvergleichlich*), and "unimaginable" (*unausdenkbar*): it defines the "sphere of the unknown" (*Sphäre des Unbekannten*). And yet, just as Helmholtz's geometry of the sphere provides a visualization of the nonintuitive character of the curvature of non-Euclidean spaces, Barth recognizes that we have some knowledge—perhaps too much knowledge—of the last things. The last things, for Barth, lie at the "foundation of things" (*in den Grund der Dinge*), such that the sphere of the unknown, unthinkable, and inscrutable constitutes our world. The given, intuitive world, Barth argues, lies between the last things, for these last things are both the end and the origin of history. This sense of in-betweenness, as in the concept of *Zwischen den Zeiten*, defines for Barth the suspension of our knowable world between the nonintuitive limits of birth and death, origin and end, *Urgeschichte* and *Endgeschichte*. For Barth, both of these liminal points, along with the tension between them, constitute the space of eschatology.

In the context of Barth's theology, the transgression of these limits becomes a possibility, if not a reality, through the figure of resurrection, which places humanity "beyond the realm of the law," as

---

76. Barth and Thurneysen, *Zur inneren Lage des Christentums*, 7–9.

Barth puts it in his translation of Romans 7:5–6, and in the space of grace.[77] But to stand in grace, Barth cautions, is not an "experience" (*Erlebnis*) that we can have but a moment of "knowledge" (*Erkenntnis*). The sudden reversal of death into life lies beyond the scope of our experience, but the overcoming of this opposition is something that we can imagine. The place at which, for Barth, we are beyond the realm of the law might be compared to Tillich's standpoint that "unsettles and suspends" every standpoint, yet it is at once a "quiet and unmoved place" in which we are released from the "net of human things":

> Still moved, shaken, and struck back and forth [*bewegt, geschüttelt und hin- und hergeschlagen*] by the peripeteias of religious life (with which we are truly somewhat familiar), we reach out for the calm, unmoved place on which the pulsating perpendicular *rests*. Still bewildered and caught up in the tangle of religious events, in which everything (everything!) is human, we already stand in *primal history* and *end history* [Urgeschichte *und* Endgeschichte], where all duality, all polarity, all both/and, all shimmering is brushed aside, because God is all in all; the temporality that we cannot escape thus confronts us as something whole, complete, bounded by the day of Jesus Christ; and we feel *released* from the net of the human, all too human, that precisely as the religiously human oppresses and strangles us most stiflingly. (R 239)

Whereas in religious experience we are thrown back and forth in sudden reversals of peripeteia, in grace we are supported by a "pulsating perpendicular" (*das schwingende Perpendikel*)—a geometrical figure for the vertical relation to God—which defines a calm, unmoved place in the midst of the turmoil of religious experience. This space represents the reconciliation of all polarity and duality, a space where the two liminal points discussed above have been transgressed and we stand already in *Urgeschichte* and *Endgeschichte*.

For Barth, this is a space that we can imagine and of which we can speak, but that we cannot occupy. Insofar as *Urgeschichte* and *Endgeschichte* define the limit of human knowledge and possibilities, we can at best stand at the threshold drawn by this limit:

---

77. See R 236–42.

We do not know what we say, and we say what we do not know when we say that the law is no longer the realm in which "we" stand, that the religious possibility is exhausted and has been left behind "us." But we say it anyway. We say it as the impossible, . . . as the arrow from the other shore on which *we* shall never set foot, yet which has struck *us*, as the truth beyond the limit [*jenseits der Grenze*] that *we* shall never transgress, but which has spoken to *us* from there. Woe to us if we did *not* say what *had* to be said where nothing remains intuitive besides—the nonintuitive [*wo nichts mehr anschaulich ist als—das Unanschauliche*]. We say it as those who are captive and yet free, blind and yet able to see, dying and yet behold, we live. *Not we* say it: Christ is the end of the law, the limit of religion. (R 239–40)

Barth negotiates the problem of a limit that cannot be transgressed as a problem of impossible representation. What lies beyond the limit is the impossible and the nonintuitive: that humanity stands no longer in the realm of the law but in grace. For Barth this is something we cannot know, an insight that we can only say as the impossible because we cannot cross this boundary. When "we" say that the realm of the law is behind us, we speak from the position of "not we," from the position of a new predication of humanity that is a negation of our predication through the flesh. As a result, Barth rejects the "unmoved place" of the "pulsating perpendicular"—with its purported resolution of our divided consciousness—in favor of an intensification of the contradictions of our dual natures: he pictures humanity as imprisoned and yet free, as blind and yet able to see, as dying and yet alive. On the boundary between we and not we, for Barth, the only thing that is intuitive is the nonintuitive.

The problem of the representability of what lies beyond the eschatological limit is a function of the incommensurability of eschatology with history. In *The Resurrection of the Dead*, his lectures on 1 Corinthians 15, Barth comments on the temporality of a *futurum resurrectionis*, describing it as a "crisis that tears open all times lengthwise" (*alle Zeiten der Länge nach aufreißenden Krisis*).[78] On the one hand, the moment of resurrection is described as taking place "*en atomoi* . . . , literally in the indivisible, thus not in a part of time, or else it could not take place simultaneously [*gleichzeitig*]

---

78. Barth, *Die Auferstehung der Toten*, 122.

for all generations, but in the *present* [Gegenwart]."[79] Yet in contrast to the indivisible contemporaneity of its presence, on the other hand, the moment of resurrection is said to interrupt the historical continuum; as Barth writes: "*en ripe ophthalmou*, in a blink of the eye [*in einem Augenblick*], hence the 'suddenness' of the breaking in of this crisis: it does not come in gradual or catastrophic developments."[80] Barth attributes a set of temporal relations to the eschatological moment of a *futurum resurrectionis* whose paradoxes extend to a breaking point the categories of historical thinking. The eschatological moment is to be found in the *present*, yet it is contemporaneous with each moment. This moment is present to all times, yet it appears in the "blink of an eye"—as Paul's Greek emphasizes—in a sudden moment of *kairos* rather than in "gradual or catastrophic developments."[81] The cut that the moment of resurrection makes in historical time is therefore negative: it does not permit an isomorphism of the historical moment with the moment of resurrection.

The cultural significance of eschatological thought for the understanding of history in the early twentieth century is therefore a function of a negative theology. At stake is neither a loss of transcendence nor a secular decoupling of history from salvation history, but rather an understanding of history as determined by an eschatological limit that is incommensurable with history. The reference of figures of history to eschatological fulfillment is unbroken, yet it is a reference to a fulfillment that exceeds objective representation. In Barth's brand of negative theology, eschatological fulfillment exceeds what can be known: the absolute crisis of the moment of resurrection cannot be represented because it involves a repredication of humanity as that which it is not, and this repredication leaves historical experience groundless. As Martin Heidegger puts it, considering Paul's Epistle to the Thessalonians, "by readily speaking of 'representation,' one fails to recognize that the eschatological is never primarily a representation" (*Indem man ohne weiteres von 'Vorstel-*

---

79. Barth, *Die Auferstehung der Toten*, 122.
80. Barth, *Die Auferstehung der Toten*, 122.
81. On the concept of *kairos*, see Tillich, *Kairos*, 8–11.

lung' redet, verkennt man, daß das Eschatologische niemals primär Vorstellung ist).[82] Insofar as the eschatological limit exceeds humanity's knowledge, it is also an epistemological limit. Eschatological thinking takes away the ground of representation because it exceeds temporal localization. History is neither dependent upon eternity understood as permanent presence nor cut loose from a transcendent origin and hence naturalized. Instead, it is grounded on the groundlessness of eschatological crisis. Eschatology therefore decenters the understanding of history in modernism and places it in relation to its negative limit.[83]

Barth's theology of history can thus be understood as a mode of nontemporal constructivism that is even more sharply antihistoricist than Troeltsch's "construction of European cultural history" (*Aufbau der europäischen Kulturgeschichte*). Barth makes use of nontemporal construction in formulating the *Gleichzeitigkeit* of historical moments, but the anchor of this construction is not a constellation of moments, but rather their extra-historical substrate. The eschatological moment requires an ultimate historical construction in which all vital forces of the past are concentrated, but the last things shatter all historical constructions and nullify all historical images by placing them before an extratemporal architecture. The last things are both the *Aufbau* of a nonhistorical vector of eternity, in its liminal intersection with historical time, and the dismantling (*Rückbau*) of all historical construction.

By moving away from constructions of history toward a figuration (*Gestaltung*) of eschatology, Barth's theological variant of constructivism draws upon and extends Spengler's philosophy of the

---

82. Martin Heidegger, *Phänomenologie des religiösen Lebens (1920/1921)*, in *Gesamtausgaube*, vol. 60 (Frankfurt am Main: Vittorio Klostermann, 1995), 110–11.

83. On the negativity of modernism's relation to history, see Ralf Frisch, *Theologie im Augenblick ihres Sturzes: Theodor W. Adorno und Karl Barth; zwei Gestalten einer kritischen Theorie der Moderne* (Vienna: Passagen, 1999), 134: "Gerade in dem für die *episteme* der Moderne des zwanzigsten Jahrhunderts charakteristischen Entsetzen vor der Geschichte der Natur und Kultur, im Entsetzen vor den destruktiven Möglichkeiten des Menschen liegt das Krisenzeitgemäße und Geistesgegenwärtige Barths."

future.[84] Their affinity lies not in a pessimistic account of the decline of history or in apocalyptic scenarios of the end of history, but rather in the rhetorical, metaphorical, and conceptual use of figures (*Gestalten*) in their thinking. But whereas for Spengler these *Gestalten* are figures of the world as history, Barth probes the figuration of an event that cannot be subsumed among historical phenomena. The eschatological possibility, which for Barth is exemplified in art, is that of a refiguration of reality in which the world of creation appears as a redeemed reality. Borrowing from Spengler's metaphorics of "figure and reality" (*Gestalt und Wirklichkeit*), Barth describes an eschatological figure that concerns both the historical present and a "new heaven and new earth." According to Barth, this eschatological figure verges on the figuration of the impossible:

> In art the human being dares ultimately not to take contemporary reality in its creaturely that-ness or even in its such-ness as a world of the Fall and reconciliation seriously, but rather to create a second reality, which as a present [*als Gegenwart*] is only possible in the most paradoxical fashion, without freeing oneself of the first. Artistic creation will of course always have a tendency toward the incredible and unprecedented, toward the figuration of the impossible [*zur Gestaltung des Unmöglichen*] and toward impossible figurations [*zu unmöglichen Gestaltungen*]. All artistic creation is in principle *futural*; but it also returns again and again to reality in order to reconfigure it, to see and show its transformation, a reality created by and reconciled with God, but this reality now as a *redeemed* reality, in its foreseen and anticipated completion, as such a transfigured and purified reality, although there is much more to it than simply transfiguration, unless one expressly means the working out of the actual, the last, and the ultimate in reality.[85]

Barth understands aesthetic creation as the eschatological figuration of a new reality that represents the redemption and completion of the

---

84. Barth's familiarity with Spengler's work can be gleaned from a reference he makes to Spengler's concept of an "eiserne[m] Zeitalter" in his preface to the second edition of *The Epistle to the Romans* (R xxiv).

85. Karl Barth, *Ethik II: Vorlesung Münster Wintersemester 1928/29, wiederholt in Bonn, Wintersemester 1930/31*, in *Gesamtausgabe*, vol. 2, ed. Dietrich Braun (Zurich: Theologischer Verlag Zürich, 1978), 440.

contemporary world. As a mode of *Gestaltung*, such aesthetic creation is a "language of forms" (*Formsprache*), as Spengler would put it, but its object is neither history nor the extension of history into the future but rather a "reality" that can never be fully present.[86] Indeed, Barth is explicit that art has significance for theology "in the context of eschatological reflection" (*in den Zusammenhang eschatologischer Betrachtung*).[87]

Barth's theology thus follows the constructivist tendency in historiography in its language and conceptual metaphors, yet it does not construct history in the mode of Troeltsch, Dilthey, and Spengler. Barth does not seek to put together, build up, and synthesize a historical structure from the material of history, but rather to give shape to an eschatological figure of a reality whose possibility belongs to each moment, yet which can neither be actualized as presence nor become history. As a result, Barth's theology of history is at its root concerned with a history that cannot be written. Such are the ends of historiography in Barth's eschatological constructivism: the historical event provides an impetus for the figuration of an eschatological moment, but in doing so, it ceases to be part of a historical narrative and points to a temporality that exceeds history.

Whereas Mannheim and Troeltsch sought to instill new life into historiography by arguing for new constructions of history in relation to the life of the present, Barth interrogates the limit of life and death in the figure of eschatology. The eschatological moment becomes the basis of a different sort of constructivism, one in which time is not simply historical time but contains moments at the limit of historical experience. The contested ground of this construct is precisely how to represent and picture such extrahistorical time,

---

86. Paul Tillich invokes a similar rhetoric of "Gestaltung" from the point of view of eternity, but he sees this mode of figuration as arising from interpretation of the present moment that is conscious of history. See Tillich, *Kairos*, x: "Ein Denken, das geschichtsbewußter Deutung der Gegenwartslage entspringt, trägt notwendig den Willen zur Gestaltung in sich. Die Dinge vom Ewigen her sehen, heißt sie verantwortlich sehen, in ihnen Gericht und Forderung anschauen."

87. Barth goes on to discuss the "eschatologische Möglichkeit" of poetry and music (*Ethik II*, 437, 441).

given that it withdraws from the intuitive visibility of historical phenomena. It is here that Barth's theology has a deep affinity with the construction of nonintuitive spaces in non-Euclidean geometry.

Barth's *The Epistle to the Romans* serves as a paradigmatic example of how modernism's spatial forms provided alternatives to the construction of history in historicism. In the time of the now, the figure of the moment is detemporalized and conceptualized through the paradoxes of spatial forms. In Barth's work, we can see how presence and nonpresence occupy the liminal spaces of points without extension and parallel lines that intersect in infinity. Indeed, the dialectical oscillation of proximity and distance comes to define the entwinement of presence and nonpresence as non-mutually-exclusive possibilities. In this way, spatial form is a key means by which modernism—from literature to theory—contended with the possibilities of the moment after historicism.

3

The Arc of History

Franz Rosenzweig's Figures of Time and Eternity

The German-Jewish philosophy of Franz Rosenzweig provides a second perspective on eschatological thought's reconfiguration of historical understanding in modernism. Rosenzweig's *Der Stern der Erlösung* (*The Star of Redemption*), composed on the Balkan Front between August 1918 and February 1919 and first published in 1921, borrows freely from both Jewish and Christian sources to offer a syncretic philosophical account of key theological concepts, such as creation (*Schöpfung*), revelation (*Offenbarung*), and redemption (*Erlösung*). Together with Rosenzweig's well-known correspondence with Eugen Rosenstock-Huessy, *The Star of Redemption* shows the fertile interplay of Jewish and Christian thought in the early twentieth century. Rosenzweig wrote his magnum opus at the same time that Barth was working on the second edition of *The Epistle to the Romans*, and despite significant differences in these works' theological perspectives, they share an important kinship in their imagination of shapes of time that exceed the historicist paradigm.[1] Indeed, Rosenzweig's concept of history emerged from a deep engagement with and critique of the nineteenth-century concept of history, Hegel's philosophy of history, and Ranke's historicism.

Rosenzweig grasps history in eschatological terms as suspended between an eternal beginning and an eternal end. He argues that the present—which he dubs the "today"—contains a moment of eternity that exceeds the narrative flow of history and points to the end of days. The metaphorical foundation for this rethinking of history is provided by a rich language of geometrical and spatial forms. Rosenzweig conceives of history as an arc suspended between two eternities rather than an unbroken circle; as a path without a focal point rather than a line of development; and as a set of lines that become curves in proximity to the points of creation, revelation, and redemption. Like Barth, Rosenzweig borrows key insights from non-Euclidean geometry in these spatial renderings of history. But for Rosenzweig, non-Euclideanism not only helps visualize the nonintuitive but also permits, via the projective geometry of Beltrami, Klein, and Poincaré, a mapping of historical time onto the space of the end of history. This mapping allows Rosenzweig to picture the "today" both as a historical present in the chain of time and as a springboard to eternity. In this way, modernist mathematics undergirds a theological perspective on history that breaks with nineteenth-century philosophies of history and their historicist commitments.

As a form of constructivism, Rosenzweig's geometrical imagination of history contributes to the production of spatial figures that are conceived as *Gestalten*. Indeed, the rhetoric of *Gestalten* is ubiquitous in *The Star of Redemption*. As Annette Simonis has noted, al-

---

1. Scholarship has drawn attention to Rosenzweig's encounter with and affinity to the Protestant "theology of crisis," including their common antihistoricism, yet the central place of eschatology in Rosenzweig's rethinking of history has been neglected. See Myers, *Resisting History*, 69–70. In a similar vein, Peter Gordon notes that "Rosenzweig shared a great deal with the Protestant crisis-theologians of the 1920s (e.g., Karl Barth, Rudolf Bultmann, and Friedrich Gogarten)." Cited in Samuel Moyn, "Is Revelation in the World?," *Jewish Quarterly Review* 96, no. 3 (2006): 396–403, at 400.

though the term came to prominence in Goethe's morphology, it experienced a renaissance in the early twentieth century, not only via Christian von Ehrenfels's Gestalt psychology but also through Georg Simmel's development of the aesthetic and poetic qualities of the concept.[2] In Rosenzweig's work, *Gestalten* provide an interface through which history is reimagined via spatial forms. Indeed, the larger trajectory of *The Star of Redemption* refigures the temporality of historical experience through a step-by-step construction of the figure of the star, proceeding from elements (points) to pathways (lines) and culminating in "the Star of Redemption itself as it finally became clear for us as a figure" (*den Stern der Erlösung selber, wie er uns endlich als Gestalt aufging*) (*S* 465). The creation of such figures, including the figure of eternity, opens up new possibilities for conceiving the temporality of the moment.[3] Rosenzweig's thinking in figures is thus a core element of his reimagination of the shape of time.

The concepts of history and eschatology that emerge from this geometrical and figural thinking can be summarized as follows: In contrast to the conventional view of the last things as the terminal endpoint of history, Rosenzweig thinks of the end of history as standing in proximity to each moment of time, thereby adopting a nonteleological concept of the end. He rejects a temporalized notion of eschatology as the terminus of history in favor of a spatial configuration of the star and constructs a curved space of history in which beginning and end are at all points equally near. Using images of curves, arcs, and intersecting parallel lines, Rosenzweig leverages key insights from non-Euclidean geometry to picture history not just as a temporal sequence en route to a *telos* but as a collection of figures that give the end in redemption a concrete shape and form.

---

2. Annette Simonis, "'Gestalt' als ästhetische Kategorie: Transformationen eines Konzeptes vom 18. bis 20. Jahrhundert," in *Morphologie und Moderne: Goethes "anschauliches Denken" in den Geistes- und Kulturwissenschaften seit 1800*, ed. Jonas Maatsch (Berlin: De Gruyter, 2014), 245–66, at 245.

3. See *S* 465: "The eternal had become a figure in the truth" (*Das Ewige war Gestalt worden in der Wahrheit*).

The Space of History: Rosenzweig's Revision of Historicism

The concept of history that Rosenzweig develops in *The Star of Redemption* has its roots in his early work, where one can observe both the influence of historicism and a subtle turn against historicism. Rosenzweig was trained as a historian under the tutelage of Friedrich Meinecke, who is widely considered to be the heir and leading twentieth-century proponent of Ranke's historicist project. Rosenzweig completed his dissertation, *Hegel und der Staat* (*Hegel and the State*) (1920), under Meinecke's direction, and he had great respect for Meinecke's *Weltbürgertum und Nationalstaat* (*Cosmopolitanism and the National State*) (1908), which was the subject of seminar discussions in which Rosenzweig took part.[4] Indeed, in many respects Rosenzweig adopted an approach to the history of ideas (*Ideengeschichte*) that was indebted to Meinecke and the historicist tradition more generally.

Nevertheless, at a very early stage in his intellectual development, Rosenzweig's opposition to the prevailing nineteenth-century concept of history, especially as embodied in Hegel's philosophy of history, was apparent. This stage was foundational to his later work. Rosenzweig confides to Hans Ehrenberg in a letter from September 26, 1910, that he was reading works such as Hegel's *Theologische Jugendschriften* (*Early Theological Writings*) (1907), Ranke's *Englische Geschichte* (*History of England*) (1859), Thomas Macaulay's *History of England* (1848), numerous works by Herder, and Schleiermacher's *Reden über die Religion an die Gebildeten unter ihren Verächtern* (*On Religion: Speeches to Its Cultured Despisers*) (1799) (*B* 53–54). Of the central questions with which he was struggling, Rosenzweig writes: "Many meta-religious issues are embedded in all these questions about the origin of evil, about God and history" (*B* 53).

The encounter with these key sources on nineteenth-century historiography and the philosophy of religion served as a foil against

---

4. See Alexander Altmann, "Rosenzweig on History," in *The Philosophy of Franz Rosenzweig*, ed. Paul Mendes-Flohr (Hanover, NH: University Press of New England, 1988), 124–37, at 124.

which Rosenzweig formulated his own view of the relationship between religion and history. Of Schleiermacher's *Reden*, Rosenzweig writes: "I got a lot out of them; ... the relationship of religion and history in the nineteenth century and today became clear to me in them" (B 54). Schleiermacher was the embodiment of the liberal Protestant view of history according to which Christ is conceived as a historical impulse and religious experience as immanent to the historical process. Drawing a connection between Schleiermacher's postulation that feeling is the center of religious praxis and Hegel's "religious intellectualism," Rosenzweig criticizes both figures for claiming that history is a manifestation of the divine: "We also refuse to see 'God in history' [*Gott in der Geschichte*] because we do not wish to see history (in its religious aspect) as an image or as being; rather, we *deny* God in *history* in order to *restore* God in the process through which history *emerges*" (B 55). In contrast to the Hegelian and Schleiermachian view that God is immanent to history—as feeling, as religious experience, as the movement of spirit toward self-realization—Rosenzweig affirms the divine origin of history, but denies, in a negative theology reminiscent of Barth's work, the presence of God in history.

Rosenzweig clarifies the relationship between religion and history in the nineteenth century and in his own time as follows: whereas for Hegel and the nineteenth century at large, world history was synonymous with theodicy, for Rosenzweig theodicy is the sole province of religion, and religion stands outside history. In his *Lectures on the Philosophy of History*, Hegel famously argued: "That world history is this course of development and the real becoming of spirit, in the midst of the varying spectacle of its histories—this is the true theodicy, the justification of God in history."[5] To this fundamental statement not only of the divinity of history but also of history as a course of development and the becoming of spirit, Rosenzweig counters:

> We see God in all ethical events but not in the completed whole, in history;—for why would we need a God if history were divine [*wenn*

---

5. Hegel, *Philosophie der Geschichte*, 540.

die Geschichte göttlich wäre], if all acts that flow into this basin were readily to become divine and justified? No, every act becomes sinful when it enters into history (the perpetrator did not want what happened) and therefore God must redeem humanity not through history, but indeed— there is no other option—as "God in religion" [*Gott in der Religion*]. For Hegel history was divine, "theodicy"; the act—as prehistorical, moral, subjective—was implicitly ungodly, "passion," "individual," "good intention," "knight of virtue." For us religion is the "only true theodicy."— The struggle against history in its nineteenth-century sense is therefore for us at once a struggle for religion in its twentieth-century sense. (B 55)

Whereas for Hegel history has a redemptive function insofar as it realizes a providential path toward reconciliation, for Rosenzweig history remains a site of fallenness, perdition, sin, and the brokenness of creaturely life; it is the site of the "Fall" and the "act of the perpetrator" (*Tat des Täters*) (B 55). Rosenzweig thus defines his struggle against "history in its nineteenth-century sense" as a "struggle for religion in its twentieth-century sense"; that is, for a concept of religion with antihistoricist undertones. For Rosenzweig, religion does not provide for redemption *in* history; it provides for redemption *from* history.

According to Stéphane Mosès, Rosenzweig's critique of Hegel's philosophy of history and of the "German historiographical school" (*deutsche historiographische Schule*) more generally was motivated by his opposition to Bismarck's founding of the German Empire "by fire and sword" and was confirmed by the catastrophe of the First World War.[6] While it is certain that Rosenzweig, especially in *Hegel and the State*, took a clear stand against the Hegelian view that the real is the rational and against the political extension of Hegel's philosophy, which enshrined the national state as the embodiment of spirit (while simultaneously excluding Judaism), Rosenzweig's opposition of religion and history has first and foremost a theological motivation. This motivation is borne out by his continued interest

---

6. Stéphane Mosès, "Hegel beim Wort genommen: Geschichtskritik bei Franz Rosenzweig," in *Zeitgewinn: Messianisches Denken nach Franz Rosenzweig*, ed. Gotthard Fuchs and Hans Hermann Henrix (Frankfurt am Main: Josef Knecht, 1987), 67–89, at 69.

in the theology of history in his correspondence and in *The Star of Redemption*.

An excellent illustration of the religious sources of Rosenzweig's opposition to the Hegelian concept of history can be found at the conclusion of his correspondence with Rosenstock-Huessy on Judaism and Christianity.[7] Here Rosenzweig brings to bear a perspective on history defined by "a final preparation for the last day" (*eine letzte Vorbereitung auf den jüngsten Tag*), one that succeeds in reversing the basic coordinates of Hegelian history:

> For "the reconciliation of the hearts of the fathers with the children" is, according to the final verse of the Prophet Maleachi, a final preparation for the last day. Without scholarship each generation would *go astray* from the preceding one, and history would seem to be a discontinuous series (as in fact it really *is*) and not (as it *ought* to appear) the parable of a single point, a *nunc stans* (as history really *is* in the final moment, but thanks to scholarship, as I have said, *appears* to be already in advance, here and now). (B 719)

Rosenzweig's interpretation of the verse by the prophet Malachi has two distinct layers. The first layer concerns the reality of history, which for Rosenzweig constitutes a "discontinuous series" (*diskontinuierliche Reihe*) that becomes the parable of a standing moment (*nunc stans*) at the end of time. The second layer concerns the appearance or semblance of history, as provided by scholarship or *Wissenschaft*: according to Rosenzweig, scholarship dispels the appearance of history as a discontinuous series and gives history the semblance of already being the parable of a standing moment.[8] This transformation of the semblance of history, which corresponds

---

7. "Franz Rosenzweig und Eugen Rosenstock: Judentum und Christentum," in Franz Rosenzweig, *Briefe*, ed. Ernst Simon and Edith Rosenzweig (Berlin: Schocken, 1935), 637–720. An English translation can be found in *Judaism despite Christianity: The 1916 Wartime Correspondence between Eugen Rosenstock and Franz Rosenzweig*, trans. Dorothy Emmet, ed. Eugen Rosenstock-Huessy (Chicago: University of Chicago Press, 2011), 77–170.

8. Myers reads this passage as evidence that "Rosenzweig maintained a deep ambivalence toward history.... On one hand, he recognized the epistemic unavoidability of ordering the world in historical terms; on the other hand, he recognized the epistemic artifice involved in such an ordering" (*Resisting History*, 86).

to an intergenerational reconciliation between fathers and children, produces not continuity or development, as in Hegel, but rather a superimposition of heterogeneous historical moments, in anticipation of the eschatological now. As Paul Mendes-Flohr notes in his commentary on this passage, the "historian's endeavor to endow history with coherence and structure is an anticipation of the meaning of history that will become manifest with redemption."[9]

Rosenzweig's mapping of a discontinuous series of historical moments onto an eschatological standpoint has an important affinity with the mathematics of projective geometry, which Eugenio Beltrami and Felix Klein used to demonstrate the consistency of non-Euclidean geometry. In his "Essay on the Interpretation of Non-Euclidean Geometry" (1868), Beltrami developed a projective model of the hyperbolic plane by asking: "Which surfaces admit a map into the Euclidean plane sending their geodesics to straight lines?"[10] It was known that a sphere could be projected onto a tangent plane in this way, but Beltrami showed that the only surfaces that allow this kind of map are surfaces with constant curvature.[11] In "On the So-Called Non-Euclidean Geometry" (1873), Klein generalized these insights by showing that geometry is fundamentally projective and that Euclidean geometry exists within projective geometry.[12] Klein's projective geometry has a natural basis in the optics of visual perception; as John Stillwell notes, "our visual world is projective rather than Euclidean. Objects constantly change shape as we shift our point of view—circles become ellipses, right angles become acute or obtuse, and so on. We are also used to seeing points at infinity; they are the points on the horizon, and they form a line. Yet it is also clear that, somehow, we abstract Euclidean geometry from the visual world of projective images."[13]

---

9. Paul Mendes-Flohr, "Franz Rosenzweig and the Crisis of Historicism," in *The Philosophy of Franz Rosenzweig*, ed. Paul Mendes-Flohr (Hanover, NH: University Press of New England, 1988), 138–61, at 158.
10. John Stillwell, *Sources of Hyperbolic Geometry* (Providence, RI: American Mathematical Society, 1996), 3.
11. Stillwell, *Sources of Hyperbolic Geometry*, 3–4.
12. Stillwell, *Sources of Hyperbolic Geometry*, 63.
13. Stillwell, *Sources of Hyperbolic Geometry*, 63.

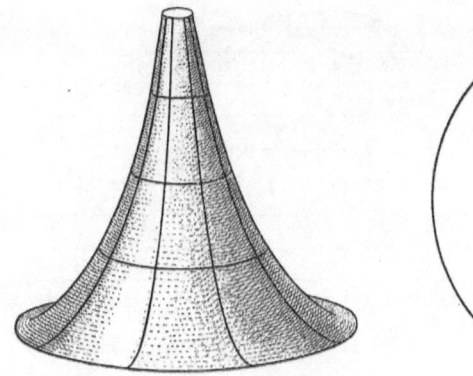 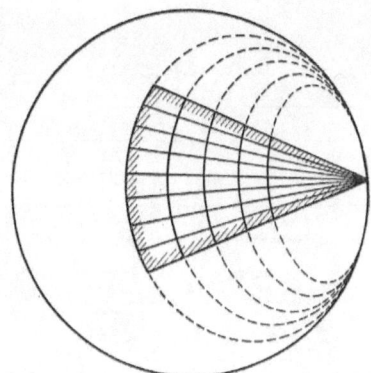

Figure 2. Map of a pseudosphere onto a disc in the Beltrami-Klein disc model. Felix Klein, *Vorlesungen über Nicht-Euklidische Geometrie* (Berlin: Springer, 1928), 286. Reprinted with permission from Springer Verlag.

In the Beltrami-Klein disc model, a surface of constant negative curvature such as a pseudosphere can be mapped onto a disc such that the whole region of the pseudosphere is projected onto the inside of a circle, with the boundary of the circle (the limit circle) representing points at infinity. As Beltrami writes, "In this map the geodesics of the surface are represented by chords of the limit circle."[14] In a surface of constant negative curvature, a geodesic—the shortest path between two points—will be a curve, but in the projection of the pseudosphere to a disc, it appears as a straight line (figure 2).

Whereas Klein and Beltrami map non-Euclidean curved space onto a model of Euclidean space, Rosenzweig proceeds in the opposite direction, mapping a set of discrete historical moments onto the nonintuitive space of the end of history. This mapping can be compared to John Paul Goode's interrupted homolosine projection of the earth, which appears discontinuous in Euclidean space but is continuous in the non-Euclidean space of the sphere (figure 3). The interruption of the projection along meridians reduces distortions in the shape and scale within the lobes by creating discontinuities

---

14. Eugenio Beltrami, "Essay on the Interpretation of Non-Euclidean Geometry," in Stillwell, *Sources of Hyperbolic Geometry*, 7–34, at 11.

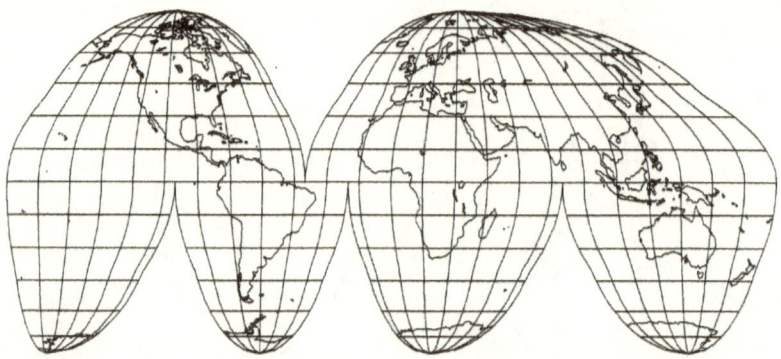

Figure 3. The Goode homolosine projection with interruptions. John P. Snyder, *Flattening the Earth: Two Thousand Years of Map Projections* (Chicago: University of Chicago Press, 1993), 197. Reprinted with permission from University of Chicago Press.

between parts of the world.[15] But when the map is applied to the non-Euclidean surface of the sphere, these discontinuities vanish. Similarly, for Rosenzweig, history appears as a discontinuous series of moments, but these discontinuities disappear when history is viewed from the standpoint of redemption.

Rosenzweig's concept of history is thus at once antihistoricist—in its recognition of the discontinuity of history—and grounded in a theological perspective.[16] Mosès speaks of Rosenzweig's new understanding of history as "that of a discontinuous, non-accumulative history [*einer diskontinuierlichen, nicht akkumulativen Geschichte*] that is not determined, for instance, by the succession of great civilizations and world events."[17] Instead, the coordinates of history, as

---

15. See John P. Snyder, *Flattening the Earth: Two Thousand Years of Map Projections* (Chicago: University of Chicago Press, 1993), 167.
16. For a related discussion of the discontinuity of time in *The Star of Redemption*, see Robert Gibbs, "Lines, Circles, Points: Messianic Epistemology in Cohen, Rosenzweig and Benjamin," in *Toward the Millennium: Messianic Expectations from the Bible to Waco*, ed. Peter Schäfer and Mark Cohen (Leiden: Brill, 1998), 363–82, at 374: "The punctuation of the hour inserts the discontinuity of the future redemption, of the messianic age, into time. Cycles are neither linear nor simply the continuity of time. Rather, they represent the interruption and break-up of continuity."
17. Mosès, "Geschichtskritik bei Franz Rosenzweig," 70.

we see in *The Star of Redemption*, are defined by the theological concepts of creation, revelation, and redemption, which are loosely coordinated with the historical concepts of past, present, and future. Accordingly, religion, and especially Judaism, defines what Mosès calls a "secret flipside" or "negative image" (*Negativ*) of history, or a "history beyond history" (*Geschichte jenseits der Geschichte*).[18] In a letter to Ehrenberg on May 9, 1918, Rosenzweig articulates an idea that will become central to *The Star of Redemption*—namely, that Judaism stands outside history: "I do not claim for Judaism any correspondence with another reality beyond itself. I completely acknowledge the worldly unliveliness [*weltliche Unlebendigkeit*] of Judaism.... Judaism is alive only by being with God. Not until the world is *also* with God will Judaism be alive in worldly terms as well. But that is again beyond history [*jenseits der Geschichte*]" (B 311–12). Judaism is beyond history because it anticipates the end of redemption, or, as Alexander Altmann puts it: "Judaism ... lives and has its being in the *eschaton*."[19] Yet Rosenzweig's antihistoricism does not merely step outside history; it also draws on theology to rethink the structure of history. In a letter to Rosenstock-Huessy dated November 30, 1916, Rosenzweig writes, "This is perhaps the essence of revelation: that it brings an absolute symbolic order into history" (B 710).

The "absolute symbolic order" that revelation gives to history has two components: it rethinks history in spatial terms and in terms of geometrical figures, and it thereby detemporalizes key historical categories. In his correspondence with Rosenstock-Huessy on "Judaism and Christianity," Rosenzweig responds to Rosenstock-Huessy's sketch of a calendar of key philosophical concepts and historical turning points with a spatial outline of key cities, places, and architectural forms.[20] Rosenzweig writes, "You philosophize

---

18. Mosès, "Geschichtskritik bei Franz Rosenzweig," 76–77.
19. Altmann, "Rosenzweig on History," 129. See also Myers, who notes that Rosenzweig and Judah Halevi had a "shared belief that the Jewish people operated outside the normal flow of historical events" (*Resisting History*, 74).
20. Rosenzweig's sketch is as follows: "1. Die Schauplätze: (Kassel, Wilhelmshöhe, Vierwaldstätter See, die Sonne Homers, Cornwall-Island-Bretagne, Harz-Frankfurt-Weimar, Göttingen, München, Freiburg.) / 2. Der Kosmos. / 3. Die Häuser.

in time; I would like to philosophize in the form of space" (*Sie philosophieren in der Zeit, ich will einmal in der Form des Raumes philosophieren*) (B 653). Just as in Paul Klee's polyphonic painting "the time element becomes a spatial element,"[21] as Zachary Braiterman notes, so too does Rosenzweig treat space as "a sensualization of time" (*eine Versinnlichung der Zeit*) (B 654). Rosenzweig further develops his historical geography of space in two essays, "Ökumene" ("Ecumene") and "Thalatta" ("Thalatta"), to which he gave the overarching title "Globus: Studien zur weltgeschichtlichen Raumlehre" ("Globe: Studies on the World-Historical Theory of Space") (1917).[22] A history of the world from antiquity to the present told via the geography of the earth and its oceans, the text bears witness to Rosenzweig's interest in writing history in spatial terms.

Indeed, in a letter to Ehrenberg from December 26, 1917, Rosenzweig notes that his "Globus" essays construct history as a spatiotemporal curve. This account of history in terms of a geometry of space allows Rosenzweig to distinguish his own mode of historiography from Ranke's historicism:

> The reduction to spiritual forces, which you found lacking in "Cannä und Gorlice," is also missing in "Ökumene und Thalatta." The special aspect of this entire approach is precisely that it does not conceive of history as a curve between the coordinates of time and *soul*, but rather between time and *space*. The former curve results in *spirit* (Ranke), mine in the *movement* of history. In Ranke, history is thus its own glorification; for me history is a bare fact. (B 273)

Here we can observe the fundamentally different driving forces of history in Rosenzweig and Ranke. Whereas for Ranke the motor of history is psychological, stemming from the spirit of nations, for Rosenzweig it is spatial and geographical, which is to say driven by

---

(Elternhaus, Schulhaus und Kaserne, fremde Häuser, eignes Haus, Rathaus.) / 4. Die Ökumene. / 5. Die Kathedrale" (B 653).

21. Braiterman, *The Shape of Revelation*, 142.

22. Franz Rosenzweig, "Globus: Studien zur weltgeschichtlichen Raumlehre," in *Der Mensch und sein Werk: Gesammelte Schriften*, vol. 3, ed. Reinhold Mayer and Annemarie Mayer (Dordrecht: Nijhoff, 1984), 313–68.

topography and the limits imposed by the seas. At the root of this differentiation, in a second layer of spatial metaphorics, is a geometrical presentation of history as a "curve between coordinates" (*Kurve zwischen Koordinaten*). The curve between "time and soul" (*Zeit und Seele*), which Rosenzweig attributes to Ranke, charts the diachronic development of spirit, whereas Rosenzweig's curve between "time and space" (*Zeit und Raum*) traces the movement of history across space. The divergence of these curves, Rosenzweig suggests, points to a more fundamental difference between his understanding of history and that of historicism: whereas for Ranke "history is its own glorification"—just as for Hegel history is theodicy—for Rosenzweig history is a "bare fact" (*nackte Tatsache*). In other words, Rosenzweig is committed to a historiography that exposes the brute realities of history: "Globe: Studies on the World-Historical Theory of Space" is above all a history of territorial expansion, conquest, and domination. As in his earlier letter on Hegel's philosophy of history, here too Rosenzweig detects no sign of "God in history," but rather conceives of history as the "Fall" and the "act of the perpetrator."

These reservations about the ideological presuppositions of Ranke's historiography notwithstanding, Rosenzweig did find some common ground with Ranke's work, which no doubt reflects his training as a historian and the influence of Meinecke. In particular, Rosenzweig was committed to a historicization of philosophy and rebelled against the age-old "claim to glory" of philosophy; namely, its claim to have knowledge "independently of time."[23] The major source for Rosenzweig's method of narrative philosophy was Friedrich Wilhelm Joseph Schelling's *Die Weltalter* (*The Ages of the*

---

23. Franz Rosenzweig, "'The New Thinking': A Few Supplementary Remarks to *The Star of Redemption*," in *Franz Rosenzweig's "The New Thinking*," ed. and trans. Alan Udoff and Barbara E. Galli (Syracuse, NY: Syracuse University Press, 1999), 67–102, at 83; Franz Rosenzweig, "Das neue Denken: Einige nachträgliche Bemerkungen zum 'Stern der Erlösung,'" in *Der Mensch und sein Werk: Gesammelte Schriften*, vol. 3, ed. Reinhold Mayer and Annemarie Mayer (Dordrecht: Nijhoff, 1984), 139–61, at 149. On Rosenzweig's "new thinking," see Reiner Wiehl, *Zeitwelten: Philosophisches Denken an den Rändern von Natur und Geschichte* (Frankfurt am Main: Suhrkamp, 1998), 177–83.

World) (1815), which Rosenzweig refers to in a letter to Rosenstock-Huessy in 1916 and again in "Das neue Denken" ("The New Thinking") (1925), his postscript to *The Star of Redemption*.[24] In "The New Thinking," Rosenzweig notes that while the first volume of *The Star* is devoted to the philosophical question "What is?," the second volume represents "reality as it is experienced," and this requires a "method of narration" (*Methode des Erzählens*) or a "narrative philosophy" (*erzählende Philosophie*) inspired by Schelling. Schelling's introduction to *The Ages of the World* begins with a programmatic statement of the temporality of philosophical knowledge: "The past is known, the present is recognized, the future is anticipated. The known is narrated, the recognized is represented, the anticipated is prophesied. Scholarship is already in the basic meaning of the word history (historía)."[25]

Given his criticism of Ranke, it is noteworthy that Rosenzweig conceives of his method of narrative philosophy in Rankean terms:

> What does it mean to tell a story (*erzählen*)? He who narrates does not want to say how it "actually" was, but how it really took place. Even when the great German historian in his well-known definition of his scientific intention uses the former and not the latter word, he means it in this way. The narrator never wants to show that it actually was entirely different . . . , rather he wants to show how such and such that is on everyone's lips as concept and name, for example, the Thirty Years' War, or the Reformation, actually happened.[26]

This passage neatly illustrates Rosenzweig's ambivalence about Ranke's historicism, or rather his own simultaneous investment in both historicist and antihistoricist methods. In distinguishing how the past "actually was" (*eigentlich gewesen*) from how it "really happened" (*wirklich zugegangen*), Rosenzweig conceives of narration in dynamic rather than static terms: it is not a question of the being of the

---

24. See *B* 711; Rosenzweig, "Das neue Denken," 148.
25. Friedrich Wilhelm Joseph von Schelling, *Die Weltalter Fragmente: In den Urfassungen von 1811 und 1813*, ed. Manfred Schröter (Munich: C. H. Beck'sche Verlagsbuchhandlung, 1993), 111.
26. Rosenzweig, "The New Thinking," 81–82; Rosenzweig, "Das neue Denken," 148.

past, but rather of its becoming. Yet Rosenzweig's reformulation of Ranke's canonical statement of the task of the historian—"he merely wants to show how it actually was" (*er will blos zeigen, wie es eigentlich gewesen*)[27]—appears to be less a matter of distancing himself from Ranke than of reframing Ranke's contribution to historiography as a narrator. For Rosenzweig, a narrative philosophy is not concerned with "what was," and hence he sees "was-sentences" (*War-Sätze*) and "substantives" (*Substanzworte*) as secondary to the importance of the "verb" (*Verbum*) or "time-word" (*Zeit-wort*), which is capable of showing the "movement" of the past.[28]

But if Rosenzweig's method of narrative philosophy is more or less compatible with a reframed version of historicism, his philosophy of religion, especially as it concerns the relationship between Judaism and Christianity—the major subject of the third volume of *The Star of Redemption*—runs up against the limits of historicism. In particular, the concept of eternity requires a mode of narration that attends to the spatial and figural qualities of images (*Bilder*) and figures (*Gestalten*). As Rosenzweig notes in "The New Thinking," his characterization of Judaism and Christianity is motivated by the question of an "existing eternity" (*seienden Ewigkeit*), and it proceeds from the "external, visible figure through which they [Judaism and Christianity] wrest their eternity from time" (*von der äußeren, sichtbaren Gestalt, durch die sie der Zeit ihre Ewigkeit abringen*).[29] In reference to the "eternal figures" (*ewigen Gestalten*) that anticipate a future of redemption, Rosenzweig writes: "The flow of events projects brilliant images onto heaven, above the temporal world, and they remain. They are not archetypes. On the contrary, they would not exist if the current of reality did not continue to break forth out of its three invisible-mysterious sources."[30] For Rosenzweig, these eternal figures bear a crucial relationship to

---

27. Ranke, "Geschichten der romanischen und germanischen Völker," vii.
28. Rosenzweig, "The New Thinking," 82; Rosenzweig, "Das neue Denken," 148.
29. Rosenzweig, "The New Thinking," 94, translation modified; Rosenzweig, "Das neue Denken," 156.
30. Rosenzweig, "The New Thinking," 93, translation modified; Rosenzweig, "Das neue Denken," 155.

the "flow of events" (*Fluß des Geschehens*) and "current of reality" (*Strom der Wirklichkeit*); that is, to the fundamental historicist understanding of time as a continuous flow, a central historicist metaphor that can be traced back to Ranke and Herder.[31] Indeed, the flow of events produces these eternal figures as "brilliant images" (*glänzende Bilder*) that remain. The eternal figures that Judaism and Christianity make visible are thus not "archetypes" (*Urbilder*) or Platonic forms but reflections of historical time: "The existing substances which are perpetual only as secret premises of the always renewed reality are superseded by figures that eternally mirror this always renewed reality [*Gestalten, die diese allzeiterneuerte Wirklichkeit ewig spiegeln*]. The third volume deals with these [figures]."[32] The eternal figures that Rosenzweig constructs in the third volume of *The Star of Redemption* are thus historicist and antihistoricist in equal measure: they are derived from the temporal flow of events yet function as spatial representations of history whose narrative character is superseded by a figural mode of thinking.

Rosenzweig's interest in figures that "wrest eternity from time" results in a dual-aspect theory of history in which temporal categories can be understood either in historical terms or as spatial configurations of eternity. As an example of how these two aspects are structured, let us consider Rosenzweig's discussion of the "today" (*Heute*) in a letter from February 5, 1917, to Gertrud Oppenheim. Glossing a Talmudic legend concerning the coming of the Messiah, Rosenzweig unpacks the competing senses of what "today" might

---

31. In "This Too a Philosophy of History," Herder uses the concept of *Strom* as an image for the continuity of historical development: "But should there not be obvious *progress* and *development* in a higher sense than one imagined? Do you see this *current* [Strom] swimming along: how it sprung from a small well, grows, breaks off there, begins here, always winding and boring further and deeper—but always remains *water*! *current*! Drops always just drops, until it rushes into the sea—if it were thus with humankind?" (Herder, "Auch eine Philosophie der Geschichte," 41). Herder focuses on the progress and development of the current from its origins to its goal, noting that it always remains itself even as its path proves uneven. Similarly, Ranke conceives of the succession of epochs as "a current [*Strom*] that channels its path in its own way" (*Über die Epochen der neueren Geschichte*, 27).

32. Rosenzweig, "The New Thinking," 93, translation modified; Rosenzweig, "Das neue Denken," 156.

signify by spatializing the concept as either a bridge or a springboard:

> I don't know whether you know the Talmudic legend: At the entrance to a cave a Rabbi meets the prophet Elias (known to be the precursor of the Messiah according to Malachi's final lines) and asks him: Where is the Messiah? Inside in the cave. So he goes inside and finds the Messiah sitting there. And he asks him: When are you coming, Lord? The Messiah answers: Today. So he leaves cheerfully and waits until evening. But when the Messiah still has not come, the Rabbi says to Elias: The Messiah lied; he said he would come today. Elias responded: He meant: (quoting Psalm 91 or 92 or 93, 94) "today, when you heed my voice." (B 157–58)

Whereas for historicism the concept of today has an unequivocal meaning as the present, as it is situated between the past and the future, for Rosenzweig it has two aspects: on the one hand, it refers to a "today that only wants to be a bridge to tomorrow" (*Heute, das nur die Brücke zum Morgen sein will*); on the other hand, it signifies an "other today that is a springboard to eternity" (*ander[es] Heute, das das Sprungbrett zur Ewigkeit ist*) (B 158). Both images, the bridge and the springboard, provide spatial representations of a temporal concept. But the contrast between these images reflects a fault line between a historicist and an antihistoricist concept of time. Today as a bridge to tomorrow reflects the historicist notion of time as a linear continuum of past, present, and future, in which one event follows upon another in a horizontal sequence. By contrast, today as a springboard to eternity represents the present as standing outside the continuum of time and historical sequence, as no longer determined by its temporal location but already in the presence of eternity. As Mosès argues, the historical time of the present as bridge follows a "linear axis of time" that is quantitative, additive, and determined by causality, while the symbolic time of the present as springboard reflects time as a "discontinuous sequence of states of quality and intensity that are each unique."[33] The discontinuous time of the present as springboard thus serves as a key

---

33. Stéphane Mosès, *The Angel of History: Rosenzweig, Benjamin, Scholem*, trans. Barbara Harshav (Stanford, CA: Stanford University Press, 2009), 60–61.

image for Rosenzweig's use of theology to critique and supplement the historicist concept of time.

But a crucial dimension of Rosenzweig's dual-aspect theory of history, which Mosès appears to overlook, is that Rosenzweig conceives of the present *both* as a bridge *and* as a springboard. Indeed, in his letter he writes: "Everything apparently contradictory can be explained by the equation of the today that only wants to be a bridge to tomorrow and the other today that is a springboard to eternity" (B 158). We can shed light on this "equation" (*Gleichung*) that explains "everything apparently contradictory" (*alles scheinbar Widersprechende*) by considering the geometrical image of parallel lines with which Rosenzweig begins his letter. As a preface to his gloss on the Talmudic legend discussed above, he writes:

> So: at the outset: parallel lines meet in infinity [*Parallelen schneiden sich im Unendlichen*]. This also applies to the realm of spirit. Just that here the infinite is not "very far away" but "indeterminably far away." That is to say: I do not know whether it will be reached and become real already in the next moment or some other time.... Every act must be carried out as if the fate of eternity depended on it. For one can never know *whether* it depends on it or not. (B 157)

In a nod to non-Euclidean geometries, which demonstrated the existence of spaces in which the parallel lines of Euclidean space may meet, Rosenzweig claims that parallel lines do in fact intersect at infinity. The today as bridge and the today as springboard can be considered as such parallel lines. Although they appear to run along parallel tracks, in the mode of an either/or, at infinity, whose distance Rosenzweig considers to be "indeterminable" (*unbestimmbar*), the either/or becomes a both/and. Since we do not know when the present of historical time and that of symbolic time will intersect, the law of our actions requires that we treat every act as if it were a springboard to eternity. This intersection of parallel lines—of the historical present and the time of redemption—is what Rosenzweig refers to as the "apparently contradictory." In the historicist concept of history, which can be compared to Euclidean geometry, the present moment cannot be eternal without contradiction. Yet a non-Euclidean geometrical imagination, in which parallel lines can indeed intersect, provides Rosenzweig with a spatial representation of a shape of time in which the present is both historical and eternal. Just as non-Euclidean geometries contain Euclidean geometry as one possible space (with a constant curvature equal to zero) among many, Rosenzweig's dual-aspect theory of history treats historicism as one of several possible configurations of time.

Circle, Arc, Star: Judaism, Christianity, and the Geometry of History

In *The Star of Redemption* Rosenzweig carries forward the construction of spatial images of history that we have observed in his letters. In his magnum opus he expounds on the relationship between time and eternity using geometrical forms such as the circle, the arc, and the star. Like a perspective on a prism whose vantage point is varied, *The Star of Redemption* stages confrontations of time and eternity, history and its eschatological other. This work challenges the reader to consider figures (*Gestalten*) of eternity in time and to conceive of eternity in terms of temporal vectors. However, like Barth, Rosenzweig rejects any isomorphism of time and eternity. His eschatological reading of history therefore runs counter to German Idealism and the Hegelian tradition, whose pantheism made a claim for the divinity of the world and for the consciousness of the I as a cipher for divine subjectivity.[34] In Hegel's philosophy of history, world history embodies the successive stages of the spirit's progression to its full recognition and consciousness of itself; the divine is thus entirely immanent in the unfolding of the spirit. In Rosenzweig's characterization of this philosophical perspective, its model is that of a circle whose end is its beginning, whose goal is a return to the origin—a circle that contains infinity

---

34. For an overview of Rosenzweig's and Hegel's competing views on history, see Ulrich Bieberich, *Wenn die Geschichte göttlich wäre: Rosenzweigs Auseinandersetzung mit Hegel* (St. Ottilien: EOS, 1990), 44–45.

within itself.[35] Rosenzweig's criticism of the Hegelian concept of history is bound up with a criticism of its geometrical model and the formulation of an alternative geometry of history.

The Hegelian concept of Being as a sphere or circle, according to Rosenzweig, implies the immanence of eternity in time and the inherence of the infinite within the finite. Rosenzweig criticizes such a "unity of a sphere that everywhere returns to itself" (*Einheit der überall in sich selbst zurücklaufenden Kugel*) because it has no genuine sense of the end; its seamless folding of the end back into the beginning denies it a true eschatological moment:

> For in its first beginnings, philosophy, with naive candor, had claimed that it wanted to regard "being" as a sphere, or at least as a circle, and this thought dominated it until its end in Hegel. Hegel's dialectic believes it can and must justify itself by leading back into itself.... That unity of running back into itself, into its own beginning, the in-finite in the sense that the end is immediately changed again into the beginning and therefore is never graspable and conceivable as an end [*nie als Ende greif- und begreifbar*], that unity was therefore situated for us at the outermost limits [*äußersten Grenzen*] of our world. (S 283)

Whereas *The Star of Redemption* is nurtured by the possibility that the end can become a new beginning, Hegel's concept of the unity of the circle tames the reversal of beginning and end in such a way that there is no proper end. Rosenzweig rejects the model of the circle because its beginning and end are so self-contained that they ultimately disappear in the infinity encompassed by the circle. For Rosenzweig, the eschatological end does not simply return to the origin but marks a threshold of death and new life that exceeds the limits of the circle.

To show the rupture of the unity of the circle and its opening to the outside, Rosenzweig constructs a different set of images of the relationship between the finite and the infinite. Picturing a "sea of

---

35. Compare Hegel, *Enzyklopädie*, 50: "Auf diese Weise zeigt sich die Philosophie als ein in sich zurückgehender Kreis, der keinen Anfang im Sinne anderer Wissenschaften hat, so dass der Anfang nur eine Beziehung auf das Subjekt, als welches sich entschließen will zu philosophieren, nicht aber auf die Wissenschaft als solche hat."

infinity" (*Meer der Unendlichkeit*) that extends "before" the beginning and "after" the end (*S* 283),[36] Rosenzweig suggests that the finitude of the world is suspended between these two infinities: "So the world that comes together for us in the ascent does not circle into itself; it breaks forth from the infinite and plunges back into the infinite, both an infinite outside of it, in relation to which the world itself is a finite reality, whereas the circumference or even the sphere had the infinite in itself, indeed was itself infinite, and therefore all apparently finite reality in it emerged out of and led into its own infinitude" (*S* 284). Rosenzweig's eschatological account of history is encapsulated in the image of an arc spanning a distance between a beginning and an end, suspended between an origin in the infinite and an end in the infinite. In contrast to Hegel's model, Rosenzweig's arc can be thought of as a broken circle, in keeping with the mathematical definition of the arc as "any portion (other than the entire curve) of the circumference of a circle."[37] Rosenzweig's effort to rethink the unity of the circle implicitly taps into and calls into question the classical emblem of God as "an infinite sphere whose center is everywhere and whose circumference is nowhere."[38] By picturing a broken circle, Rosenzweig stands in a tradition that began with Nicolaus Cusanus and Blaise Pascal and was further developed in the twentieth century by writers such Jorge Luis Borges and Samuel Beckett.[39] In Rosenzweig's image of the arc, the world bears a relation to the infinite only as to that which

---

36. A similar metaphorics is at work in Barth's *Epistle to the Romans*: "Wir sehen dann die *Zeit*, in der wir leben, charakterisiert, als 'die Zeit des *Jetzt*,' d.h. als das Meer der gegebenen Wirklichkeit, das die submarine Insel des 'Jetzt,' der göttlichen Offenbarung, der Wahrheit vollständig überflutet, unter dessen anschaulicher Oberfläche sie aber nichtsdestoweniger vollständig intakt vorhanden ist" (*R* 313).

37. Margherita Barile and Eric W. Weisstein, "Arc," *MathWorld—A Wolfram Web Resource*, https://mathworld.wolfram.com/Arc.html.

38. On the history of the metaphor of the infinite sphere, see Karsten Harries, "The Infinite Sphere: Comments on the History of a Metaphor," *Journal of the History of Philosophy* 13, no. 1 (1975): 5–15.

39. See Jorge Luis Borges, "The Fearful Sphere of Pascal," in *Labyrinths, Selected Stories and Other Writings* (New York: New Directions, 1964), 189–92; and Samuel Beckett, "Watt," in *The Grove Centenary Edition*, vol. 1, ed. Paul Auster (New York: Grove Press, 2006), 169–379, at 272–73.

is outside itself.[40] Like Barth, Rosenzweig articulates a relationship between time and eternity but denies any commensurability or identification between the two. Whereas the cyclical model is fundamentally Hellenic (pantheistic, Hegelian), Rosenzweig's arc represents a Judeo-Christian eschatological understanding of history.

In conceiving of history as an arc extending between an eschatological beginning and end, Rosenzweig pictures an open geometry that is not only decentered but also irreducible to a set of focal points of that sort that define a hyperbola. Insofar as the eternity of the beginning and end is open to the temporality of the world and the arc it traces, the eschatological account of history embodies an open infinity. The open infinity of the eschatological model may be imperfect or "bad" in comparison with its Hegelian counterpart, yet for Rosenzweig the decentered nature of this model is its virtue:

> In order to make visible this infinity that did not curve back on itself, hence precisely the "bad" infinity according to the philosophical view, we therefore had to shatter idealism's infinity that curved back on itself; since, that is to say, instead of the circumference completely determined by the relation of one of its own points to a point of reference, we set single points against each other, none of which could be clearly taken as a point of reference for the others, we forced the construction of the line [*Konstruktion der Linie*] through these points, without there being a law of construction [*Konstruktionsgesetz*] that set an "absolutely mentally" valid relation between "any arbitrary" point of the line and a common point of reference. Through such a relation, namely in the formula made possible through it, even the infinity that is "bad" in itself, namely, the non-enclosed [*ungeschlossene*] infinity of, say, a hyperbola [*Hyperbel*],

---

40. Silvia Richter arrives at a similar conclusion in her discussion of eschatology in the work of Emmanuel Levinas: "The function of eschatology lies exactly in the realization of this relationship of the same to the other, manifested in the experience of being faced by the other which relates the Same to a beyond of history: 'Eschatology institutes a relation with being *beyond the totality* or beyond history'" ("Language and Eschatology in the Work of Emmanuel Levinas," *Shofar: An Interdisciplinary Journal of Jewish Studies* 26, no. 4 [2008]: 54–73, at 56). Peter Gordon's reservations about reading Rosenzweig through Levinas notwithstanding, the comparison emphasizes the extent to which Rosenzweig breaks with Hegel's understanding of the totality of history, as embodied in the figure of the circle. See Peter Eli Gordon, *Rosenzweig and Heidegger: Between Judaism and German Philosophy* (Berkeley: University of California Press, 2003), 9–12.

becomes a "good" infinity, namely one that can be formulated as closed [*geschlossen*]. (S 284)

Whereas the circle has a definite center by means of which it can be constructed through a formula, the arc cannot be formulated in advance because it has no "common point of reference" (*gemeinsamen Beziehungspunkt*). This is consistent with the more general mathematical definition of the arc as "any smooth curve joining two points."[41] In Rosenzweig's example, the points of the arc have no inherent relation to one another; a line can be constructed through these points, but there is no center. Rosenzweig therefore extends his critique of the closed infinity of the Hegelian circle to any formulaic relation of points on a line, such as that of a hyperbola. For once a law of construction can be posited by which any arbitrary point on the line can be determined in relation to a focal point, the infinity to which the line tends is no longer truly open. The contingency of the arc is decisive for Rosenzweig's eschatological understanding of history, and it grounds his opposition to linear, hyperbolic, and cyclical models of history.

Having established an alternative geometry for thinking about the relationship between time and eternity, Rosenzweig proceeds to consider Judaism and Christianity as two distinct embodiments of the way the present moment relates to the last things. In each case, his aim is to reconfigure the end of days as a spatial matrix rather than as the terminus of a historical teleology. In the case of Judaism, Rosenzweig argues for an anticipation of the end that lifts the Jewish people out of history, placing them on the threshold of the end of days by collapsing the space between the present and a future of redemption. In the case of Christianity, he argues that the in-between character of the Christian path defines a set of midpoints that are at all points equidistant to the beginning and end. Ultimately, Rosenzweig resolves the apparent opposition of Judaism and Christianity in the figure of the star of redemption, which provides a spatial representation of his nonteleological concept of eschatology.

---

41. Barile and Weisstein, "Arc."

152   Chapter 3

In his reading of Judaism, Rosenzweig detects a relation to historical experience defined by detemporalization. Inasmuch as the law is the point of reference for the Jewish people in each moment, he claims, Judaism is not entangled in history; instead, it lives in an eternal present.[42] Rosenzweig's thesis should be read, not as a statement about the disposition of Judaism, but rather as a reflection on the mode of temporality that it makes manifest. And this temporality, he claims, echoing a familiar antihistoricist trope, entails being "lifted out of" (*herausheben*) history:

> Because the teaching of the Holy Law—for the appellation Torah comprises both, teaching and law in one—therefore lifts the people out of all temporality and historicity of life [*das Volk aus aller Zeit- und Geschichtlichkeit des Lebens heraushebt*], it also deprives them of their power over time. The Jewish people do not calculate the years of their own chronology. Neither the memory of their history nor the official times of their lawgivers can become their measure of time; for historical memory is not a fixed point in the past that becomes more past every year by one year, but rather a memory that is always equally near, really not at all past, but eternally present. . . . The people's chronology thus cannot be here the calculation of their own time; for it is timeless, it is without time [*es ist zeitlos, es hat keine Zeit*]. (*S* 337–38)

Rosenzweig gives the *topos* of being lifted out of historicity, which we have seen in Nietzsche, Kierkegaard, and Barth, a unique spin: for him, the mode of being without time, of being lifted out of the historical continuum, takes place as an eternal present for which the past and future are always close at hand.[43] Whereas in Christianity the law comes to stand for a relation of tradition and a grounded-

---

42. Compare Altmann, "Rosenzweig on History," 137, who argues that for Rosenzweig, "God remains wholly outside of history. . . . Eschatology in its absolute sense falls outside of history. Neither God nor the eternal have a history." Similarly, Bieberich argues that Rosenzweig's opposition to Hegel's philosophy of history allows an "unhistorical" Judaism to stand for the "end of history": "Rosenzweig gibt also Hegel und Schelling geschichtlich recht: das Judentum steht wirklich außerhalb der Geschichte. Aber gerade darin erblickt er den Wert des Judentums. Nur als ungeschichtliches kann es für das Ende der Geschichte stehen" (*Wenn die Geschichte göttlich wäre*, 44–45).

43. Compare Myers, who interprets Rosenzweig's claim that the Jewish people are outside history as a function of his antihistoricism (*Resisting History*, 102–3).

ness in the past, for Judaism the law is eminently present in its mode of being. Indeed, Rosenzweig claims, because the Jewish people are bound to an immutable law, they are an eternal people that are not measured by changing historical epochs. This is not to say that the people are without historical consciousness, but rather that the historical memory of the people is grounded in the present of remembering, in which past events are as near as if they had taken place yesterday. According to Rosenzweig, the Jewish people live with an "eternally present memory," and yet for just this reason they stand apart from time (S 337). In what is certainly the most controversial aspect of his argument, Rosenzweig claims that the Jewish people do not actively take part in the world-historical life of peoples, but rather maintain an everlasting existence above and beyond time, not in the eternal afterlife of the individual, but in the endless persistence of the people in their relation to the law.

Yet as much as the confrontation of history with eschatological time carves out a space for an extrahistorical presence in history, Rosenzweig claims that the timelessness of this position is again and again inscribed and reinscribed in the calendrical movement of history. The "eternal people" that are lifted out of history in the "fully compact present" (*ganz dichte Gegenwart*) (S 352)[44] of their relation to the law are thrown back into history insofar as their present cannot fully embody eternity. For Rosenzweig, there is a tension in Judaism's spatial proximity to eschatology: on the one hand, he claims that the Jewish people live their own redemption and are conscious of themselves as being at the "end" or "goal" (*am Ziel*); on the other hand, he suggests that the Jewish people live on the threshold of the eschatological end because redemption has not yet been attained:

> It was the cycle of a people. A people was at its goal in it and knew it was at the goal. . . . It anticipated eternity for itself. In the cycle of its year the future is the motive power; the circular movement does not arise as it were by push but by tug; the present elapses not because the past shoves it forward, but because the future drags it along. The holidays of creation and revelation also somehow lead to redemption. That the consciousness

---

44. Franz Rosenzweig, *The Star of Redemption*, trans. Barbara E. Galli (Madison: University of Wisconsin Press, 2005), 336.

of the still unattained redemption again breaks forth [*wieder hervorbricht*] and thereby the thought of eternity again foams over the chalice of the moment [*Becher des Augenblicks*] into which it appeared to have already been filled—this gives to the year the power to begin anew and place its beginningless and endless link [*anfangs- und endelosen Ring*] properly into the long chain of times [*lange Kette der Zeiten*]. But the people still remain the eternal people [*das ewige Volk*]. For their temporality, this fact that the years recur, is considered only as a waiting, perhaps as a wandering, not as a growing. (S 364–65)

History, Rosenzweig claims, is too limited a vessel to contain the fulfillment of eschatological time. The anticipation of eternity places the Jewish people in the midst of redemption, yet as a historical people stranded in time, the redemption that eternity promises escapes their grasp. The "chalice of the moment" (*Becher des Augenblicks*) brims over because it cannot contain the measureless, boundless infinity of eternity. Rosenzweig's spatial image of the chalice of the moment overflowing with eternity suggests that redemption exceeds the bounds of the moment. Hence, redemption remains unattained, not because the chalice can never be filled but because the moment is too finite to hold the energies of eternity with which it might be fulfilled.

As a result of this deficiency of the moment, in Rosenzweig's view, the Jewish people are cast back into historical time and returned to the progression of time that passes from year to year. The end of the calendar year marks the threshold of eternity that cannot be crossed, and hence it must return to its origin in a gesture of repetition. Yet for Rosenzweig this disjunction of eternity and the "long chain of times" (*lange Kette der Zeiten*) does not imply a negation of eternity; on the contrary, it results in a concept of the eschatological now—what Rosenzweig calls the "today" (*Heute*)—that suspends the temporal gap between the present and the future.[45] There is "completion" (*Voll-*

---

45. This dialectic of the temporalization of theological concepts and the detemporalization of historical concepts is crucial. The Rosenzweig scholarship generally falls into two camps: those who emphasize his challenge to the metaphysical tradition in the temporal emphasis of the "new thinking" (Wiehl, Bieberich), and those who emphasize a resistance to history and historicism in Rosenzweig's thesis of the eternal people as "outside of time and history" (Myers, Kaplan, Hollander). This chapter claims that the tension between these perspectives is at the heart of the confrontation that Rosenzweig stages between history and eschatology.

endung) for the eternal people, he claims, just not completion in a temporal mode. Rosenzweig thus articulates a concept of eternity that spatializes the relationship of the present moment to completion: "For eternity is precisely this, that between the present moment and completion time may no longer claim a place, but rather already in the today every future is graspable" (*Denn Ewigkeit ist grade dies, daß zwischen dem gegenwärtigen Augenblick und der Vollendung keine Zeit mehr Platz beanspruchen darf, sondern im Heute schon alle Zukunft erfaßbar ist*) (S 365). Rosenzweig's concept of eternity, which Braitermann has called "the new anti-historicist figure par excellence,"[46] entails a detemporalization of historical time in which the space between the present moment and the future of completion no longer admits any temporal extension. This figure of eternity rests on a shape of time in which the gap between present and future becomes infinitesimally small, yet without resulting in their identification. Like Barth's notion of contemporaneity, Rosenzweig's figure of eternity implies a liminal superimposition of distant temporal moments that, because it cannot be conceived in historicist terms, avails itself of the rhetoric of spatial proximity.

In his interpretation of Judaism, Rosenzweig recognizes the potential of the present to bear a relation to its eschatological other, even as its temporality can never contain eternity. The figure of completion is the crossroads of this dual determination of the present. Completion, by definition, can take place only at the end of time—that end in which time is finally fulfilled—and yet when time is placed in relation to eternity, its end no longer has a temporal index. If completion remains unattained in time, then the historical can have no claim to eternity and will be cut off forever from it. Yet completion implies that time has come to an end and is finished, and hence that the end of time is essentially futural. Rosenzweig's concept of eternity circumvents this tension by conceiving eternity such that there is no deferral of completion into a distant future: time cannot insert itself between the present moment and the future of completion. This view implies that the present is excerpted out of time and placed in direct relationship to completion, without thereby

---

46. Braiterman, *The Shape of Revelation*, 140–41.

suggesting that time has come to its terminus. The eternity of such a present, for Rosenzweig, does not deny temporality but rather carves a space for an eschatological now within the chain of time.

In his interpretation of Christianity as "the eternal way" (*der ewige Weg*), which complements his theory of Judaism as "the eternal people" (*das ewige Volk*), Rosenzweig extends his spatial account of history and eschatology with the help of geometrical figures of paths, rays, and midpoints. His discourse is not concerned with a Christian position as such but with the specific spatial and temporal dynamics for which Christianity stands in his writing. The figure of Christianity that Rosenzweig generalizes and falsifies has its roots in the nineteenth-century Protestant theology of Schleiermacher and in Hegel's philosophy of history. The eternal way, which Rosenzweig describes as a historical "path" or "line" (*Strecke*) that extends outwardly following an originary pulse, reflects the historicization of Christianity in the wake of Schleiermacher and Hegel. At the same time, Rosenzweig's account of Christian temporality constructs the eternal way in relation to eschatology, thereby subverting Schleiermacher's and Hegel's historicization of Christianity. In Rosenzweig's work, the eternal way embodies the eschatological arc that breaks open the Hegelian circle and intensifies the tension of beginning and end. The spatial metaphors of "paths" (*Strecken*) and "rays" (*Strahlen*) thus have a dual function: from the perspective of history, they suggest linearity and continuity, yet from an eschatological perspective, they are curves representing an alternative shape of time.

The chapter of *The Star of Redemption* titled "Die Strahlen oder der ewige Weg" ("The Rays or the Eternal Way"), in which Rosenzweig lays out his reading of Christianity, returns to the eschatological problematic by delineating the temporal quality of an event that enters time but has its beginning and end outside time. In taking on this problematic, Rosenzweig steps back from the symbiosis of time and eternity that he worked out in his reading of Judaism. The concept of a today in which the future is already anticipated in the present, as we saw above, suggested a figure of eternity that nestles within the movement of time. The repetition and renewal of the moment in the cycle of the calendar year entailed an eternity that is completely contained within time, one whose beginning and end are temporalized in the mode of presence.⁴⁷ The event that structures the eternal way, by contrast, has a beginning and end that resist temporalization:

> If an event confronted [the rhythm of time] that had its beginning and its end outside of time, then the pulse beat of this event could regulate the striking of the hours of the world clock. Such an event would have to come from beyond time and flow into a beyond of time. In each present it would of course be in time; but because it knows that it is independent of time in its past and in its future, it therefore feels strong against time. Its present stands between past and future; yet the moment does not stand but vanishes as swift as an arrow and as a result is never "between" its past and its future, but before it could be between anything, it has already vanished. (S 374)

The "event" (*Geschehen*) conceived here has the shape of an eschatological arc, in contrast to the Hegelian circle that contains infinity within itself. The eschatological arc is suspended between an eternal beginning whence it comes and an eternal end toward which it tends. Since this event is not contained within time, the rhythm of historical time has no power over it. On the contrary, Rosenzweig suggests that the "pulse beat" (*Pulsschlag*) of such an event, insofar as its past and future are independent of time, has a world-historical intensity whose pulse provides a measure for historical time. Rosenzweig may well have Schleiermacher's concept of the inner-historical force of the figure of Christ in mind, yet his claim that the present of the event stands "between" past and future in a

---

47. For a nuanced reading of Rosenzweig's comparison of Judaism and Christianity, see Gregory Kaplan, who notes: "Insofar as the Jewish People stand in the present as the actuality (or, from a historical viewpoint, prospective fulfillment) of redemption, their ritual practice stands apart from the ordinary history which Christianity not only inhabits but, even more, conducts" ("In the End Shall Christians Become Jews and Jews, Christians? On Franz Rosenzweig's Apocalyptic Eschatology," *Crosscurrents* 53, no. 4 [2004]: 511–29, at 514). From the point of view of eternity, however, the situation is reversed: Rosenzweig thinks that Judaism articulates the presence of eternity in time, whereas Christianity places eternity outside historical time as its eschatological beginning and end.

more emphatic sense than the fleeting moment also resonates with Dialectical Theology's postulation of humanity as "between the times" (*zwischen den Zeiten*).[48]

The temporality of the eternal way entails a radical revision of historical time. In the historiography of world history, past epochs are preceded and succeeded by other eras, such that one can delineate points in time that structure relations of before, between, and after.[49] At stake in the eternal way, by contrast, is a between on a different scale: here the entirety of time is conceived as a present that is set off not by a before and after of historical time but by eternity:

> If the present too should thus be elevated to the baroness of time, it too would have to be a between [*ein Zwischen*]; the present, every present, would have to become epoch-making. And time as a whole would have to become an hour—this temporality; and as such stretched into eternity [*eingespannt in die Ewigkeit*]; eternity its beginning, eternity its end, and all time only the between that is between that beginning and that end [*das Zwischen zwischen jenem Anfang und jenem Ende*]. (S 375)

In this remarkable shape of time, the entirety of historical time is thought of as a moment of presence, one for which the true past and future lie in eternity. Time is held in tension between two eternities, an eternity of the beginning (creation) and an eternity of the end (redemption). The entirety of time between the coming of Christ and his return in the moment of *parousia* is thus conceived as a single epoch, *Christuszeit*, with the historical past and historical future be-

---

48. While it is unlikely that Rosenzweig had read Barth's *The Epistle to the Romans* when he wrote *The Star of Redemption*, since he composed his work from the front during the First World War, a letter from Rosenzweig to Siegfried Kracauer dated June 5, 1923, makes it clear that Rosenzweig was familiar with Barth's work and approached it with a certain degree of skepticism: "Dass also Gott kein Mensch und der Mensch kein Gott ist, sagt so wenig etwas gegen beide, dass es vielmehr die 'Beziehung' erst möglich macht, jedenfalls die absolute Beziehung. . . . Gott greift doch nicht als ein 'fremder Gott' in diese Welt ein,—wäre es so, dann wäre es freilich zum Verzweifeln oder zum Professor-für-reformierte-Theologie-in-Göttingen-werden, sondern in seine Welt, in seine Schöpfung" (Stephanie Baumann, "Drei Briefe: Franz Rosenzweig an Siegfried Kracauer," *Zeitschrift für Religions- und Geistesgeschichte* 63, no. 2 [2011]: 166–76, at 176).

49. Rosenzweig refers to "Epochen, Haltepunkte in der Zeit" (S 374).

ing nothing more than aspects of this extended presence between the times (S 375).

This figure of time has implications for the concept of eschatology, because it displaces the end from its position as a terminal point in time and places it in direct relation to the present at each moment. Extending his spatial representation of the temporality of the eternal way, Rosenzweig stretches the limits of the imagination of this space by proposing that each point in time is equally near the beginning and the end and is hence a midpoint:

> Time is now mere temporality. As such, it can be surveyed entirely from each of its points; for beginning and end are equally near to each of these points; time has become a single path, but a path whose beginning and end lie beyond time, and hence an eternal path; whereas on paths that lead from out of time into time only the next section can be surveyed. On the eternal way, on the other hand, because of course beginning and end are equally near regardless of how time advances, each point is a midpoint [*Mittelpunkt*]. (S 375–76)

From the perspective of Euclidean space, Rosenzweig's geometry of midpoints on the eternal way appears paradoxical, since movement along a line increases or decreases the distance to a given starting point or endpoint. This is because the scale of distance is at all points uniform in a system of Cartesian coordinates in Euclidean space. In non-Euclidean geometries, by contrast, the scale of distances can change as one approaches a boundary. For example, consider Klein's construction of a scale of elliptically equidistant points on the line (figure 4). Here we can observe that points that are elliptically equidistant on the circumference of a circle become infinitely distant when projected onto a line. Similarly, we saw above that points on the limit circle in Beltrami's disc model appear to have a finite distance but are in fact infinitely distant. This discrepancy points to a basic problem of non-Euclidean geometry; namely, that hyperbolic and Euclidean distances are not necessarily the same.[50] The reason is that the projections conceived by Beltrami, Klein, and Poincaré

---

50. See Riku Klén and Matti Vuorinen, "Apollonian Circles and Hyperbolic Geometry," *Journal of Analysis* 19 (2011): 41–60.

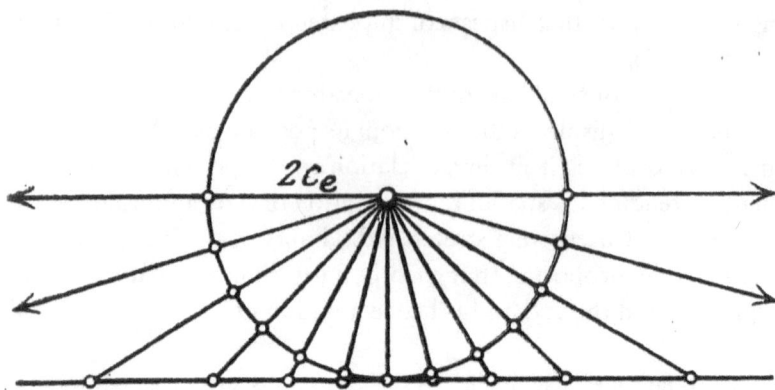

Figure 4. Construction of a scale of elliptically equidistant points on a line. Felix Klein, *Vorlesungen über Nicht-Euklidische Geometrie* (Berlin: Springer, 1928), 171. Reprinted with permission from Springer Verlag.

are not isometric. Analogously, Rosenzweig's construction of the eternal way entails a projection of points from the space of time to the space of eternity. In the former, there can be only a single midpoint between two points defined as starting and endpoints, but in the space of eschatology, all points on the path are equidistant to the beginning and the end.

While the metaphors of the way (*Weg*) and path (*Strecke*) suggest that the points along the way precede and succeed one another, in Rosenzweig's image the beginning and end are not of the same order as points in time. Time, because it is fixed between two eternities, becomes "mere temporality" that is devoid of development. If time is conceived as an arc spanning two eschatological extremes, this arc has no high point. Instead, the high point is the event that transforms the beginning and the end into eternal beginnings and eternal ends. In such an event, the "notches [*Kerben*] that are struck [into time]" are so deep, Rosenzweig claims, that there is no time between the present and the eternal beginning and eternal end (*S* 375). Here we can observe a striking similarity to Rosenzweig's account of eternity in Judaism in which there is no space between the present and completion. In Christianity, he suggests, eschatology does not mark a horizon that looms in the distance, but is equally close to each moment on the eternal path: "Both beginning and end are for [the Christian] at every moment equally near, because both are in eternity; and only for this reason does he know himself at each moment to be a midpoint. A midpoint not of a horizon that he surveys, but rather of a line that consists of nothing but midpoints, indeed which is entirely middle [*ganz Mitte*], entirely between, entirely path" (*S* 376–77). The "line" (*Strecke*) that defines this path has a distinctly nonteleological character, because there appears to be no motion along the line from start to finish but only a collection of midpoints. Because each point on the line is equally a midpoint, there is no development—instead, there is only a single moment of presence. By emphasizing the proximity of each moment to the eschatological beginning and end, Rosenzweig's account of Christianity shares an important kinship with Barth's Dialectical Theology.

In Rosenzweig's work, Jewish and Christian temporalities approach the eschatological tension of beginning and end with different emphases: Christianity is driven by an eternal beginning and dwells in the present of revelation, while Judaism is motivated by confidence in a messianic future and is therefore characterized by its hope of redemption.[51] Yet in the final analysis, Rosenzweig overcomes this typology and shows how its elements exist in a state of mutual relation and dependence. The ultimate perspective from which such a dependency of temporal moments upon one another can be grasped is an end that surpasses any end that might be associated with a messianic future. In the final book of *The Star of Redemption*, Rosenzweig introduces a vantage point from which the opposition of Christianity's outward-directed path, gathered in a present that repeats the beginning, and Judaism's inward-directed life, focused always on an imminent messianic future, is finally overcome.[52] These modalities exhaust themselves in "the truth" (*die*

---

51. See *S* 385–86: "Therefore Christian consciousness, absorbed entirely in faith, pushes toward the beginning of the path, to the first Christian, to the crucified one, just as Jewish consciousness, gathered entirely in hope, toward the man of the last days [*zum Manne der Endzeit*], to David's royal scion."

52. See *S* 386: "Rootedness in the deepest self [*Verwurzelung ins tiefste Selbst*], this had been the secret of the eternity of the people. Expansion through all

Wahrheit), and the truth, according to Rosenzweig, has an eschatological figure. From the perspective of eschatology, the Jewish and Christian temporalities converge, and their polarity dwindles. To develop this claim, Rosenzweig considers an eschatology of a second order that features not an eternal end that sustains time in tension with an eternal beginning, but the end of all ends.

In a section titled "Die Gestalt der Bewährung: Eschatologie" ("The Shape of Verification: Eschatology"), Rosenzweig traces the eternal nature of the Christian path and Jewish life to the point where they confront the figure of death despite their eternity:

> The rays of the star that refract outwardly in this way, the fire that glows inwardly—neither rest until they have arrived at the end, at the outermost, at the innermost [*ans Ende, ans Äußerste, ans Innerste*]. Both draw everything into the circle that their activity fills. Though for the rays it is by branching off in the outside, dispersing, going separate ways that unite again only beyond the fully traversed outer space of the primordial world; for the fire, however, it is by gathering and collecting in itself the rich diversity of existence into oppositions of inner life within the flickering play of its flame: oppositions that likewise find their unity only where the flame may be extinguished, because the burnt out world does not offer it any more combustible material and the flickering life of the flame dies away in what is more than human-worldly life: the divine life of truth. For this, the truth is our concern here, no longer the splitting of the path in the visible world, no longer the inner opposition of life. But the truth always appears only at the end. The end is its place. (S 442–43)

The truth of the eternal way of Christianity (i.e., the rays) and of the eternal people of Judaism (i.e., the fire), Rosenzweig suggests, arises only from the perspective of their ultimate end. The approach to the eschatological end has a different quality and texture for Christianity and Judaism, yet they meet at this limit: the metaphors of "radiance" (*Strahlen*) and "fire" (*Feuer*), the two dimensions of the star of redemption, exhaust and consume themselves so that the star can emerge as the eschatological figure *par excellence*. The temporalities of radiance and fire, according to Rosenzweig, are each constituted

---

that is outside [*Ausbreitung durch alles Außen hin*]—this is the secret of the eternity of the path."

in relation to an end at the extreme, whether it be an outermost or an innermost extreme, and they take no rest until they reach this end. In doing so, however, they break apart, precisely because they leave no ground uncovered: the radiance disperses, splits apart into separate ways, and thus finds no unity in the externality of space, whereas the fire collects and pulls together the diversity of life, holding together its contradictions. These contradictions can be united only when the flame is extinguished. The final dispersion of the radiance and the dying down of the fire therefore stand for that end in which the divided is unified. This eschatological end, embodying the metaphorical space of death by exhaustion and depletion, is for Rosenzweig "the divine life of truth." Such truth, Rosenzweig claims, can be glimpsed only by following the splitting of the path and the contradictions of life to their death at the limit.

Commentators on Rosenzweig's "dual-covenant theology" often claim that Judaism and Christianity will be united at the end of time. In doing so, they project this end into the future. For example, Jacob Taubes writes:

> Until the end of history, Rosenzweig argues, the "eternal life" (the suprahistorical fate of Israel) and the "eternal path" (the historical fate of the gentile peoples in Jesus Christ) are separated. Only at the end of time, when the son of God hands over the "Kingdom" to God and gives up his power and authority, will eternal life and eternal path come together; Israel will cease to live in the eternal, divine present as a sacred people when Christ ceases to rule over the eternal path of the peoples.[53]

Similarly, Gregory Kaplan concludes that "the Christian approach to and the Jewish accomplishment of living with God are coeval, structurally equivalent positions. The end of time (merely) 'restores' their coincidence following a provisional separation."[54] The problem with these interpretations is that they depend on a temporalized notion of eschatology as the terminus of history, whereas the

---

53. Jacob Taubes, *Der Preis des Messianismus: Briefe von Jacob Taubes an Gershom Scholem und andere Materialien*, ed. Elettra Stimilli (Würzburg: Königshausen & Neumann, 2006), 14–15.

54. Kaplan, "Rosenzweig's Apocalyptic Eschatology," 514.

metaphorics surrounding Rosenzweig's star are spatial. Ultimately, both Judaism and Christianity are oriented toward an end that cannot be conceived by analogy to the temporal moments of past, present, and future. The end of the star of redemption, for Rosenzweig, is at once "the outermost extreme" (*das Äußerste*) and "the innermost extreme" (*das Innerste*). In Rosenzweig's dual-covenant theology, Christianity and Judaism converge—keeping in mind the root meaning of the term *Eschaton* as the extreme—at the eschatological limit at the heart of his figure of the star.

From Curves to Shapes: The Mathematics of Figural Thinking

Rosenzweig develops his spatial and geometrical account of the end of history, finally, with the help of mathematics, which he famously referred to, following Hermann Cohen, as an "organon of thought" (*Organon des Denkens*) (*S* 23).[55] In his correspondence with Ehrenberg and Rosenstock-Huessy, he uses irrational numbers and differential calculus to support his reading of a dual-covenant theology in which Judaism and Christianity are both contrapuntal and correlative. And in *The Star of Redemption*, he draws on non-Euclidean geometry to show how the lines connecting the fundamental theological concepts of God, human being, and world can be conceived as curves. Mathematics thus facilitates Rosenzweig's development of nonintuitive geometrical figures and what he calls "supra-mathematical" shapes (*Gestalten*), culminating in the figure of the star.

In a letter to Ehrenberg from May 10–11, 1918, during the time when he was working on *The Star of Redemption*, Rosenzweig elab-

---

55. On the importance of mathematics for Rosenzweig's work, see Smith, "The Infinitesimal as Theological Principle," 562–88; and Handelman, *The Mathematical Imagination*, 104–44. Smith and Handelman explore Rosenzweig's use of differential calculus, building on Hermann Cohen's work on the infinitesimal, as a tool to generate something out of nothing. Whereas Smith detects an affinity between Rosenzweig and Barth in sketching a liminal relationship between God, world, and humanity, Handelman aligns Rosenzweig with Newtonian mechanics and an Enlightenment tradition of messianism that emphasizes the agency of the subject.

orates his thoughts on the difference between Judaism and Christianity using a mathematical analogy of rational and irrational numbers. Arguing here, as he does in *The Star of Redemption*, that Christianity is oriented toward the beginning and Judaism toward the end, Rosenzweig claims that Christianity affirms the "intermediate realm" (*Zwischenreich*)—a concept denoting the space between the arrival of the Messiah and the inception of the Kingdom of God—while Judaism negates it. Christianity's affirmation of the "between" (*Zwischen*) posits the beginning as positive and the end as negative, whereas Judaism's denial of the *Zwischen* does just the opposite: "The beginning of the intermediate realm, the arrival of the Messiah, has not yet transpired; the end, the Kingdom of God, has already begun, is already here, is already given today for every Jew in the immediate, ultimate relation to God himself, in the daily 'shouldering of the yoke of the kingdom of heaven' through the fulfillment of the law" (B 316). Rosenzweig expounds on these inverse polarities of Judaism and Christianity by comparing Christianity to rational numbers and Judaism to irrational numbers:

> In the mathematical analogy you can see at once how Judaism appears from the perspective of Christianity. The irrational number—what does it mean for the rational number? For rational numbers the infinite would be the limit that is eternally unattainable [*ewig unerreichbare Grenze*], both upwardly and downwardly, something eternally *improbable*, although it is certainly and always true. Only in the irrational number does this limit of the rational order of numbers strike physically, numerically, and presently against each of its individual points, redeeming it from the "linear" abstraction and uncertainty of its one-dimensionality and becoming a "spatial" totality [*räumlichen Ganzheit*] and thereby a self-assured reality. As an infinitesimal number, the infinite is the secret, eternally invisible driving force of the "visible" reality of the rational number. As an irrational number, by contrast, the infinite is overt, visible, and yet eternally foreign—a number and yet not a number, a "non-number" [*Unzahl*].... Only with non-numbers, real visible non-numbers, and not with invisible operative primal numbers, the infinitesimal numbers, is the world of numbers complete. (B 316–17)

On the basis of this analogy, Rosenzweig develops two relationships to infinity, characteristic respectively of Judaism and Christianity. From the standpoint of *rational* numbers (Christianity), infinity is a magnitude conceived in quantitative terms as an "eternally unattainable limit." This infinity is one-dimensional and the product of a "linear abstraction" in two respects. First, in the direction "nach oben," for every natural number "n," we can posit an "$n + 1$" to infinity. Second, in the direction "nach unten," the set of rational numbers contains all fractions of integers (e.g., p / q, where p and q are integers), to the infinitesimal limit of $1 / (n +1)$, where n is taken to infinity. For this reason, Rosenzweig considers the *Infinitesimalzahl* to be the "invisible driving force" of the "visible reality" of rational numbers. This quantitative infinity of rational numbers corresponds to what Rosenzweig, in *The Star of Redemption*, conceives of as Christianity's "expansion through all that is outside" (*Ausbreitung durch alles Außen hin*) (*S* 386).

By contrast, the infinity of *irrational* numbers, which for Rosenzweig provides an analogy for Judaism, is "manifest, visible, but nevertheless eternally foreign." Consider, for example, the irrational number $\sqrt{2}$, which can be represented as the hypotenuse of a right triangle whose sides are each 1, or the mathematical constant π, which defines the ratio of a circle's circumference to its diameter. In such irrational numbers, the one-dimensionality and "linear abstraction" of rational numbers becomes a "spatial totality"—that is, embodied and visible in the geometric figure, thus very much real, yet as a "number" something foreign to the set of rational numbers, hence a "non-number" (*Unzahl*). Mosès describes the infinity of irrational numbers in Rosenzweig's letter as a "qualitative infinity" or "abundance infinity."[56] Whereas the set of rational numbers is an accumulation of fractions, irrational numbers are indivisible: they cannot be placed on the line of rational numbers, yet they can be represented spatially in figures. If the quantitative infinity of rational numbers can be compared to Christianity's "expansion through all that is outside," then the qualitative infinity of irrational numbers corresponds to Judaism's "rootedness in its own innermost" (*Verwurzelung ins tiefste Selbst*) (*S* 386). The incommensurate *Unzahl* of the irrational number thus provides a metaphor for Rosen-

---

56. Mosès, *The Angel of History*, 53.

zweig's turn from the linear model of historical development to a spatial model of religion in terms of figures.[57]

Already in his letters to Rosenstock-Huessy dating from late 1916, Rosenzweig distinguishes between Christianity and Judaism along an axis of exteriority and interiority. Yet in contrast to the polarity observed above, in these letters he insists that in terms of their "content of piety" (*Gehalt der Frömmigkeit*), Judaism and Christianity are coeval. To make this point, Rosenzweig uses a metaphor taken from differential calculus to assess the relative status of beginning and end in these two religions:

> Christianity has its soul in its externals; Judaism, on the outside, has only its hard protective shell, and one can speak of its soul only from within.... You must take my word for it that the abstract character, as it were, of the content of piety is the same with us [the Jews] and with you [the Christians]. Beginning and end, if I may so express it, are the same with us and with you; to use Newton as an assistant in metaphor, the "continually emerging" [*eben entstehende*] and the "continually vanishing" [*eben verschwindende*] curves have the same formula for both of us, and you know that one can define the whole curve from such differential quotients, but you and we choose different points on the path of the curve in order to describe it, and therein lies our difference. (B 688–89)

As in the letter to Ehrenberg, Rosenzweig argues for a correspondence of beginning and end in Christianity and Judaism. But here the accent falls on the similarities between the two religions rather than on their incommensurable differences. Beginning and end, in this spatialization of religion, are fixed points connected by a curve. Using Newtonian calculus as a metaphorical aid, Rosenzweig describes the determination of this curve on the basis of a differential quotient (i.e., $\lim (\Delta x \to 0) \Delta y/\Delta x$), in which, as the distance between points becomes infinitesimally small, the curve can be approximated by its tangent. It makes no difference, Rosenzweig claims, whether

---

57. Rosenzweig's claim that it is only with "non-numbers" that the "world of numbers is complete" can be compared to the mathematician David Hilbert's claim that the theory of real numbers provides a foundation for all geometries, both Euclidean and non-Euclidean. See Stillwell, *Sources of Hyperbolic Geometry*, 65.

one analyzes the "emerging curve" (i.e., the curve that arises from the beginning, in Christianity) or the "vanishing curve" (i.e., the curve that is oriented to the end, in Judaism); in each case, the curve has the same formula. In other words, Judaism and Christianity describe the same curve, but use different points on the curve for their analysis.

In his response to this letter, Rosenstock-Huessy contests Rosenzweig's image of a common curve of piety in Judaism and Christianity. His critique has two prongs. First, he grants that the curves have the same shape, but he denies that they occupy the same space. Invoking the geography of Romania, with which Rosenzweig was familiar from his wartime post on the Balkan Front, he writes: "You can of course say that the Carpathian province of Siebenbürgen is equal in surface area [*flächeninhaltlich gleich*] to the Wallachian Plain, but have you then achieved anything? In the same way, it is true that piety is piety, and remains so, and in this sense it is found among Jews and Christians" (*B* 705). The regions of Transylvania (*Siebenbürgen*) and Wallachia (*Walachei*), the argument goes, may be comparable in terms of their surface areas, but they have very different elevations and are separated by the Carpathian Mountains. Similarly, Jewish piety and Christian piety may have similar curves, yet they are by no means identical. Second, Rosenstock-Huessy challenges the mathematical basis of Rosenzweig's use of Newtonian calculus: "But there is still one nut to crack: the always identical piety for Jew and Christian.... I can't follow you there, unless you only mean it as a hyperbole; namely, if for you [A – B] [A ~ B] are two equal curves just because they both touch the same two points" (*B* 704). Two curves, Rosenstock-Huessy suggests, need not be identical simply because they each touch two identical points. As his shorthand shows, a straight line passing through points A and B and a curve passing through these same points are by no means identical, even if, as in Rosenzweig's scenario, they contain the same points.

Yet despite this skepticism, Rosenstock-Huessy admits a potential point of agreement on the condition that Rosenzweig intends his postulate of an "always identical piety for Jew and Christian" as a "hyperbole" (*Hyperbel*). The term *Hyperbel* is both a concept in rhetoric (hyperbole), signifying the use of exaggeration as a rhetorical device, and a mathematical concept (hyperbola), defined as a special curve consisting of two symmetrical branches, each of which extends into infinity. In a first reading of the passage, Rosenstock-Huessy seems to intend "hyperbole," since he proceeds with a counterexample. But a reading of *Hyperbel* as "hyperbola" also yields important insights. As in the curve representing the function $f(x) = 1/x$, the two branches of a hyperbola share a common set of asymptotes and appear as mirror images of one another. In this reading, the piety curves of Judaism and Christianity would be at once identical, and indeed determined by the same function, and diametrically opposed—that is, oriented toward fundamentally different kinds of infinity. A common formula of piety in Judaism and Christianity, Rosenstock-Huessy suggests, may yield two radically different curves.

Rosenzweig has the final word in this exchange, and in his rejoinder he grants that Rosenstock-Huessy is correct that piety (*Frommsein*) has an entirely different reality in Judaism and in Christianity. Indeed, he is willing to grant that these two conceptions of piety are "contrary to one another" (*etwas Entgegengesetztes*) (B 716). However, in keeping with his larger aim of showing the common root of Judaism and Christianity, despite their apparent differences, Rosenzweig immediately adds the caveat that Jewish and Christian piety are "correlatively opposed" (*korrelativ entgegengesetzt*):

> You are certainly right that piety, when it has become a living reality, that is to say, the pious human being ... is something entirely different in Jew and Christian, and even an opposition, though admittedly a correlative opposition like a suture of two bones dovetailed together (thus before God they are the same, but before men they are directly opposed). But behind the images on these two coins is hidden the same metal. The forms of sanctity themselves are different, but ... this piety is what they share in common. (B 716)

Whereas earlier in his exchange Rosenzweig used mathematical metaphors, here he turns to anatomy, comparing Jewish and Christian piety to "two interlocking sutures" (*zwei ineinanderverzahnte Knochennähte*) to illustrate their correlative opposition.

The metaphor is well chosen, since a suture (*Knochennaht*) refers to a joint in which "the jagged edges of two bones are tightly interlocked."[58] In human anatomy, sutures are a prominent feature in the skull; whereas in youth these cranial sutures admit movement to accommodate the growth of the brain, in old age they fuse together and become ossified. Thus, while he grants that Jewish and Christian forms of piety are "fundamentally different" (*etwas ganz Verschiedenes*), Rosenzweig reconfigures the terms of this opposition to suggest an underlying unity, which is ultimately a unity before God: "before God they are the same, but before men directly opposed" (*vor Gott einerlei, vor Menschen schnurstracks entgegengesetzt*) (*B* 716). Insofar as the diametrical opposition of these pieties is also correlative, the jagged interlocking teeth of the fault line become fused as a unity of differences. To underscore this point, Rosenzweig notes that these pieties are made of up the "same metal" (*ein gleiches Metall*), thus pointing to their common materiality. In transposing the anatomical metaphor of the suture into a metallurgical metaphor, Rosenzweig suggests the welding together of two materials, which, despite their divergent appearances, are fundamentally of one substance.[59]

Rosenzweig's turn from mathematics to anatomy and metallurgy in his exchange with Rosenstock-Huessy anticipates a similar turn from geometry to supra-mathematical spatial forms in *The Star of Redemption*. A key example of Rosenzweig's interest in supra-mathematical figures can be found in his reflections on the arc versus the circle. As discussed above, the circle, for Rosenzweig, is emblematic of a Hegelian concept of infinity in which the end is a return to the beginning. The arc, by contrast, expresses a Judeo-Christian eschatological concept of history. Rosenzweig takes pains to distinguish his model of the arc from the formulaic concept of a hyperbola, argu-

---

58. *Meyers Großes Konversations-Lexikon*, vol. 11 (Leipzig: Bibliographisches Institut, 1907), 186: "die zackigen Ränder zweier Knochen dicht ineinander greifen."

59. Rosenzweig goes on to argue that the correlative opposition of Jewish and Christian piety can be traced back to a common objective origin in revelation or *Offenbarung*. He further clarifies that while for the poet (*Dichter*) the opposition is visible, the common "potentiality" of Jewish and Christian piety is apparent to the thinker (*Denker*) (*B* 716).

ing that the arc, unlike the hyperbola or the circle, does not have a "point of reference" that anchors the figure. Instead, a line is constructed through a series of points without a "law of construction" (S 284).

In tracing an arc that bursts forth from infinity and plunges back into infinity, Rosenzweig insists on the "impossibility of formulating the course of its path" (*Unformulierbarkeit des Bahnverlaufs*) and notes that the relationship of the three points through which the arc is constructed is by no means geometrical: "If a relationship existed between the singular points, obviously this could not be in the manner of a geometric relationship" (S 284). Similarly, he notes that a set of three lines connecting three points (that is, a triangle) should not be considered geometrical lines:

> And the three lines to which, in the three books of this part, we joined the three points discovered in the first part, are really not lines in the geometric sense nor the shortest connections between two points. Rather, these lines sprang from these points through an act of reversal [*Akt der Umkehrung*] grounded in the history of the emergence of these three points, yet in itself groundless—therefore real lines, and not mathematical ones [*wirkliche, nicht mathematische Linien*]. (S 284)

The points that Rosenzweig mentions here designate the elemental categories of "God" (*Gott*), "world" (*Welt*), and "human being" (*Mensch*). He describes the lines between these points as real lines insofar they are grounded in the genesis of the points, rather than as mathematical lines that are constructed as the shortest path between two points. Pointing to Rosenzweig's knowledge of differential calculus and his interest in Hermann Cohen's work on the infinitesimal, John H. Smith argues that Rosenzweig is concerned here not with "straight lines connecting fixed points" but with "curves emerging from the non-points or the limit of the tangent."[60]

The idea that lines can be conceived as curves was foundational for Riemann's differential geometry. As Gray notes, Riemann measured length at the limit by using an "infinitesimal ruler," allowing one to find "the geodesic between two points and compute the

---

60. Smith, "The Infinitesimal as Theological Principle," 574–75.

curvature of the surface as Gauss had shown. If the surface is a plane, a geodesic is a straight line, and on a sphere it is a great circle."[61] This means that a straight line is the shortest route between two points only in Euclidean space, but given a space of constant curvature other than zero, the most direct path between two will be defined by a curve. Together with Gauss's work on the curvature of surfaces, Riemann's differential geometry showed that curved space could have "genuine geometric ... properties even if it was not embedded in a Euclidean space," thus destroying "the paramount status of Euclidean geometry and Euclidean space as the source of geometry."[62] Just as non-Euclidean geometry showed the consistency of spaces that exceed our capacity for visualization, so too does Rosenzweig construct a figure of the relationship of God, world, and human being that explicitly undermines the foundations of mathematical intuition—at least in its commonplace, Euclidean form. Indeed, he claims that the reality of the lines between God, world, and human being can be adequately described only under the condition "that one expressly and intuitively disrupts the concept of the mathematical line according to which it is the shortest connection between two points" (*daß man den Begriff der mathematischen Linie, wonach sie eben kürzeste Verbindung zweier Punkte ist, ausdrücklich und anschaulich stört*) (S 285). This disruption of the intuitive visibility of the mathematical line has an affinity with non-Euclidean geometries, which, when constructed on a surface of constant curvature other than zero, entail that the shortest route between two points will be a curve.

Just as mathematicians such as Helmholtz, Beltrami, and Poincaré produced analogies and projections enabling the visualization of nonintuitive and multidimensional spaces, Rosenzweig aims to give intuitive visibility to his disruption of the intuitive concept of the line. He thus introduces a set of three additional points to demonstrate the curvature of the lines between the points of God, world, and human being:

---

61. Jeremy Gray, *Worlds Out of Nothing: A Course in the History of Geometry in the 19th Century* (London: Springer, 2011), 199.

62. Gray, *Worlds Out of Nothing*, 308–9.

If this process is itself to take place with a kind of mathematical intuitiveness [*mathematischer Anschaulichkeit*], it would have to take place in such a way that the line itself, although it is already sufficiently determined as a mathematical line by the two points, would also be designated by a point of its own, and certainly if those first three points correspond to the elements God, world, man, these three new points that stand for the paths creation, revelation, redemption would have to be situated in such a way that the triangle formed by them would not be inside the preceding triangle; . . . on the contrary, the joining of a point to two others must itself pass through the line of the original triangle, such that the two triangles overlap by intersecting. (S 285)

The superimposition of these triangles, of course, results in the eponymous star of *The Star of Redemption*. The line between God and human being, in this figure, is constituted, not as the shortest path between these points, but through a third point—namely, revelation—which does not lie on the line between God and human being, yet gives rise to two lines that intersect this line. The line between God and human being, one might say, is bent toward revelation (*Offenbarung*), just as the line between God and world is indicated through creation (*Schöpfung*), and the line between human being and world tends toward redemption (*Erlösung*) (figure 5). Rosenzweig appears to be aware of the cognitive dissonance that results from his claim that the star does not consist of mathematical lines, given that the lines connecting the points of the star do indeed appear to follow the shortest path between these points. What, then, is at stake in Rosenzweig's claim that the reality of these lines explicitly and intuitively disrupts the concept of the mathematical line?

In a comment that goes to the core of his interest in figures and spatial forms, Rosenzweig clarifies that while the elements of the star of redemption are geometrical, the figure itself is a supra-mathematical figure:

But really a figure thus arises that may be geometrically constructed, but is itself foreign to geometry, namely not a "figure" [*Figur*] at all—but a Gestalt. For a Gestalt is differentiated from a figure by the fact that while it may be composed of mathematical figures, in truth its composition did not take place according to a mathematical rule, but according to a

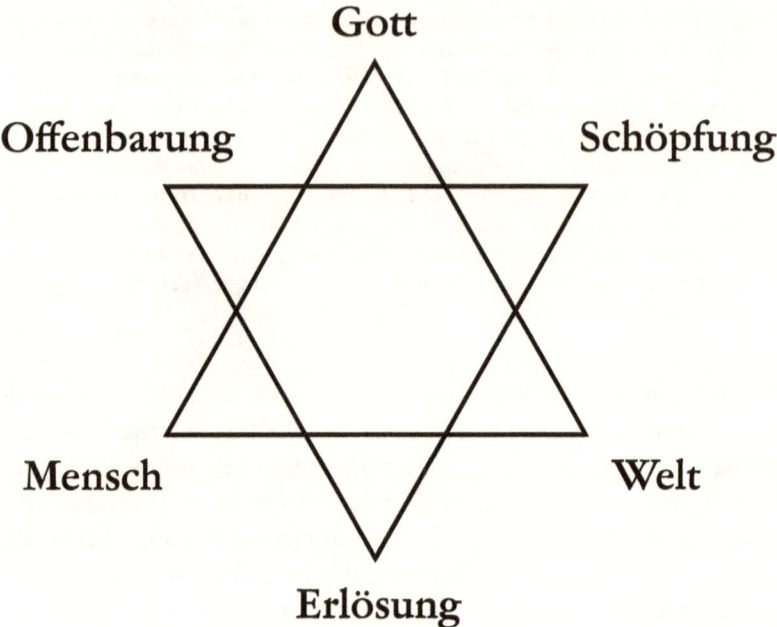

Figure 5. The star of redemption.

supra-mathematical principle; here this principle was provided by the idea of characterizing the connections of the elementary points as symbols of a real event instead of as realizations of a mathematical idea. (S 285)

Echoing his earlier criticism of the formulaic construction of the hyperbola, Rosenzweig distinguishes between a *Figur* that is constructed according to a mathematical rule and a *Gestalt* that has a "supra-mathematical" (*übermathematischen*) ground. The star of redemption, he claims, is in this sense a *Gestalt* rather than a *Figur*: although it is constructed in geometrical terms—as an assemblage of lines and triangles—it is ultimately a figure utterly foreign to geometry. As a *Gestalt*, the star of redemption has a symbolic value representing a set of real relations; it is not merely the realization of a mathematical idea. In short, Rosenzweig works with geometrical and mathematical figures, but he insists that their genesis is not in mathematics but in history, such that these spatial forms serve as representations of a philosophy of religion.

Indeed, it is as a representation of the relations among elementary religious categories such as God, world, and human being that Rosenzweig's star reconfigures geometrical figures such as points and lines into a supra-mathematical *Gestalt*. The transformation of *Figuren* into a *Gestalt*, Rosenzweig claims, lifts these figures out of the realm of mathematics: "But simple geometric forms, like points and lines, can acquire Gestalt only by being lifted out of the vital element of mathematics; this vital element is universal relativity" (*S* 285). The rhetoric of being "lifted out" (*herausgehoben*) of the realm of mathematics in this passage echoes the antihistoricist trope of being lifted out of history. Like history, mathematics, according to Rosenzweig, is defined by the relativity of its figures: "the direction of a line depends on the direction of a line of reference [*Beziehungslinie*] originally adopted arbitrarily, the position of a point on a point of departure of a system of coordinates [*Koordinatenausgangspunkt*] originally established arbitrarily" (*S* 286). By contrast, the star of redemption is defined by its absolute position and direction and is in this sense a "demathematicized figure": "If the points and the lines of the two triangles with which we are dealing are to become demathematicized figures [*entmathematisierten Gestalten*], they must be given an absolute position and direction" (*S* 286). Rosenzweig thus situates his figure of the star in a theological context that allows for a thinking in *Gestalten* that leaves behind the relative and formulaic constructions of mathematical logic.

The trajectory that Rosenzweig traces here marks the culmination of a larger set of transpositions in *The Star of Redemption*, in which Rosenzweig's methodology undergoes a series of revisions in accordance with the development of his argument. He begins with mathematics as an "organon of thought" and introduces the differential as a means of producing something from nothing. In the later sections of the book, an elemental mathematics is reformulated and transposed into the domains of spatial geometry, language and grammar, and finally liturgy.[63] Whereas a "mathematical language of signs"

---

63. Peter E. Gordon, drawing on the work of Amos Funkenstein, comes to a similar conclusion, noting that, for Rosenzweig, "mathematics cannot itself be the governing 'principle' of reality" but rather has a "metaphorical function" in *The*

(*mathematische Symbolsprache*) is capable of expressing a reversal of the negative into the positive (of the *Nein* into the *Ja*), geometrical symbols, Rosenzweig notes, have the advantage of being able to represent "the reciprocal influences of the points in space ... through which the various concepts are symbolized" (*S* 138).[64] Geometry refines the elemental language of mathematics, in other words, because of its more precise ability to capture the spatiality of the way concepts relate to one another. Similarly, while mathematics provides the organon for conceptualizing the "primordial world" (*Vorwelt*)—the elemental sphere in which the world emerges out of nothingness—Rosenzweig argues that the world of revelation requires a different organon: "Instead of a science of mute signs, there has to come a science of living sounds; instead of a mathematical science, there has to appear the doctrine of linguistic forms, grammar" (*S* 139). Yet Rosenzweig does not view grammar as having simply supplanted mathematics; on the contrary, he describes language as "the truly 'higher' mathematics" (*S* 167). If mathematics provides the symbolism of the "elements" (the subject of the first part of *The Star of Redemption*), language allows for a description of the "course" (*Bahn*) of these elements (the subject of the second part of Rosenzweig's book). Analogously, the third part of *The Star*, titled "Die Gestalt oder die ewige Überwelt" ("The Figure or the Eternal Supra-World"), is the province of liturgy: "For this third part, therefore, liturgy will assume a similar role as organon, just as mathematics did in the first part and grammar in the second" (*S* 327). In this succession of organons, mathematics and geometry provide the metaphorical foundation for a set of instruments geared toward the production of figures.

Rosenzweig's *Star of Redemption* is constructed, in each of its parts and stages, as a language of forms, figures, and *Gestalten*. In a letter to Margrit Rosenstock-Huessy from August 22, 1918, in which the figure of the star of redemption is sketched and described for the first

---

Star of Redemption ("Science, Finitude, and Infinity: Neo-Kantianism and the Birth of Existentialism," *Jewish Social Studies* 6, no. 1 [1999]: 30–53, at 46). Serving as a means of describing the interactions of the primal elements of creation, revelation, and redemption, mathematics, Gordon claims, is for Rosenzweig a "helpful yet ultimately provisional tool" (46).

64. Rosenzweig, *The Star of Redemption*, trans. Galli, 136.

time, Rosenzweig writes: "I think in figures" (*Ich denke in Figuren*).[65] These figures begin as elemental mathematical figures, are grasped in terms of their spatial relations as geometrical figures, become rhetorical figures or figures of language, and finally are treated as religious or liturgical figures. While Rosenzweig leaves behind the formulaic qualities of mathematics, and indeed experiments with figures that probe the limits of intuitive visibility, the fundamental spatial and figural parameters of *Gestalten* remain the core element of his thought.

In the third book of *The Star of Redemption*, Rosenzweig directs his attention to a "future of redemption" as it is anticipated in "eternal figures" (*ewigen Gestalten*).[66] Here Rosenzweig is finally able to grasp the moment as a *Gestalt*—that is, to discern the coordinates of the eschatological moment as a configuration of spatial forms. The emergence of the *Gestalt*, for Rosenzweig, depends upon a standstill of the moment as an eternal moment: "Only that which lasts for more than a moment can be seen with the eye; only the moment that has been brought to a halt by becoming eternal permits the eye to see the Gestalt in it" (*S* 328). Playing on the literal meaning of the term *Augenblick*, Rosenzweig claims that, paradoxically, the eye (*Auge*) is only able to catch sight (*erblicken*) of what lasts for more than a moment (*Augenblick*). Rosenzweig is concerned here with the perception of a *Gestalt* within the moment, and this is impossible if the moment is ephemeral or fleeting. To apprehend the *Gestalt*, Rosenzweig configures the moment as "the moment that has been brought to a halt by becoming eternal" (*der durch seine Verewigung stillgelegte Augenblick*), an expression that at once detemporalizes the moment by halting the flow of time and articulates its "becoming eternal" (*Verewigung*). By endowing the moment with the qualities of a *Gestalt*, Rosenzweig underscores his central insight that history is defined by not only the movement of time but also its spatial configuration.

Rosenzweig's "thinking in figures" demonstrates how, as Braiterman puts it, "meta-physical pictures saturate this peculiar

---

65. Franz Rosenzweig, *Die "Gritli"-Briefe: Briefe an Margrit Rosenstock-Huessy*, ed. Inke Rühle and Reinhold Mayer (Tübingen: Bilam, 2002), 124.
66. Rosenzweig, "Das neue Denken," 155.

philosophical thought."[67] As Braiterman has shown, aesthetics and aesthetic theory play a key role in Rosenzweig's conception of the star as a *Gestalt*, and indeed in his larger conception of religion.[68] Rosenzweig's rhetoric and the philosophy of *Gestalten* equally draw sustenance from a range of spatial concepts taken from mathematics and geometry, especially in their nonintuitive and supra-mathematical varieties. As philosophical pictures, however, Rosenzweig's *Gestalten* have as their ultimate horizon a theory of redemption and a figure of the eschatological moment. Here, beyond the purview of the historicist concept of time, Rosenzweig's figural thinking offers a prism for considering eschatology in spatial terms.

A comparison of Rosenzweig's work with that of his teacher Hermann Cohen, a neo-Kantian philosopher, underscores how the spatial and figural rendering of eschatology breaks with a developmental and teleological concept of history. Whereas Rosenzweig brings the moment to a standstill in order to grasp eternity as a *Gestalt*, Cohen posits a messianic concept of history in which the end of days embodies the ideal goal toward which all historical development strives.[69] In Cohen's teleological understanding of messianism, the messianic idea signifies a historical period at the end of days. In his essay "Die Messiasidee" ("The Idea of the Messiah") (1892), Cohen describes the development of the idea of the

---

67. Braiterman, *The Shape of Revelation*, 5.

68. Braiterman shows that Rosenzweig's philosophy has a deep affinity with the aesthetics of Wassily Kandinsky, Paul Klee, and Franz Marc, especially with regard to their "reordering of creation ... from individual points, lines, spots, and dabs" (*The Shape of Revelation*, 47, 59–60).

69. Cohen's late work brought his systematic philosophy to a close by turning to Judaism. In *Die Religion der Vernunft aus den Quellen des Judentums* (*Religion of Reason: Out of the Sources of Judaism*) (1919) and in essays such as "Monotheismus und Messianismus" ("Monotheism and Messianism") and "Die Messiasidee" ("The Idea of the Messiah"), Cohen set into motion a revival of Jewish messianic thought that both nurtured Rosenzweig's work and stood in contrast to it. Rosenzweig took part in Cohen's seminars and was familiar with draft material from *Religion of Reason* when he began writing *The Star of Redemption*. On these historical connections, see Wolfdietrich Schmied-Kowarzik, "Cohen und Rosenzweig zu Vernunft und Offenbarung," in *Die Gottesfrage in der europäischen Philosophie und Literatur des 20. Jahrhunderts*, ed. Rudolf Langthaler and Wolfgang Treitler (Vienna: Böhlau, 2007).

messiah from a person (from "the Anointed One" as priest, to king, to prophet, to a "people") to a historical idea, and then to a key concept of history: "The Messiah becomes, to put it bluntly, a calendar concept. Instead of the person of the Messiah one later refers to 'the days of the Messiah.' The Anointed One is idealized as the idea of a historical period of humankind."[70] Cohen's messianism thus entails progress and development toward a historical ideal. Indeed, Cohen's description of the future messianic age is saturated with these key historicist concepts and metaphors:

> The creator of heaven and earth is not sufficient for this being of the future [*Sein der Zukunft*]. He must create "a new heaven and a new earth." The being of history up to now is also inadequate for nature; *development* [Entwicklung] is required for the course of things. And development presupposes a goal toward which it strives. Hence, progress [*Fortschritt*] is required in the history of humankind. This is the meaning of the future as the establishment of the true, divine being on earth: the future, this idea of being, represents the *ideal* of history. Not a golden age, not paradise. Both are past.[71]

In this passage, Cohen extends the Kantian philosophy of history, for which the slow march of history toward an ideal society is guided by the hand of providence, by grounding its theodicy in the eschatological notion of new creation.[72] In doing so, he gives a theological grounding to the secular concepts of *Entwicklung* and *Fortschritt*, which derive from the Enlightenment and nineteenth-century philosophies of history. For Cohen, the teleological narrative of history as development and progress toward a goal is buttressed by the expectation of a messianic future.

---

70. Hermann Cohen, "Die Messiasidee," in *Hermann Cohens Jüdische Schriften*, vol. 1, ed. Bruno Strauß (Berlin: C. A. Schwetschke & Sohn, 1924), 105–24, at 108.

71. Cohen, *Die Religion der Vernunft aus den Quellen des Judentums* (Leipzig: Gustav Fock, 1919), 294.

72. In his essay "Idea for a Universal History from a Cosmopolitan Point of View," Kant argues for an "Aussicht in die Zukunft..., in welcher die Menschengattung in weiter Ferne vorgestellt wird, wie sie sich endlich doch zu dem Zustande emporarbeitet, in welchem alle Keime, die die Natur in sie legte, völlig können entwickelt... werden" and claims that such a view of history amounts to a "Rechtfertigung der Natur—oder besser *der Vorsehung*" ("Idee zu einer allgemeinen Geschichte," 18).

In Rosenzweig's work, by contrast, redemption is found in the eschatological now or *Heute* in which all future is contained. Cohen's commitment to a future history of messianic fulfillment stands in contrast to Rosenzweig's claim that the Jewish people are outside history and already with God. This antihistoricist dimension of Rosenzweig's thought calls into question the common reading of his philosophy of history as a form of messianism.[73] Rosenzweig supports neither the narrative of otherworldly eschatological completion nor that of inner-historical progress toward a messianic future.[74] Instead, his critique of history depends on figures of eternity that are placed squarely in the present while reconfiguring the historicist categories of time.[75] Rosenzweig thus rejects a teleological concept of eschatology whose end is a terminal point of history. Indeed, at the conclusion of *The Star of Redemption*, he proposes that the end is a beginning: "This inversion [*Verkehrung*] of temporal sequence, in which awakening precedes being for the world, establishes the life of the eternal people. Its eternal life, that is to say, constantly anticipates the end and thus makes it into the beginning" (*S* 467). Just as Barth understands *Endgeschichte* as coextensive with *Urgeschichte*, Rosenzweig reverses the teleological trajectory of history by turning the eschatological end into a start-

---

73. Dana Hollander, for example, refers to the "messianic philosophy of history that is suggested by Rosenzweig's *Star of Redemption* and related writings" ("On the Significance of the Messianic Idea in Rosenzweig," *Cross Currents* 53, no. 4 [2004]: 555–65, at 556).

74. Compare to Gregory Kaplan, "Rosenzweig's Apocalyptic Eschatology," 511: "What Scholem articulates and, I aim to show, Rosenzweig illustrates, is a tension within the messianic idea of Judaism between this-worldly and otherworldly, temporal and eternal foci of redemption."

75. Despite Rosenzweig's persistent critique of historicism, he refuses to view eschatology and history as strictly separate, instead using eschatology to inform a new understanding of history. Peter Gordon reaches a similar conclusion about Rosenzweig's interest in history in terms of his concept of revelation: "Revelation, Rosenzweig concludes, thus finds its highest 'certainty' not as a phenomenon incommensurable with history, but rather, precisely 'in' our Geschichtlichkeit, or 'historicity.' This is an intriguing claim, and it is evidence of a thinker who can abandon neither religious transcendence nor the historicist doctrine that threatens to dissolve it" (cited in Moyn, "Is Revelation in the World?" 401).

ing point for a new experience of history as the anticipation of eternal life.[76]

The metamorphosis of the concept of the end as a teleological goal (*Ziel*) in Cohen to a concept of the end as a beginning (*Anfang*) in Rosenzweig is at the heart of the new concept of history in modernism. For Cohen this end is still temporal in form; it is not the repetition of the origin but a future inception of a unified humanity, a messianic epoch that defines the goal of world history: "The end of days, the future of humanity, this is the fruit the life of peoples, this is the epoch of the Messiah. The end lies neither close at hand nor in the blue yonder [*weder in der Nähe, noch in blauer Ferne*]; it is the goal of world history."[77] By contrast, Rosenzweig's concept of eschatology detemporalizes the end: the end of history is no longer conceived as a historical period, or its terminal cessation, but as a space of history configured in relation to the eschatological now. History does not strive toward an ideal at the end of time, but rather begins from the situation of the end.

In Rosenzweig's work, history is given shape in images of curves, arcs, and parallel lines that meet at infinity; in the space of irrational numbers; and in lines that are bent by the curvature of creation, revelation, and redemption. Rosenzweig's spatial construction of history, culminating in the *Gestalt* of the star of redemption, projects the eschatological end into the space of time, just as Beltrami, Klein, and Poincaré produced projections of non-Euclidean space into Euclidean space. The result is a shape of time that appears nonintuitive, one in which history is at all times equally close to its beginning and end, in which the present is both a link in the chain of time and an embodiment of eternity. Geometry, mathematics, and a unique mode of figural thinking are thus central to Rosenzweig's critique of nineteenth-century historicism and philosophies of history.

---

76. Hollander arrives at a similar conclusion, noting that "Rosenzweig, with his distinction between Christian history and Jewish eternity, has broken with the idea that Jewish existence should be understood as a linear historical process, along the lines of Kant's view of history as a teleological progression" ("Messianic Idea in Rosenzweig," 561).

77. Cohen, "Die Messiasidee," 117.

4

Temporal Exterritoriality

Siegfried Kracauer and the Shape of History

In *History: The Last Things before the Last* (1969), Siegfried Kracauer offers a critique of the historicist paradigm of time as a homogeneous medium that serves as a vehicle for the chronological flow of events. Instead, he argues for the "temporal exterritoriality" of historical phenomena, which "must be thought of as lying both within and outside flowing time" (*L* 199–200). The elements of history, that is, are located in a historical context and are simultaneously part of time sequences to which they have no temporal proximity. For Kracauer, history is therefore a collection of diverse "shapes of time" (*L* 144) whose spatial and morphological characteristics take priority over their temporal indexes. This understanding of history in terms of shaped times is connected to a spatialized eschatology. Kracauer defines history as an "anteroom area" (*L* 191) that stands in spatial proximity to the last things. Eschatology is figured not as the culmination of the historical process but as a disturbance in the smooth surface of history where the elements of history begin to dissolve. Here he observes "pockets and voids" in the "temporal currents" of history that demarcate negative spaces in which figures of the last things may materialize (L 199). In Kracauer's work, the shape of the end of time is the form given to the elements of history in a state of dissolution.

Kracauer's theory of shaped times is defined by a dialectic of form and formlessness, of the configuration of shapes and their disintegration. Whereas Barth and Rosenzweig turn to calculus and projective geometry to describe the relation of history and eschatology, Kracauer finds an impetus for his approach in the mathematical discipline of topology, whose development was closely connected to the idea of curved spaces in non-Euclidean geometry.[1] In the work of mathematicians such as Leonhard Euler, Johann Benedict Listing, Bernhard Riemann, and Henri Poincaré, topology addresses the relations of points or elements in terms not of their metrical properties but of their possible connections. In topology, objects can be deformed and change shape as long as they retain the fundamental qualities of their invariants, such as the number of their holes. A circle is topologically equivalent to an ellipse, a donut to a teacup. In the mathematics of topology, objects do not exist in space but rather are themselves unique spaces. As Riemann showed, objects with variable curvature do not need to be thought of as embedded in Euclidean space; instead, they can be understood as defining their own intrinsic space.[2] The parameters of the relations between topological elements are flexible and can be deformed, dissociated, and reassociated. Similarly, Kracauer shows how the elements of

---

1. On Kracauer's relationship to mathematics, see Handelman, *The Mathematical Imagination*, 145–85. While we have little record of the details of Kracauer's familiarity with modern mathematics, Handelman draws attention to a letter to Margarete Susman dated January 11, 1920, in which Kracauer states: "More and more, my thinking is approaching higher mathematics" (*The Mathematical Imagination*, 145). This interest in mathematics is a perennial, though typically unspoken, feature of Kracauer's work. Even in his late work in *History*, Kracauer compares the histories of different time curves to "mathematical incommensurables" that have "different orders of magnitude" (L 145).

2. See Celia Lury, Luciana Parisi, and Tiziana Terranova, "Introduction: The Becoming Topological of Culture," *Theory, Culture & Society* 29, no. 4/5 (2012): 3–35, at 21.

urban space, photography, and history break down, disintegrate, and are reconfigured in new shapes and forms. In Kracauer's work, the geometries of urban spaces, the remnants of the photographic archive, and the shapes of history admit transformations and distortions that can be understood topologically.

These distortions become legible in Kracauer's accounts of the deformed geometries of cities such as Marseille and Berlin. They also appear in voids and hollow spaces on the surface of the material world, both in the afterlives of photographic images and in disturbances that unsettle the flow of chronological time. In each case, the topological rendering of such deformations is connected to an underlying eschatological account of the disintegration of a fallen world. Indeed, this combination of a materialist approach to modernist cultural and aesthetic forms with an underlying theological framework is a hallmark of Kracauer's approach. As he puts it in a letter to Ernst Bloch dated May 27, 1926, "One would have to encounter theology in the profane, into whose holes and fractures, which would have to be exposed, truth has descended" (*Man müßte der Theologie im Profanen begegnen, dessen Löcher und Risse zu zeigen wären, in die die Wahrheit herabgesunken ist*).[3] Similarly, in his essay "Zwei Arten der Mitteilung" ("Two Forms of Communication") (1929), Kracauer argues for a program of critical negativity in which theology emerges from the hollow spaces of material history: "But theology exists, and I attribute eternal reality to the word, just as you do. Very well, one should conceive of revolutionary negativity [*revolutionierende Negativität*] such that spaces (hollow spaces) [*Hohlräume*] remain open for the unspoken positive."[4] As much as he eschewed any attachment to theological or philosophical systems that posit timeless truths, Kracauer was attuned to the

---

3. Bloch, *Briefe, 1903–1975*, 1:274.
4. Siegfried Kracauer, "Zwei Arten der Mitteilung," in *Werke*, vol. 5, no. 3, ed. Inka Mülder-Bach and Ingrid Belke (Frankfurt am Main: Suhrkamp, 2006), 180–87, at 180–81. As Inka Mülder-Bach notes, Kracauer's approach is "materialistically conceived with a characteristic theological 'coloring'"; it arises from a "critical strategy that abstains from theological language in order to preserve its truth content" (*Siegfried Kracauer—Grenzgänger zwischen Theorie und Literatur: Seine frühen Schriften, 1913–1933* [Stuttgart: J. B. Metzlersche Verlagsbuchhandlung, 1985], 50).

manifestations of theology within the materiality of contemporary reality.[5]

By bringing together a topological account of the deformation of figures and an attention to the eschatological undercurrents of the modern world, Kracauer developed a framework for a concept of history defined by asynchronous and noncontemporaneous shaped times. A reading of Kracauer in terms of this convergence

---

5. Kracauer's frequent crossing of boundaries between theology and the profane has given rise to scholarly interpretations that typically characterize his thought as primarily either religious or secular. Dagmar Barnouw, for example, emphasizes the secularity of Kracauer's project by contrasting the theological meaning of redemption in Walter Benjamin's work with the "open-ended secularity" and "sense of salvaging from cultural oblivion" in Kracauer's use of the term (*Critical Realism: History, Photography, and the Work of Siegfried Kracauer* [Baltimore: Johns Hopkins University Press, 1994], 31–32, 260–61). Barbara Thums, in an insightful reading of Kracauer's *Der Detektiv-Roman* (*The Detective Novel*) (1925), points to key theological figures of thought in Kracauer's work, but insists that his demand for a "theology in the profane" entails a radical rejection of religion and the category of faith ("Kracauer und die Detektive: Denk-Räume einer 'Theologie im Profanen,'" *Deutsche Vierteljahrsschrift für Literaturwissenschaft und Geistesgeschichte* 84, no. 3 [2010]: 390–406, at 391). By contrast, Michael Kessler compares Kracauer's appropriation of theology to Bloch's concept of "religion in inheritance" (*Religion im Erbe*), a term that designates the afterlife of religion in modernity ("Entschleiern und Bewahren: Siegfried Kracauers Ansätze für eine Philosophie und Theologie der Geschichte," in *Siegfried Kracauer: Neue Interpretationen*, ed. Michael Kessler and Thomas Y. Levin [Tübingen: Stauffenburg, 1990], 105–28, at 122). Seeking to mediate between these positions, Gertrud Koch notes a characteristic "vacillation between phenomenological concretism and theology" in Kracauer's work and suggests that he regarded the "loss and collapse of the major systems of religion" with "mixed feelings" ("'Not Yet Accepted Anywhere': Exile, Memory, and Image in Kracauer's Conception of History," trans. Jeremy Gaines, *New German Critique* 54 [1991]: 95–109, at 97; and Koch, *Siegfried Kracauer: An Introduction*, trans. Jeremy Gaines [Princeton, NJ: Princeton University Press, 2000], 118). Similarly, Harry T. Craver notes that even in his later works, theological concepts remain "stowed away as contraband close to the core of his critical project," and argues that Kracauer's "critical vocation derives from the conflict between religious revival and secularism" (*Reluctant Skeptic: Siegfried Kracauer and the Crises of Weimar Culture* [New York: Berghahn, 2017], 19). Finally, Miriam Hansen detects an affinity in Kracauer's work with "the discourse of secular Jewish Messianism" but argues that it "emerges less in his conceptual constructions than in recurring motifs and interpretive tropes" ("Decentric Perspectives: Kracauer's Early Writings on Film and Mass Culture," *New German Critique* 54 [1991]: 47–76, at 53). Hansen suggests that Kracauer's attempt to distance himself from theology as a matter of "overt construction" need not exclude the presence of theological figures of thought in his work that can "take on a theoretical life of their own" (53).

of topology and eschatology demonstrates that there are fruitful intersections between two traditions of scholarship on Kracauer that often follow divergent paths: those commentaries that read Kracauer, who was trained as an architect, as a theoretician of modern urban spaces, in which the geometries of spatial form are entwined with the architectures of the metropolis and new strategies of literary representation,[6] and those that read Kracauer as a theoretician of history, often with philosophical and theological undertones, who sought alternatives to the progressive model of history championed by historicism.[7] The theory of time curves that emerges in Kracauer's work offers an alternative to the historicist paradigm in which history develops along the axis of chronological time. At the same time, it redefines that status of the end of history: in his analysis of the legendary figure of Ahasuerus, Kracauer shows that the shape of history is constantly changing and never takes on an ultimate form. Given that a temporal end of history is unimaginable, the space of eschatology emerges negatively in the rifts and fractures of historical time.[8]

---

6. See Andreas Huyssen, "Modernist Miniatures: Literary Snapshots of Urban Spaces," *PMLA* 122, no. 1 (2007): 27–42; Huyssen, "The Urban Miniature and the Feuilleton in Kracauer and Benjamin," in *Culture in the Anteroom: The Legacies of Siegfried Kracauer*, ed. Gerd Gemünden and Johannes von Moltke (Ann Arbor: University of Michigan Press, 2012), 213–25; and Esther Leslie, "Kracauer's Weimar Geometry and Geomancy," *New Formations: A Journal of Culture/Theory/Politics* 61 (2007): 34–48.

7. See Vincent P. Pecora, "Benjamin, Kracauer, and Redemptive History," in *Secularization and Cultural Criticism: Religion, Nation, and Modernity* (Chicago: University of Chicago Press, 2006), 67–100; Barnouw, *Critical Realism*, 20–52; and Kessler, "Entschleiern und Bewahren."

8. By contrast, critics such as Mülder-Bach and Hansen discern a messianic and apocalyptic dimension of Kracauer's thought in which the end of history is a tangible possibility. Mülder-Bach comments on Kracauer's diagnosis of the decline of religious certainty in his time by remarking that an "apocalyptic mood of end times [*Endzeitstimmung*] is palpable here" (*Siegfried Kracauer*, 20). In her discussion of "The Mass Ornament" (1927), Mülder-Bach glosses Kracauer's perspective as follows: "In the messianic 'institution of truth in the world,' which is 'intimated' by historical reason, history would find its end" (*Siegfried Kracauer*, 58). Similarly, Hansen argues that "a distinctly apocalyptic undercurrent continued to characterize [Kracauer's] observations of contemporary life, specifically, a perception of modernity as traumatic upheaval which will lead to catastrophe. Like Benjamin and Bloch, he could not envision change as immanent in history, as in bourgeois-liberal

Constructions of Urban Space and the Disintegration of the Figure

In *Soziologie als Wissenschaft* (*Sociology as Science*) (1922), Kracauer outlines an approach to working with history that entails a construction of space. Noting that an encounter with history is vital for the discipline of sociology, he argues that history provides sociology with a series of points that can be connected as a construction. In doing so, Kracauer contrasts the "continuous flow" of history with an "empty space" in which sociology constructs a relationship among these points:

> Sociology emerges point by point [*Punkt für Punkt*] from the current of events [*Fluß des Geschehens*] in order to wrest from historical time a piece of the events that have taken place therein. But it does not pull the continuous flow [*kontinuierlichen Fluß*] itself over into the empty space [*in den leeren Raum*], regardless of how densely packed the points may be. For indeed material sociology would be at the end of its knowledge if it restricted itself to building its constructions [*Konstruktionen*] in thin air, without a plain assessment of what occurs, if it did not incorporate history.[9]

The empty space of sociological reflection, according to Kracauer, is populated with points that are excerpted from the continuous flow of time. These points provide, in piecemeal fashion, the material for the construction of sociological insights. In this act of spatial construction, the flow of time is interrupted, while the fragmentary points are reconstituted in new structures. Yet Kracauer clarifies that history does not merely provide sociology with the material for its constructions; it also assists in the "establishment of connections between all the points" (*Herstellung der Verbindung zwischen all den Punkten*) and indicates the direction these constructions are to take.[10]

This insight that history can be assembled and constructed by mapping points from the continuous flow of time onto a space of

---

notions of progress and reform, but only as a total break" ("Decentric Perspectives," 53).

9. Siegfried Kracauer, *Soziologie als Wissenschaft*, in *Werke*, vol. 1, ed. Inka Mülder-Bach (Frankfurt am Main: Suhrkamp, 2006), 9–101, at 91.

10. Kracauer, *Soziologie als Wissenschaft*, 91–92.

ideas provides the germ cell of Kracauer's theory of history. Indeed, the imagery and conceptual language of this early account of sociology has a strong affinity with the approach to historiography that Kracauer went on to develop in *History: The Last Things before the Last*. The constructivist dimension of Kracauer's approach to history, however, differs in important respects from the historiography of Troeltsch and Spengler. Kracauer's constructivism is unique in that it not only assembles new structures but also draws attention to the fundamental deformability of their elements. To understand how deformation and disintegration are central to Kracauer's spatial construction of history, a comparison with Kracauer's texts on the geometry of urban space is illuminating.

In his urban miniatures from the Weimar period, Kracauer gives us renderings of the metropolitan landscape that express the central insight of modernist mathematics—that space is not given but constructed.[11] Drawing on his training as an architect, Kracauer shows how the geometries of spatial form are entwined with the architectures of the metropolis and new strategies of literary representation. That is, Kracauer constructs geometrical images of the city in the act of writing. These images are derived from the architectural forms that Kracauer encountered in cities such as Berlin and Marseille, yet they are also images that estrange the perception of the city. Imagery and motifs taken from the vocabulary of non-Euclidean geometry enable the construction of images of urban space that border on the nonvisualizable.

The collection of Weimar feuilleton pieces published as *Das Ornament der Masse* (*The Mass Ornament*) (1963) begins with a section that Kracauer calls "Natürliche Geometrie" ("Natural Geometry"). A close examination of these texts, especially "Zwei Flächen" ("Two Planes") (1926), however, reveals that the interplay of geometry and social space that Kracauer depicts is anything but natural. Indeed, Kracauer's textual rendering of urban space is defined by a geometry of distortion that challenges Renaissance perspective and points to

---

11. On the genre of the urban miniature as a unique modernist textual form, see Huyssen, *Miniature Metropolis*.

the curvature of space. The aim of these texts is neither to trace the natural lines of empirical space nor to glorify the clean lines and symmetry of cities meticulously laid out by urban planners. Instead, a geometrical presentation of the metropolis heightens the uncanniness of urban space.

The first of two pieces in "Two Planes," titled "Die Bai" ("The Bay"), likens the city of Marseille to an amphitheater constructed around the "rectangle" (*Rechteck*) of the old harbor (O 11). The main thoroughfare, the Cannebière, which connects the harbor with the interior of the city, serves as the central axis of a Renaissance perspective that Kracauer glosses as "the vanishing point of all perspectives" (*den Fluchtort aller Perspektiven*) (O 11). In contrast to the monumentality of this urban space, in which the city is gathered, like a crowd of spectators, around the arena of the bay, stands the reality of contemporary life in Marseille. With its glory days in the past, the city has lost its luster; the bay is no longer a shifting kaleidoscope of activity but merely a deserted rectangle. The result is a spatial inversion in which the bay is transformed from the focal point of a visual spectacle into a "hollow space" (*Hohlraum*) whose "emptiness straddles a distant angle" (*ihre Leere spreizt sich im fernen Winkel*) (O 12). In geometrical terms, the image of the bay is difficult to visualize. This difficulty is intentional on Kracauer's part. The bay defines the empty center of the city, a center that distorts urban space rather than serving as its focal point. Backing up into this void, the "streets terminate here as dead ends" (*laufen sich die Straßen tot*), while the bay "bends straight lines into curves" (*biegt die Graden zu Kurven um*) (O 12). In these images, the vanishing point of Renaissance perspective has been supplanted by a space of negative curvature, as in the geometry of a pseudosphere, in which lines follow the curvature of space.

In "Das Karree" ("The Quadrangle"), the companion piece to "The Bay," Kracauer provides a second example of an urban space in which we can observe both the overwhelming power of Cartesian geometry and a subtle distortion of its principles. Amid an urban labyrinth characterized as a "tangle of pictorial alleys" (*Knäuel der Bildergänge*),[12] Kracauer discovers a deserted square that takes on a rigid geometrical form: "The horizontal lines are drawn with a ruler, dead-straight" (*Die Horizontalen sind mit dem Lineal gezogen, schnurgrad*).[13] Accordingly, the square is described as a "quadrangle" (*Karree*) or "quadrilateral" (*Quadrat*) that has been stamped into the urban landscape, a shape that pushes those it traps into its center. But at the same time, the quadrangle defies the basic laws of perspective. Indeed, the walls that surround the square do not appear to decrease in size as they become more distant: "Other walls of equal length foreshorten like railway tracks, but not this one. Its vanishing points diverge" (*Andere Mauern von gleicher Länge verkürzen sich wie Eisenbahngleise; diese nicht. Ihre Fluchtlinien laufen auseinander*).[14] These perceptual disturbances suggest that the "Zwei Flächen" at stake here are not "planes" at all but rather "surfaces" whose curvature can be defined by non-Euclidean geometry.

In "Aus dem Fenster gesehen" ("View from a Window") (1931), a short piece that appeared in the collection *Straßen in Berlin und anderswo (Streets in Berlin and Elsewhere)*, Kracauer provides a textual rendering of the city of Berlin that evokes the unruly geometries of unplanned urban spaces. Whereas the architecture of a city square such as the Pariser Platz creates perspectival effects through the layout of groups of buildings and views, a different set of urban geometries is at work in those "creatures of chance" that emerge when buildings and streets built for diverse reasons come together.[15] At the outset, Kracauer both draws attention to the position of spectatorship from which his view of the city emerges and relativizes this position as a placeless place. The view from the window can be found "high above" a square that is said to have the ability to make itself invisible—not because it is deserted, but because it is the intersection of several major avenues, crossed by thousands every day in busses

---

12. See Siegfried Kracauer, "Two Planes," in *The Mass Ornament: Weimar Essays*, trans. Thomas Y. Levin (Cambridge, MA: Harvard University Press, 1995), 37–39, at 39; Kracauer, "Zwei Flächen," O 11–13, at 13.

13. Kracauer, "Two Planes," trans. Levin, 39; O 13.

14. Kracauer, "Two Planes," trans. Levin, 39; O 13.

15. Siegfried Kracauer, "Aus dem Fenster gesehen," in *Straßen in Berlin und anderswo* (Frankfurt am Main: Suhrkamp, 2009), 53–55, at 53.

and trams who take little note of the square. In the midst of the hustle and bustle of the city, the square lives an "incognito" existence.[16] It is open to all sides, yet no one notices the view it affords. Kracauer therefore apprehends the city from a point without extension, a standpoint on which it appears impossible to stand given the perpetual motion of the city.

Kracauer's picture of the city is shot through with geometrical images and motifs. At its center stands a surface comprised of railway tracks near the Charlottenburg train station:

> The city-image itself, which begins from this little square, is a space of extraordinary expanse [*ein Raum von außerordentlicher Weite*] pervaded by a metal field of iron [*Eisenacker*]. It resounds with railroad tracks. They emerge from the direction of the Charlottenburg train station behind a larger-than-life tenement building wall, run in bundles side by side [*laufen bündelweise nebeneinander*], and finally disappear behind ordinary houses. A swarm of radiant parallels [*Ein Schwarm von glänzenden Parallelen*] that lies low enough under the window to be able to survey its entire extension. With its numerous signal masts and sheds, the surface almost gives the impression of a mechanical model that a boy, kneeling invisibly somewhere, uses for experiments.[17]

Kracauer's image transforms the scene of an industrial metropolis into a model of geometrical space that is riven by acoustical and visual effects such as echoes and glares. The image is constructed as a space that is expansive and yet bounded, surveyable in its entirety and yet on the verge of a certain disorder, as evident in the "swarm" of railroad tracks. Indeed, the phrase "a swarm of radiant parallels" couples the image of an orderly layout of parallel tracks with the suggestion of a teeming abundance of lines. Kracauer's image of bundles of parallel lines evokes the non-Euclidean geometry of the pseudosphere, in which a bundle of lines can be drawn through a point without intersecting with other lines. All of this suggests that the surface of this geometrical model can be best understood as a curved space.

Kracauer underscores the disturbance of visual perception in this scene with the description of a "roaring underpass" (*rauschende*

---

16. Kracauer, "Aus dem Fenster gesehen," 54.
17. Kracauer, "Aus dem Fenster gesehen," 54.

Straßenunterführung) that is stretched "perfectly straight" (*schnurgerade*) under the entire railroad level. An endless progression of automobile traffic traverses this tunnel, its speed doubled in a kind of time-lapse photography, producing what Kracauer calls a "running cross belt" (*laufende Querband*) whose incessant motion clashes with but cannot disturb the composure of the "iron surface" (*eisernen Fläche*). To these horizontal and lateral axes, finally, Kracauer appends a vertical axis in the figure of the Berlin radio tower, which is described as "a vertical line thinly drawn with a ruling pen through a piece of the sky" (*ein senkrechter Strich, der mit der Reißfeder dünn durch ein Stück Himmel gezogen ist*).[18] Here the technical drawing of the engineer is combined with the cinematographer's manipulation of time and motion, resulting in a dynamic set of spatial relations in which a curved surface appears pinned to a set of coordinate axes and yet somehow apart, floating in a space of its own. In a final projection of this urban landscape, Kracauer considers the same scene at night, when the tracks, masts, and buildings disappear such that "a single field of light shines in the darkness" (*ein einziges Lichterfeld glänzt in der Dunkelheit*).[19] Here the lights appear distributed across a space that has lost its dimension of depth: they mark singular points that remain still or are stretched into cords by movement. Whereas "Two Planes" demonstrated the warping of perspective and the inversion of spatial relations, "View from a Window" shows how dynamic movement—the swarm of parallel tracks, the roaring lateral movement of the underpass—creates an unstable geometry of urban space that is perpetually in flux, a geometry that is hardly visualizable as an image, even when it is reduced to two dimensions.

In Kracauer's texts on the geometry of space, it is possible to discern both the construction of geometrical figures and an interest in the disintegration and dissolution of these figures. Kracauer's geometrical figures are highly mobile and are subject to processes of transformation and decay. As such, they resemble the shapes of the mathematical discipline of topology, which can be stretched and bent

---

18. Kracauer, "Aus dem Fenster gesehen," 54–55.
19. Kracauer, "Aus dem Fenster gesehen," 55.

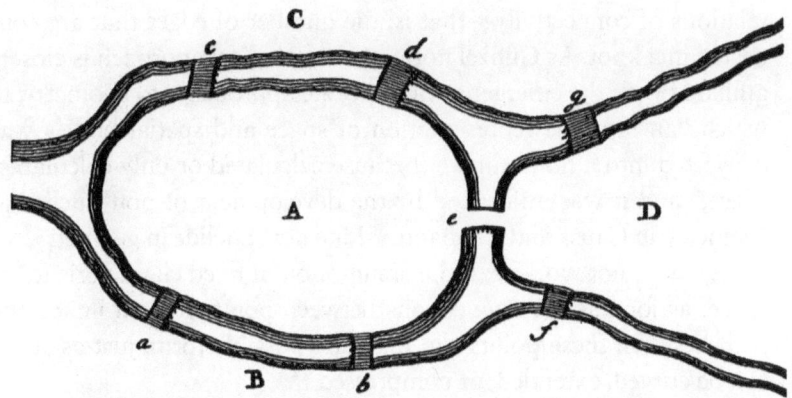

Figure 6. Euler's diagram of the bridges of Königsburg (1736). Rob Shields, "Cultural Topology: The Seven Bridges of Königsburg, 1736," *Theory, Culture & Society* 29, no. 4/5 (2012): 44.

as long as they are not cut or torn. The mathematical field of topology was established in the late nineteenth century by Johann Benedict Listing, a student of Gauss, who called topology the theory of "modal relationships of spatial figures" (*modaler Verhältnisse räumlicher Gebilde*).[20] The origins of topology can be found in Leibniz's sketch of a theory of "Analysis situs," in which he proposed that space can be described on the basis of the relation of bodies or points to one another—without regard to metrical distance or physical properties.[21] Euler developed these ideas by showing that a "geometry of position" could be used to solve the famous Königsburg bridge problem, which asked whether it was possible to go on a walk in Königsburg and cross each of its seven bridges only once (figure 6).[22] Euler recognized that it is not the distances of physical space that are at stake in the Königsburg bridge problem, but only topological

---

20. See Stephan Günzel, "Raum—Topographie—Topologie," in *Topologie: Zur Raumbeschreibung in den Kultur- und Medienwissenschaften*, ed. Stephan Günzel (Bielefeld: Transcript, 2007), 13–29, at 22.
21. See Günzel, "Raum—Topographie—Topologie," 22.
22. See Rob Shields, "Cultural Topology: The Seven Bridges of Königsburg, 1736," *Theory, Culture & Society* 29, no. 4/5 (2012): 43–57.

relations of connectivity—that is, the number of edges that are connected in a knot. As Günzel notes, the topological approach is closely affiliated with the emergence of algebraic approaches to geometry, in which "an intuitive representation of space and spatial bodies was converted into a nonintuitive (because calculated or only calculable) space," and it was influenced by the development of non-Euclidean geometry in Gauss and Riemann.[23] Like non-Euclidean geometry, topology does not work with the assumption of fixed characteristics of space: as long as the connections between points remain intact, the path between these points can take any possible form, just as space can be curved, extended, or compressed.[24]

An important consequence of the "modal relationship of spatial figures" is that topological spaces can be reshaped by continuous stretching and bending without changing their fundamental invariant properties. For example, a circle is topologically equivalent—or homeomorphic—to an ellipse, just as a torus is homeomorphic with any other object with one hole. In topology, object-spaces can be submitted to homeomorphic deformations while retaining their basic spatial properties, such that one can speak of a mathematically rigorous "shape consistency under deformation."[25] Such deformations can be used to show that Euclidean geometry is just one case among any number of possible geometries: while a triangle on a flat plane will have angles that add up to 180 degrees, the same triangle subject to deformation on the surface of a sphere will have angles greater than 180 degrees.[26] Similarly, Riemann showed that objects can have variable curvature and define their own intrinsic space rather than being embedded in Euclidean space. In Riemann's manifold geometry, "spatial structures may change from point to point on a curve," resulting in a "continuously changing curvature of n-dimensions."[27]

Critics such as Handelman, Huyssen, and Leslie have shown how Kracauer uses geometry in texts such as "Two Planes" and "The

---

23. Günzel, "Raum—Topographie—Topologie," 21.
24. See Günzel, "Raum—Topographie—Topologie," 22–23.
25. See Shields, "Cultural Topology," 47.
26. Shields, "Cultural Topology," 47.
27. Lury, Parisi, and Terranova, "Introduction: The Becoming Topological of Culture," 10.

Mass Ornament" to represent the dominance and control of rationalization in modern capitalism.[28] These texts demonstrate that urban spaces and mass ornaments have been stamped with the straight and parallel lines of Euclidean geometry and submitted to rigid forms of order and organization. In "Das Ornament der Masse" ("The Mass Ornament") (1927), for example, Kracauer describes dancing troupes such as the Tiller Girls as "indissoluble complexes of girls whose movements are mathematical demonstrations" (*unauflösliche Mädchenkomplexe, deren Bewegungen mathematische Demonstrationen sind*) that give us "performances of ... geometrical precision" (*Darbietungen von ... geometrischer Genauigkeit*) (O 50–51). Similarly, he notes that the mass ornaments produced in stadium performances are "comprised of straight lines and circles as they can be found in the textbooks of Euclidean geometry" (*besteh[en] aus Graden und Kreisen, wie sie in den Lehrbüchern der euklidischen Geometrie sich finden*) (O 53).

At the same time, however, Kracauer points to the deformation of these geometrical forms—such as the curvature of the straight lines in "The Bay"—in a manner comparable to the topological bending and warping of space. Kracauer is interested not only in the overwhelming dominance of rationality in modern culture but also in how this rationality breaks down through the deformation of its geometrical figures. These two aspects of Kracauer's thought—the construction of figures and an attention to their deformation and dissolution—are given a theoretical grounding in the essay "Gestalt und Zerfall" ("Figure and Disintegration") (1925), which analyzes the interplay of empty formlessness and moments of configuration (*Gestaltung*) and construction (*Aufbau*) in contemporary

---

28. For example, Handelman notes that Kracauer uses geometry "as a metaphor to describe the rationalized spatial forms produced by modern capitalist society that embodied in material objects the troubling spirit of the age" (*The Mathematical Imagination*, 149). Similarly, Huyssen remarks that in "The Quadrangle," "abstract perspectival space exudes power and domination, disciplining and surveillance" ("Modernist Miniatures," 35). Leslie argues that the abstract geometrical principle of the mass ornament can be understood as an aesthetic translation of the historical forces of industrialization and militarization ("Kracauer's Weimar Geometry and Geomancy," 36).

culture.[29] In Kracauer's view, the construction of figures is fundamentally bound up with the dissolution of an older stratum of social forms. The resulting dialectic of configuration (*Gestaltung*) and disintegration (*Zerfall*) has an eschatological edge in which construction is predicated on a prior moment of dissolution.[30]

Kracauer develops his theory of disintegration through a sociological analysis of contemporary life, but this theory also provides a foundation for his understanding of the geometry of urban space, and, as we shall see, for his analysis of photography and for his concept of history. Kracauer perceives a "will to configuration" (*Willen zur Gestaltung*) and a desire for new constructions of life in a wide range of contemporary groups, including both secular and religious movements, from young Socialists to Nationalists, from the circles of Reformed Protestants to the Catholic students of Max Scheler and Romano Guardini.[31] As Kracauer showed in his essay "Die Wartenden" ("Those Who Wait") (1922), regardless of their political or religious affiliations, these movements responded to the vacuum left by the absence of religious belief and the resulting set of unsatisfied needs. They were united by a "wish to find a way out of chaos, this search for a positively suitable form of life, . . . this entire striving for construction [*Streben nach Aufbau*]."[32] Kracauer cautioned that movements that aim for positive constructions must be careful not to be naive about the realities of a rationalized and functionalized economic system, the desertedness of the soul in a realm dominated by the intellect, and "the elimination of the realm of the

---

29. Siegfried Kracauer, "Gestalt und Zerfall," in *Werke*, vol. 5, no. 2, ed. Inka Mülder-Bach (Frankfurt am Main: Suhrkamp, 2004), 283–88, at 283. On the rhetoric of *Gestaltung* and *Aufbau* in the context of constructions of history, see chapter 1.

30. Similarly, Hansen argues for "Kracauer's affinity with the discourse of secular Jewish Messianism" and notes that "Kracauer's account of the disintegration of the world also resonates with the legacy of Jewish Gnosticism, although, as a doctrine, Jewish Gnosticism was just as suspect to him as other variants of religious mysticism" ("Decentric Perspectives," 53–54). Likewise, the claim made here that there is an eschatological edge in Kracauer's thought implies, not that he subscribed to a particular religious worldview, but simply that an eschatological dynamic informs his concept of history.

31. Kracauer, "Gestalt und Zerfall," 284.

32. Kracauer, "Gestalt und Zerfall," 284.

last things" (*das Ausgeschaltetsein der letzten Dinge*) from human language and relationships.[33] Instead of challenging the realities of an atomized and machine-like existence, movements of reconfiguration often simply posit counter-images that seek to compensate for the absence of meaning and form. Concrete thought is put forth as an antidote to abstract rationalism, or the restoration of community as a cure for the functionalization of society, but these juxtapositions do not confront or overcome what they oppose. As Kracauer puts it with a note of irony, "in the midst of disintegration one catches sight of meaning and figure" (*mitten in dem Zerfall erblickt man Sinn und Gestalt*).[34]

Kracauer's critical intervention is to suggest that movements that seek new constructions fatally miss the realities of contemporary social life. Drawing a contrast between the actuality (*Realität*) of a mechanized, atomized world and the yearned-for reality (*Wirklichkeit*)[35] of the religious sphere, Kracauer elaborates: "Each circle in its own way is intent on leaving the scorned actuality [*die verpönte Realität*] behind and beneath itself and directly embodying the fancied reality [*die gemeinte Wirklichkeit*]."[36] In other words, the failure to adequately criticize and overcome *Realität* dooms any premature effort to bring about a new *Wirklichkeit*. Or, as the conclusion of the essay puts it, without a proper measure of "disintegration" (*Zerfall*), one cannot begin the task of "configuration" (*Gestaltung*).

"Figure and Disintegration" thus depends on an eschatological view of history: the fallenness of the present, captured vividly in the rhetorical intensity of the term "disintegration" (*Zerfall*), must complete its disintegration before new constructions are possible. Here we can observe how the concepts of actuality (*Realität*) and reality (*Wirklichkeit*) are deeply intertwined, despite their oppositional

---

33. Kracauer, "Gestalt und Zerfall," 285.
34. Kracauer, "Gestalt und Zerfall," 285.
35. As Michael Schröter notes, "Der Begriff *Wirklichkeit*" functions for Kracauer "als Symbol eines Defizits und damit zugleich, wie rudimentär auch immer, als Chiffre einer eschatologischen Hoffnung" ("Weltzerfall und Rekonstruktion: Zur Physiognomik Siegfried Kracauers," *Text + Kritik* 68 [1980]: 18–40, at 22).
36. Kracauer, "Gestalt und Zerfall," 286.

character. On the one hand, Kracauer notes that the actuality of mechanized, atomized existence is fundamentally *unwirklich* and yet *real*: "For the unreal reality [*unwirkliche Realität*] is at hand and more real [*realer*] than ever."[37] On the other hand, he claims that the yearned-for "real life" (*wirkliche Leben*) must don the "mask of the unreal [*Entwirklichten*] and lowly" if it is to have any chance of striking against contemporary reality in the sense of *Realität*.[38] The actualities of a rationalized and functionalized world, Kracauer suggests, must be confronted in their "own medium," and accordingly, the construction of a community that might embody reality in the sense of *Wirklichkeit* can only be advocated from a position of "exile," in the midst of an "environment that fails to recognize such community."[39]

Kracauer's dialectical oscillation of the competing concepts of *Realität* and *Wirklichkeit* is closely connected to the spatial dynamics of modernist eschatology. The path to the reality of the religious sphere, whose articulation is connected to a spatial metaphorics of figuration, configuration, construction, and shaping, must pass through the realm of the "unreal" (*Entwirklichten*) in which existence is emptied of shape and form. This is a realm characterized by what Kracauer calls the "facticity of emptied existence."[40] The spatial construction of figures—with its promise of access to the sphere of religious reality—has as its condition of possibility the disintegration of fallen figures: "For the constructed cannot be lived unless the disintegrated is collected and carried along" (*Denn das Gestaltete kann nicht gelebt werden, wenn das Zerfallene nicht eingesammelt und mitgenommen wird*).[41] Kracauer thus insists that the path to new configurations of reality must pass through processes of disintegration: "Only by passing through disintegration [*Zerfall*], only by way of this detour, can reality [*Wirklichkeit*] be won."[42] These remarks provide a key insight into the way Kracauer

---

37. Kracauer, "Gestalt und Zerfall," 286.
38. Kracauer, "Gestalt und Zerfall," 287.
39. Kracauer, "Gestalt und Zerfall," 287.
40. Kracauer, "Gestalt und Zerfall," 287.
41. Kracauer, "Gestalt und Zerfall," 288.
42. Kracauer, "Gestalt und Zerfall," 288.

constructs history in his later work: namely, that the figuration of shapes of time is bound up with the dissolution and deformation of the homogeneous flow of chronological time.

Photography and the Eschatological Void of History

In his essay "Die Photographie" ("Photography") (1927), Kracauer shows how the photographic medium, through its inevitable disintegration into its base elements, can reveal lacunae and hollow spaces in the surface of time and history. A topological approach to the transformation and deformation of the elements of the photograph provides Kracauer with a model for thinking of history beyond the historicist paradigm of a reconstruction of a temporal continuum whose shape is fixed. "Photography" offers a set of reflections on an eschatological void that hangs over the memory-scape of modernity, a void that is uniquely revealed by photography. In the spatial configuration of the photograph, that is, Kracauer discerns a nothingness that gives shape to the end of history. In "Photography," Kracauer introduces a schematic distinction between spatial and temporal categories to develop an analogy between photography and historicism, yet this schematism begins to break down as the spatial continuum offered by photography confronts the passage of time. At a distance from its referent, the photograph is reduced to a series of elements that have become estranged from their context of meaning. The aging of the photograph demonstrates the precarious status of the spatial continuum that it preserves. In this way, photography serves as a medium through which the eschatological dimension of Kracauer's criticism of historicism becomes apparent.

"Photography" begins with a juxtaposition of two photographs whose contrast allows Kracauer to develop his ideas about photography. The first shows a twenty-four-year-old film diva on the cover page of an illustrated magazine. In a series of staccato phrases and sentences, Kracauer sketches the details of the photograph. What emerges is a picture full of contradictions. Everything about the appearance of the photograph suggests immediacy: its highly contemporary place on the cover of a magazine, the liveliness of the diva, the ease with which she is recognized by all readers. Kracauer comments laconically: "Time: the present" (*Zeit: Gegenwart*) (O 21). This immediacy is matched by the precision of the photograph, to which Kracauer ascribes complete spatial continuity: "The bangs, the seductive pose of her head, and the twelve eyelashes on the right and left—all of these faithfully enumerated details sit properly in space, a gapless appearance [*eine lückenlose Erscheinung*]" (O 21).

Yet Kracauer's commentary undercuts the immediacy and continuity of the photograph. He puts the photograph under a magnifying glass, revealing the "grid" (*Raster*) and "network of points" (*Punktnetz*) that make up its print reproduction (O 21). He draws attention to the mediation of the diva via her cinematic appearance, a mediation that calls into question the indexical relation of the photograph to its referent: "Everyone recognizes her, for all have already seen the original on the big screen" (O 21). Finally, Kracauer points out the dependence of the image on its paratext: it is only through the "accompanying text" (*Begleittext*) that the diva becomes "our demonic diva" (O 21). To specify the temporal location of the photograph, Kracauer must intervene with a gesture that suggests both authority and arbitrariness: "The date is September" (*Wir schreiben September*) (O 21).[43] The English translation "The date is September" evokes the recording of minutes for a transcript. Significantly, however, the date itself is not part of the photograph, and the expression "Wir schreiben September," which Kracauer repeats twice for emphasis, might also be translated as "Let's say it's September," suggesting that the month is less significant than the ability to agree upon a date, however arbitrary it may be. Bearing in mind the metaphorics of "Figure and Disintegration," one might say that Kracauer both establishes the photograph of the film diva as a figure and subtly draws attention to its looming disintegration.

Just as the "gapless" spatial continuity of the photograph is fragile, so too is its temporal presence. Already in the very first paragraph of the essay, Kracauer begins to develop an analogy between the spatial and temporal aspects of the photograph. The formula

---

43. Kracauer, *The Mass Ornament*, trans. Levin, 47.

"Time: the present" is ambiguous, for it establishes the photograph as contemporary but raises the question of whether the photograph would function as it does outside the moment in which it was taken. In short, the questions of the future history of the photograph and the possibilities of its aging and afterlife are already posed here. And indeed, Kracauer pursues these very questions in his analysis of a second photograph, of a grandmother shown as a girl of twenty-four years. The examination of the second photograph is staged from a position of temporal distance; the grandchildren of the grandmother behold the photograph, now sixty years old, and, unable to recognize its subject, ask themselves: "Did *grandmother* look like that?" (O 21). Kracauer stages these doubts and allows them to linger: "In the end the photograph does not portray the grandmother but rather her friend who resembled her" (O 22). With the passage of time, the frame of reference for the photograph begins to recede, as does the "resemblance" (*Ähnlichkeit*) upon which the indexicality of the photograph depends:

> Contemporaries no longer exist, and the resemblance? The original image [*Urbild*] has long since decayed. The darkened figure has so little in common with the remembered features that the grandchildren, taken aback, must compel themselves to encounter their fragmentarily handed down ancestress [*fragmentarisch überlieferten Ahnfrau*] in the photograph. Very well, the grandmother, but in reality, it is an arbitrary young girl in the year 1864. (O 22)

There is nothing in the photograph itself, Kracauer notes, that can be connected to the memories of the grandchildren. If the original of the diva is mediated by her screen appearance, here the "original image" (*Urbild*) of the grandmother no longer exists. The photograph has emancipated itself from its referent but thereby lost any lingering connection to life.[44]

---

44. As Hansen writes, "In the measure that the photograph ages and outlives its referential context, the objects or persons depicted appear to be shrinking or diminishing in significance" ("Kracauer's Photography Essay: Dot Matrix—General (An-) Archive—Film," in *Culture in the Anteroom: The Legacies of Siegfried Kracauer*, ed. Gerd Gemünden and Johannes von Moltke [Ann Arbor: University of Michigan Press, 2012], 93–110, at 100).

Over time, the photograph becomes (or perhaps was from the beginning) a death mask of the figure it represents, however lively that figure may have been. One aspect of the photograph's death mask is the frozen appearance of the face, which leads Kracauer to liken the grandmother to a mannequin: "The girl smiles continuously, always the same smile, the smile stands still, no longer referring to the life from which it has been taken.... Mannequins in the hair salon smile so rigidly and perpetually" (O 22). The lifelessness of the grandmother's frozen smile stands in contrast to the liveliness of the film diva. Though each photograph represents a twenty-four-year-old girl, the diva embodies the present moment in all its demonic glamor, while the grandmother's youthful vitality has receded into the frozen smile of a no-longer-recognizable face. The viewer of the aged photograph thus witnesses the fragmentation of embodied presence into an untimely material residue: "In the eyes of the grandchildren the grandmother dissolves into fashionably old-fashioned details" (O 22).

The dissolution of the figure into its most ephemeral parts, those fashionable items of clothing that paradoxically make a stronger claim to duration than the grandmother herself, leads Kracauer to a series of reflections on the relationship between photography and time. Whereas the only temporal aspect that attaches to the spatial continuity of the diva's gapless appearance is the aura of her presence, the picture of the grandmother allows for the perception of a time that has passed away. The grandchildren laugh at the grandmother's old-fashioned dress—her chignons, cinched waists, crinolines, and Zouave jacket—but they are also overcome by a "shudder" (*Gruseln*), for there is something uncanny about the image, despite the absurdity of the costume (O 23).[45] As Kracauer writes, "For behind the ornamentation of the costume, in which the grandmother has disappeared, they think they catch sight of a moment of elapsed time, of time that passes away without return" (O 23). The picture of

---

45. Hansen comments astutely on the uncanniness of the photograph: "What makes the grandchildren giggle and at the same time gives them the creeps, Kracauer suggests, is that the photograph amalgamates these remnants with the incongruous assertion of a living presence" ("Kracauer's Photography Essay," 99).

the grandmother bears witness to death and to the irretrievable passage of time. It is precisely this effect that accounts for Kracauer's choice of the photograph of the grandmother as a counterpoint to the immediacy and presence of the image of the film diva.

Kracauer is careful to give a nuanced account of the relationship between photography and time. Although one can discern the passage of time in an old photograph, it does not follow that photography represents time. Kracauer attributes this conclusion to the grandchildren, but he offers a corrective of his own: "Time is of course not photographed like the smile or the chignons, but photography, it seems to them, is a representation of time. If photography alone could provide these details with duration, they would not be preserved beyond mere time; rather, time would create images out of them [*die Zeit schüfe aus ihnen sich Bilder*]" (O 23).[46] In what manner, Kracauer asks, do the grandmother's smile and her fashion choices, as captured by the photograph, have a claim to permanence? Is it the photograph itself, as a representation of time— of a time that has passed and will never return—that secures an afterlife for a present that no longer exists? Here Kracauer answers in the negative: if the features of a picture could be made permanent by the medium of the photograph alone, they would not outlast mere time, for the time to which they belong would have ceased to exist. This is the crux of Kracauer's argument: photography preserves the details captured in the photograph not as the things represented but rather as images. Time itself—even considered negatively as that which one day will no longer exist—is not represented by the photograph. Kracauer intervenes by reversing the terms of the relationship between photography and time. Photography does not represent time; rather, the passage of time makes images out of the figures represented in the photograph. The photographed objects have no permanence as things, but they survive as images of the past. Far from capturing a time that has ceased to exist, the photograph itself is transformed with the passage of time into a set of images. The detachment of the photographic image from its referent could not be stronger.

---

46. Compare Kracauer, *The Mass Ornament*, trans. Levin, 49.

Kracauer's confrontation of the medium of photography with the onslaught of time serves to sharpen a fundamental contrast between the completeness of the spatial continuum that photography represents and the lack of traction that it obtains in temporal terms. As a medium of modernity, the photograph makes visible—negatively, in what it does not capture—the blind spots, lacunae, and hollow spaces of time and history.[47] The spatial continuum of the photograph is haunted by the death mask of the historical process, which runs through Kracauer's essay as a powerful subtext.[48] At stake in this account of the temporal aporias of the photograph is a critique of historicism with eschatological undertones. Kracauer develops this critique by drawing an analogy between photography and historicism. Just as all the details in the photograph of the diva sit properly in space, producing an "appearance without gaps" (*eine lückenlose Erscheinung*) (O 21), so too does historicism "reconstruct a series of events in their temporal succession without gaps [*lückenlos*]" (O 24). The absence of vacant spaces, voids, or gaps in photography and historicism defines for Kracauer the continuity of these media: "Photography offers a spatial continuum [*Raumkontinuum*]; historicism attempts to fulfill the temporal continuum [*Zeitkontinuum*]" (O 24). The continuity of the temporal continuum required by historicism, Kracauer suggests, depends on the completeness of the course of history it reconstructs: "The complete mirroring [*vollständige Spiegelung*] of the intratemporal course of events contains, according to historicism, the meaning of everything that has taken place in time" (O 24).

---

47. Similarly, Hansen describes the temporal afterlife of the photograph in terms of an "irreversible disjuncture and dissociation into dissimilarity" ("Kracauer's Photography Essay," 101).

48. The death mask of history becomes manifest not in the photograph itself, but in the disjuncture between the aged photograph and the present. As Hansen comments, "The photograph of the young grandmother-to-be does not return the gaze across generations. For Kracauer, the chilly breeze of the future that makes the beholder shudder conveys not only intimations of his own mortality but also the liberating sense of the passing of a history that is already dead, depriving the bourgeois social order of its appearance of coherence and continuity, necessity and legitimacy" ("Kracauer's Photography Essay," 101).

The visual dimension of the metaphor of mirroring invoked in this characterization of historicism sets the tone for Kracauer's comparison of its historiographical ambitions with the medium of photography. At first he considered these media schematically, with one presenting a spatial continuum and the other a temporal continuum, but now Kracauer borrows from the concept of photography to define the aims of historicism: "Historicism is concerned with the photography of time. The equivalent of its temporal photography would be a giant film depicting the temporally interconnected events from every vantage point" (O 24).[49] The clear delineation of realms in the analogy is broken down by a metaphorical crossing of boundaries. Historicism's treatment of time is described as a kind of photography: the temporal continuity that it seeks takes as its operative metaphor the spatial continuity of the photograph. The result of this crossing of boundaries is a curious hybrid, a film capable of depicting everything that takes place from "all sides" (*allseitig*). In his analysis and critique of historicism, Kracauer treats time as space, adopting spatial categories to describe the operative principles of its historiography.[50]

Kracauer's strategy for challenging the spatial and temporal continuums of photography and historicism is two-pronged. In a first pass, he deploys the concept of the memory image as a counterpoint to the photographic image in order to underscore the limitations of its spatial continuity. Then, in a second pass, by staging a confrontation between photography and death, he develops the critical potential of photography to reveal the provisional and unredeemed character of the reality it records. The horizon for these reflections is a view of photography that at once points to a tear in the fabric of history and suggests possibilities for the reconfiguration of history.

---

49. Kracauer, *The Mass Ornament*, trans. Levin, 50.
50. Similarly, Hansen notes that Kracauer's interpretation of film makes manifest the spatialization of time in modernity. See Miriam Bratu Hansen, *Cinema and Experience: Siegfried Kracauer, Walter Benjamin, and Theodor W. Adorno* (Berkeley: University of California Press, 2012), 9: "With its emphasis on fragmentation and discontinuity, the film visualizes the spatialized experience of time typical of modernity: 'the moment, which is only a point in time, becomes visibility.'"

The photographic image allows Kracauer to explore in concrete terms the key ideas developed in "Figure and Disintegration."

While Kracauer clearly does not aim to set up a "mere opposition of the photographic image to the 'memory image,'"[51] the contrast between photographic and memory images, and the critical commentary *on* photography that this contrast entails, lays the groundwork for the account of the critical potential *of* photography at the conclusion of his essay. The crux of the difference turns on the fragmentary quality of the memory image in contrast to the totality and continuity of the photograph. Because the memory image is concerned only with the significance of things and events, it is fragmentary and "full of gaps" (*lückenhaft*), whereas the spatial continuum of the photograph is "without gaps" (*lückenlos*) (O 24, 21). As Kracauer writes:

> Photography captures what is given as a spatial (or temporal) continuum; memory images preserve what is given insofar as it has significance. Because significance is realized neither in a purely spatial nor in a purely temporal context, memory images stand askew to photographic reproduction. If memory images appear as fragments from the perspective of photography—as fragments, however, because photography does not comprehend the meaning to which they are related and in relation to which they cease to be fragments—, so too does photography appear from the perspective of memory images as a mixture or jumble that is composed in part of refuse. (O 25)

According to Kracauer, neither the photograph nor the memory image is able to comprehend the function of the other. Because photography operates through spatial and temporal continuities, it cannot grasp that the memory image stands outside spatial and temporal contexts. And because the memory image operates through relations of meaning and significance, it views much of what the photographic image depicts as extraneous and unworthy of preservation.[52]

---

51. Hansen, "Decentric Perspectives," 54.
52. Yet this sense of the photograph as a collection of discarded refuse makes the medium attractive to Kracauer. As Hansen comments: "In the tension between history and that which history has discarded, photography begins to occupy the intermediary zone that appeals to Kracauer the ragpicker, the intellectual seeking

This account of the memory image critiques not only the photographic image but also the historicist notion of temporal continuity. Invoking the eschatological motif of a "last image," Kracauer argues that memory images provide the history of a person:

> The image in which such characteristics can be found is distinguished from all other memory images; for it does not preserve, as they do, an abundance of opaque recollections, but rather contents that concern what has been recognized as true. To this image, which quite rightly may be called the last image, all memory images must be reduced, for it is only in the last image that the unforgettable endures. The last image [*Das letzte Bild*] of a human being is his actual *history*. (O 25)

Crucially, the concept of history that Kracauer puts forward here does not align with the temporal continuum of historicism, which discerns the significance of a person's history only in the totality of its sequential unfolding. Nor does the concept of the last image have the teleological sense of a final image at the end of a temporal sequence. Instead, the last image embodies a concept of history that is fragmentary and selective. As a condensation or reduction of all memory images, the last image is both outside time and present in each moment of time.

The history of a person, in Kracauer's view, is not a function of his "natural constitution" or of the "illusion of his individuality" but is unequivocally a construction (O 26). As a history that is represented, the last image of a person contains only those fragments of his or her constitution or supposed individuality that are meaningful. In keeping with his larger approach of spatializing time, Kracauer emphasizes the graphic and figural qualities of the construction of a person's history: "This history is akin to a *monogram* that condenses the name into a polyline which has meaning as an ornament" (O 26). Kracauer thus defines history as a spatial construction. Just as the last image is a reduction of a collection of memory images, so too is the monogram a condensation of a name into a "polyline" (*Linienzug*). Of the metaphors that Kracauer invokes here (monogram, polyline,

---

to gather the refuse and debris, the ephemeral, neglected, and marginal, the no longer functional" ("Kracauer's Photography Essay," 100).

ornament), the mathematical metaphor of the *Linienzug* is the most precise but also the most overlooked. The translation of this word as a "single graphic feature"[53] may strike readers as intuitive, but it lacks the precision of the mathematical term, which defines a curve connecting a set of points. A standard late nineteenth-century textbook on graphical statics defines the *Linienzug* as a curve described by a point that moves from one position to another without leaving the plane.[54] Whereas a constellation connects a series of points to make a figure, finding within space a set of connections, the polyline, as the condensation of the history of a person, provides a spatial representation of a temporal path or move (*Zug*). Kracauer's metaphor implies the compression of history into a spatial figure.

In the second part of "Photography," Kracauer shows how photography can reveal fissures and breaks in history by staging a confrontation of photography and death.[55] According to Kracauer, because photography is a "function of flowing time," its meaning changes depending upon whether it belongs to the present or the past (O 29). Because the film diva is among the living, her contemporary photograph not only represents her external features but also refers to a memory image that is still in circulation. In its present moment, the photograph of the diva is both a photographic image and an "optical sign" that points to her memory image. In turn, the memory image pierces the surface effects of the photograph

---

53. Kracauer, *The Mass Ornament*, trans. Levin, 51.
54. The polyline may be open or closed; it may intersect itself at nodal points; and its segments may be straight lines or curves. See Luigi Cremona, *Elemente des graphischen Calculs*, trans. Maximilian Curtze (Leipzig: Quandt & Händel, 1875), 11: "Wir wollen die Linie, welche ein Punkt beschreibt, der sich in einer Ebene aus einer Lage (der Anfangslage) bis zu einer andern (der Endlage) ohne Sprünge, d.h. ohne jemals die Ebene zu verlassen, bewegt, einen *Linienzug* nennen."
55. On the complex association of photography and death, see Jessica Nitsche, "Dem Tod ins Auge (ge)sehen: Protagonistinnen der Fotografietheorie bei Döblin, Kracauer, Barthes und Benjamin," in *Blick.Spiel.Feld*, ed. Malda Denana et al. (Würzburg: Königshausen & Neumann, 2008), 93–109. As Nitsche writes, "Der Augenblick, in dem fotografiert wird, wird zu einem tot stellenden, tödlichen Unterfangen. André Bazin bringt diesen Aspekt in Anschlag, indem er die Fotografie als ein *Einbalsamieren der Zeit* beschreibt" (95).

and lends it a heightened sense of visibility that Kracauer describes as "transparency" (O 29–30).

The passage of time, however, severs the immediacy of the connection between the photograph and its referent. When the living figure dies, its life recedes from the photograph as well: "Life, whose spatial appearance had covered over the mere spatial configuration [*bloße räumliche Konfiguration*], has drained out of the old photograph" (O 30). As the aura of the living figure vanishes, all that remains is the externality of the mere spatial configuration that the photograph records. Kracauer thus compares the photograph to a "sediment" (*Bodensatz*) or "remnant" (*Restbestand*) that has been deposited by history and whose "semiotic value" diminishes over time (O 30). The photograph of the grandmother is exemplary of this process of decay. When the grandmother ceases to signify and becomes nothing—as Kracauer puts it, "the nothing of the grandmother" (*dem Nichts der Großmutter*)—her image disintegrates into its elements: "When the grandmother in the photograph can no longer be found, the image taken from the family album must disintegrate into its details [*in seine Einzelheiten zerfallen*]" (O 30). Kracauer's use of the term *zerfallen* (to disintegrate) resonates with "Figure and Disintegration" and suggests that the photograph can reveal the fallenness of history.

It does so, however, not by representing death, but by exhibiting the unstable boundary between life and death. While the living figure of the grandmother can no longer command presence, the details of her fashion lay claim, through the photograph, to a kind of spectral afterlife: "In the photograph the grandmother's costume is recognized as a discarded remnant that continues to assert itself. In the sum of its details the costume rises up like a corpse and conducts itself amply as though there were life in it" (O 31). As in "Figure and Disintegration," here Kracauer views the disintegration of the living figure in the photograph as a prelude to new configurations. However, the afterlife of the photograph very clearly bears the traces of its disintegration: the photographic image of the grandmother's costume is conceived not as her resurrection but merely as the self-assertion of her fashionable corpse.

These reflections lead Kracauer to the theological crux of his photography essay, namely his thesis that the photographic image presents an unredeemed reality. Kracauer's motivations for introducing the concept of redemption into his reflections on photography are complex. First, he returns with increasing emphasis to the idea that a "liberated consciousness" is one that is able "to escape the bonds of nature" (O 31). Yet the photographic image, by producing a mask or costume of its objects, reproduces such natural bonds. Second, he emphasizes the uncanniness of old photographs, which becomes manifest in the realization that the photograph, although it contains the "refuse" (*Abfall*) of the past, was once present (O 31). The figure in the photographic image thus has a ghostly or spectral presence, although what survives the passage of time is the figure's costume, not the figure itself. However, even as it preserves images of the past in their external shells rather than reconstituting their elements, the photograph has the virtue of revealing the unredeemed character of the reality it conjures up:

> The playing of old hits or the reading of letters written long ago, just like the photographic image, conjures up the disintegrated unity anew. This spectral reality is *unredeemed*. It is made up of parts in space whose connection is so little necessary that one could conceive of the parts arranged differently. It clung to us once like our skin, and our property still clings to us in this way today. We are contained in nothing, and photography collects fragments around a nothing [*Wir sind in nichts enthalten, und die Photographie sammelt Fragmente um ein Nichts*]. (O 32)

The subtle eschatological horizon of this passage demands attention. It expresses less an eschatological longing of the sort that Michael Schröter detects in Kracauer's work than a sober recognition of the nothingness of history from the perspective of the last days.[56] The claim that "we are contained in nothing" points to an eschatological void beneath the stratum of surface expressions in which we move. The specifically modernist element in Kracauer's appropriation of the eschatological motif is that he conceives of photography as a medium that brings to light and comments on this eschatological

---

56. See Schröter, "Weltzerfall und Rekonstruktion," 25: "Eine Intensität der Betroffenheit ... wird greifbar in vielen der leuchtkräftigen Metaphern, die Kracauers Texte durchziehen und die so oft ihr eigenes Leben entwickeln. Sie verleiht seinen Schlüsselbegriffen eine Aura eschatologischer Sehnsucht."

situation. It does so in several ways. First, it invites a view of the past as that which was once present, yet with a sense of estrangement, because the connection between the photographic image and its referent has broken down.[57] Second, it reveals that the energy of life in the present is bound to become a specter of death with the passage of time. Finally, to this disjunctive conjunction of past and present, life and death, Kracauer adds the disintegration of the photograph into its elements and the capacity of the photography to reconstitute the disintegrated unity. As a medium that represents this dialectic of past and present, life and death, disintegration and configuration, photography points to the eschatological situation by revealing the unredeemed quality of reality.

Crucially, however, just as photography points to the unredeemed status of reality, it also suggests the possibility of its redemption.[58] The reality that the photograph reveals, according to Kracauer, is utterly contingent. It consists of parts in space that have no necessary connection and could be ordered and arranged differently. The critical distance offered by the separation between the present of the viewer and the present of the photograph shows that what was taken for granted as reality could be conceived otherwise. That the order of things presented by the photograph is not necessary is both an index of the unredeemed status of reality and an intimation that a different order, the contours of which Kracauer does not spell out, might offer possibilities for its redemption.

These redemptive possibilities are made explicit in the conclusion of "Photography," which completes the turn from a critical take *on* photography to an investigation of the critical potential *of* photography.[59] Here the productivity of Kracauer's reflections on the negative

---

57. On the estrangement produced by the photograph, compare Hansen, "Kracauer's Photography Essay," 100–101.

58. In his later work, Kracauer extends this claim about the redemptive qualities of photography to the medium of film, arguing that film redeems physical reality by rendering visible a world that we could not see before its advent. See Siegfried Kracauer, *Theory of Film: The Redemption of Physical Reality* (Princeton, NJ: Princeton University Press, 1997), 300.

59. For a different account of this shift in Kracauer's account of photography, see Steve Giles, "Making Visible, Making Strange: Photography and Representation in Kracauer, Brecht and Benjamin," *New Formations: A Journal of Culture/*

potential of photography come to a head. Describing the totality of photographs as a "general inventory" (*Generalinventar*), a "collected catalog" (*Sammelkatalog*), and a "main archive" (*Hauptarchiv*), Kracauer underscores once again the spatial continuity of photographic images: the photographic catalog collects "all appearances that offer themselves in space" (*O* 37, 38). Significantly, however, photography not only reproduces the spatial continuity of the natural order, but also reveals possibilities for its perceptual reconfiguration. His larger insight is that photography makes it possible for spatial configurations that are not determined by the habits of human perception to be added to the archive: "All spatial configurations [*alle räumlichen Konfigurationen*] are incorporated into the main archive in unfamiliar combinations [*Überschneidungen*] that distance these configurations from human proximity" (*O* 38). As the memory image that grants recognizability to the photographic image recedes, the photograph is drained of life and becomes a cipher of a "world of death" (*O* 38). The elements of the photograph thus begin to emerge in new and unfamiliar combinations; in the absence of human proximity, the natural and material base of these elements is revealed, and new possibilities for their spatial configuration arise.

The photographic archive, in Kracauer's view, is home to an inventory of images whose human proximity gradually gives way to its spatial and material residue. Paradoxically, the archive preserves the elements of the photograph just as they are falling away from the contexts that originally gave them meaning.[60] In Kracauer's suggestive imagery, the holdings of the collection crumble, but the archive

---

Theory/Politics 61 (2007): 64–75, at 65–66. Whereas the account of photography in the first part of the essay is "unremittingly negative," the final section "suddenly change[s] direction": "Hitherto, photography had been seen as a mere reflector of surface appearance, whereas now it is suddenly invested with the power to make visible the as yet unseen" (65–66). Giles notes that Kracauer criticizes a realistic concept of photography that operates in mimetic or naturalistic terms, but recognizes a "utopian dimension of avant-garde photography ... in its ability to mirror the sheer negativity of contemporary life" (66).

60. As Hansen notes, photography functions as both the symptom and the agent of this process of disintegration: "As both symptom and agent of the petrification of the world, photography also gathers the detritus of history and reveals it in all its negativity" ("Decentric Perspectives," 54–55).

preserves these remains in the state of their disintegration: "The holdings crumble [*zerbröckeln*] because they are not held together. The photographic archive assembles in the image [*im Abbild*] the last elements [*letzten Elemente*] of nature alienated from meaning" (O 38). Photography captures the dialectic of disintegration and figuration in the medium of the image. Its elements crumble because they are no longer held together by relations of meaning and significance. As images, they are reduced to their basic natural and material elements. Yet this disintegration allows the image to acquire figural qualities of its own. Just as the last image of a person takes on eschatological qualities as his or her history becomes condensed into the spatial features of a monogram, so too does the photographic archive preserve images that have been reduced to their "last elements." The last elements of the photograph, however, are not condensed as a singular, unforgettable unity of meaning, but remain open to new configurations. In the photographic archive, history is confronted with the eschatological situation in a state of thorough dissolution.[61] The elements of the photograph in such a state are images of unredeemed reality, but because they have been reduced to their basic natural elements, they are also eminently capable of redemption. As Hansen notes, Kracauer conceives of the collection of the remnants of history in the photographic archive as connected to an unrepresentable possibility of redemption in "a utopian restoration of all things past and present as implied in the cabbalist concept of *tikkun*."[62] The prospect of a restoration of the past, however, implies the transformation and reconfiguration of its disintegrated elements.[63]

Kracauer famously called photography "the go-for-broke game of the historical process" (*das Vabanque-Spiel des Geschichtsprozesses*)

---

61. Stressing the anarchic dimension of the photographic archive, Hansen describe it as an "*an*-archive": "This archive, however, is anything but easy to access and navigate; it is rather an *an*-archive—a heap of broken images—that lends itself to the task precisely because it lacks any obvious and coherent organizational system" ("Kracauer's Photography Essay," 103).
62. Hansen, *Cinema and Experience*, 22.
63. Kracauer sums up his theology of history in a letter to Bloch cited by Hansen as "the postulate that nothing must ever be forgotten and nothing that is un-forgotten must remain unchanged" (*Cinema and Experience*, 23).

(O 38). The key to understanding this claim is the idea that the photographic archive, by collecting images that have been reduced to the eschatological zero-point of meaning, reveals the provisional character of the natural order. In the photographic image, nature has been dissolved into its basic elements, but what is to be done with these elements is an open question. They have been given over to consciousness to make with them what it will:

> The images of the inventory of nature dissolved into its elements have been handed over to consciousness to dispose of freely. Their original order is gone, and they no longer cling to the spatial context that connected them with an original out of which the memory image was selected. But if the natural remnants do not point to the memory image, then the order mediated by the image is necessarily a provisional arrangement. It is therefore up to consciousness to demonstrate the *provisionality* of all given configurations, and perhaps even to awaken an intimation of the true order of the inventory of nature. (O 38–39)

Two important aspects of Kracauer's theory of photography are evident in this passage. First, as an archive of images dissolved into their elements, the spatial continuum that Kracauer had defined as the essential feature of the photograph is cast in a different light. The ordering principle of the photograph is dissociated here from the "spatial context" (*räumlichen Zusammenhang*) in which its original was embedded. All of the remnants of the natural order are present in the image, but their arrangement is revealed as provisional.[64] Kracauer's insight that the elements of the photograph move freely with regard to their original spatial context resonates with the characteristics of topological objects, which, as Lash notes, are not located *in* space but rather define their own intrinsic modes of space.[65] Second, the provisionality of the natural order, separated from historical

---

64. Hansen extends Kracauer's insights into the negative potential of photography to include the destabilization of social and cultural orders ("Kracauer's Photography Essay," 104). The condition for the disintegration of these orders, this chapter argues, is the collapse of their "spatial context."

65. See Scott Lash, "Deforming the Figure: Topology and the Social Imaginary," *Theory, Culture & Society* 29, no. 4/5 (2012): 261–87, at 265: "Topological objects are not located *in* space at all. Topological objects *are* spaces. Topological objects do not move in space. They are instead spaces of movement. Topographical

consciousness and memory, places the image at the eschatological limit. It is drained of meaning and life, yet it is fully capable of being given new meaning and new life.

Kracauer's gesture of awakening an intimation of the true order of the inventory of nature expresses the key turning point of modernist eschatology at which the liminal situation of the end flips over into possibilities for new configurations and new beginnings. Kracauer famously assigns the task of realizing these possibilities to film: "The lack of order in the refuse reflected by photography cannot be made any clearer than through the suspension [*Aufhebung*] of all familiar relations between the elements of nature. To stir up these elements is one of the possibilities of film. It realizes these possibilities whenever it associates parts and excerpts to form foreign structures" (O 39).[66] The suspension of the natural order, its reduction to its basic elements and parts, and the reassembly of these elements in new configurations constitute a process that has as its heart an eschatological dynamic of disintegration (*Zerfall*) and reconfiguration (*Neugestaltung*). The elements are dissolved and await their redemption in new constructions. For Kracauer, photography points to the last days by showing the dissolution of history into refuse and detritus. In "Photography," eschatological motifs shatter the spatial and temporal continuities of history, while the redemptive possibilities of new configurations of history are predicated on the disintegration of its elements. The spatial orientation of eschatology in Kracauer's work is concerned with the recognition of gaps, voids, and hollow spaces in what appears to be the continuous surface of history.[67] The shape of time that Kracauer gives us in "Photography" can therefore be best characterized by its shapelessness, by the looming disintegration of its form.

---

objects are forms that move in space. Topological objects are not forms, but themselves spaces: spaces of deformation."

66. Compare Kracauer, *The Mass Ornament*, trans. Levin, 62–63. The term "to stir up" has been borrowed as a translation of *umtreiben*.

67. On Kracauer's problematization of the concept of the surface, see O 27: "In order for history to be represented, the mere surface relation [*Oberflächenzusammenhang*] that photography offers must be destroyed."

"The Last Things before the Last": Kracauer's Anteroom Area of History

Kracauer's late work on history, beginning with the essay "Time and History" (1963)[68] and culminating in the posthumously published *History: The Last Things before the Last* (1969),[69] combines a penetrating critique of historicism with the formulation of a new spatial paradigm for thinking about history in terms of shapes of time.[70] Calling into question the notion of historical time as an empty container for a homogeneous flow of events, Kracauer argues that the historian forms and molds history, developing sequences of historical ideas that he describes as time curves. Historical ideas are thus displaced from the flow of chronological time and are defined, like the historian, by a state of exile and exterritoriality. Kracauer redefines history as a curved surface containing pockets and voids in which time folds on itself, deforms, and takes on new shapes. These manifold shapes of history, for Kracauer, are always on the threshold of the end of time— they are the last things before the last. But the end of history itself remains unthinkable and unrepresentable: history can never take on an ultimate shape, Kracauer argues, because shape itself, understood topologically, is always subject to deformation and disintegration.

Although more than thirty years separate Kracauer's Weimar essays from his late work on history, there is a strong connection between the key concepts and approaches in *History* and in texts such as "Photography" and "Figure and Disintegration."[71] To be

---

68. Kracauer, "Time and History," in *Werke*, vol. 4, ed. Inka Mülder-Bach (Frankfurt am Main: Suhrkamp, 2009), 377–93.

69. As Paul Oskar Kristeller notes in the foreword to *History*, although the book was unfinished, Kracauer completed the majority of its chapters and left detailed synopses for those he had not completed (*L* xi). The essay "Time and History" laid the groundwork for an important chapter in *History*.

70. As Stephanie Baumann argues, in his concept of history Kracauer aimed for "die Ablösung der Vorstellung eines linearen Geschichtsverlaufs durch ein räumlich determiniertes Geschichtsdenken" (*Im Vorraum der Geschichte: Siegfried Kracauers "History. The Last Things before the Last"* [Konstanz: Konstanz University Press, 2014], 141).

71. On the continuities and discontinuities between Kracauer's early and late works, see Inka Mülder-Bach, "Schlupflöcher: Die Diskontinuität des Kontinuierlichen im Werk Siegfried Kracauers," in *Siegfried Kracauer: Neue Interpretationen*, ed.

sure, Kracauer's sense of redemption was altered by the aporias of the history of the Holocaust, and the experience of exile in France and the United States provided a crucial new framework for thinking about the position of the historian. Nevertheless, Kracauer recognized that his theory of photography provided a germ cell for his work on history: "I realized in a flash the many existing parallels between history and the photographic media, historical reality and camera-reality. Lately I came across my piece on 'Photography' and was completely amazed at noticing that I had compared historism with photography already in this article of the 'twenties" (L 3–4). Just as "Photography" shows the provisionality of the spatial order found in the archive of photographic images, *History* aims to show the provisionality of the temporal order of history. Likewise, Kracauer notes that "history resembles photography in that it is, among other things, a means of alienation" (L 5). The photograph induces alienation by producing a perspective from which the familiar can be viewed from a position of distance.[72] What it accomplishes in terms of spatial configurations, history accomplishes in temporal terms: as Kracauer puts it, history "enable[s] us to look at the contemporary scene from a distance" (L 5).[73]

Michael Kessler and Thomas Y. Levin (Tübingen: Stauffenburg, 1990), 249–66. Noting a subtle "process of theoretical reorientation" between the Weimar essays and Kracauer's later works, Mülder-Bach argues that terms such as "nonsimultaneousness, heterogeneity, [and] discontinuity," which in the 1920s referred solely to the periphery, come to define in *History* "the entirety of historical reality" ("History as Autobiography: The Last Things before the Last," *New German Critique* 54 [1991]: 139–57, at 152).

72. Elena Gualtieri suggests that the meaning of "alienation" in Kracauer's conceptual universe undergoes an important shift between the Weimar essays and his late work: "Still understood as the malaise of modern life in *Theory of Film*, alienation will become in Kracauer's last book the condition of possibility for a modern utopia" ("The Territory of Photography: Between Modernity and Utopia in Kracauer's Thought," *New Formations: A Journal of Culture/Theory/Politics* 61 [2007]: 76–89, at 85). Yet Gualtieri overstates the difference between "Photography" and *History* by failing to recognize that already in "Photography" there is a sense of the utopian possibilities of photography.

73. Despite these continuities between "Photography" and *History*, the character of Kracauer's analogy between photography and history undergoes an important shift in *History*. In "Photography," the comparison of photography and historicism emphasizes the limitations of their production of spatial and temporal continuums

The ability of photography to reveal the provisional status of the order of things it mediates is a key aspect of its critical function that Kracauer formulates in the conclusion of "Photography." In *History*, he similarly describes the "intermediary area of history" as a "provisional insight into the last things before the last" (*L* 16). In a later definition of his key concept of the "anteroom," Kracauer makes this comparison explicit: "One may define the area of historical reality, like that of photographic reality, as an anteroom area.... They share their inherently provisional character with the material they record, explore, and penetrate" (*L* 191). In "Photography," Kracauer argues that the provisionality of a given spatial configuration is predicated on a kind of temporal decay. When the remains of an image no longer maintain any connection to a memory image, the provisional status of the image becomes apparent: "But if the natural remnants do not point to the memory image, then the order mediated by the image is necessarily a provisional arrangement" (*O* 39). In *History*, similarly, an insight into the provisionality of history depends on a rethinking of its spatial qualities, as suggested by the spatial metaphor of the anteroom.[74]

Kracauer's theory of history both criticizes historicism and attempts to conceive of a new set of historical concepts that exceed the scope of historicism.[75] According to Kracauer, historicism forms

---

without meaningful relations. In *History*, by contrast, the comparison of photography and history emphasizes their critical functions and their capacity to unsettle existing orders. As Mülder-Bach argues, "the positive and negative signs" of the analogy between photography and history are "reversed" in *History*. In the essay on "Photography," these media produce a "reduction of reality to juxtaposition and succession, to the spatial and temporal relations among its elements," whereas in *History*, "self-alienation" becomes "the precondition of genuine knowledge" ("History as Autobiography," 142).

74. The reading proposed here emphasizes a continuity between "Photography" and *History* insofar as both reveal the provisionality of the order of things. Mülder-Bach, by contrast, sees a stronger shift in Kracauer's theory of a history, arguing that in the later work, the provisionality of history is taken as given, whereas in the earlier work it had to be produced by consciousness ("History as Autobiography," 153).

75. To Kracauer's credit, it should be noted that he was not blind to the important achievements of historicism. Indeed, he criticizes Nietzsche for "[shutting] his eyes to the enormous achievements of historicism" out of an "inordinate and rather juvenile infatuation with 'life'" (*L* 75). Kracauer aims not to "liquidate"

the backbone of the modern concept of history understood as "an immanent continuous process in linear or chronological time which on its part is thought of as a flow in an irreversible direction, a homogeneous medium indiscriminately comprising all events imaginable" (L 139). This homogeneity of chronological time in historicism provides the basis for its concept of the continuity of the historical process in which all events can be incorporated:

> Under the spell of the homogeneity and irreversible direction of chronological time, conventional historiography tends to focus on what is believed to be more or less continuous large-scale sequences of events and to follow the course of these units through the centuries.... The underlying idea is that, in spite of all breaks and contingencies, each such inclusive unit has a life of its own—an individuality, as Meinecke puts it. (L 142)

Historicism does not need to deny discontinuities, breaks, and ruptures in history as long as they can be subsumed within a larger sense of continuity. While these assumptions about the linearity and continuous flow of chronological time may appear to describe solely the formal qualities of history, they also have implications, as Kracauer notes, for the way the content of history is understood: "Uncritical acceptance of the conception of flowing time kindles a desire to translate the formal property of an irreversible flow into content—to conceive, that is, of the historical process as a whole and to assign to that whole certain qualities; it may be imaged as an unfolding of potentialities, a development, or indeed a progress toward a better future" (L 142). Historicism thus provides the theoretical grounding for philosophies of history that view the historical process as a totality defined by progress and development.

Kracauer's aim is not to suggest that philosophies of history are the necessary result of historicist principles, but to show how a variety of theories of history share the basic premise that the medium of history is chronological time. In his discussion of natural history, Kracauer notes that for Comte and Marx, "the historical process is tantamount to a linear movement—a necessary and meaningful

---

historicism but to grasp both what historicism can to contribute to the study of history and what exceeds its conceptual framework.

succession of periods along a time continuum indefinitely extending into the temporal future" (L 38). Comte and Marx presume that calendric time is an "all-powerful medium" that serves as the "carrier of all significant historical forces and developments" (L 38). By contrast, the "present-interest" theory of the history developed by Croce and Collingwood makes no such assumptions about the lawfulness of history. Rather, it views history as contemporary history in which the historian "re-enacts the past out of his immersion in present-day concerns" (L 63). Yet the present-interest theory of history also proceeds from the premise that "the flow of chronological time is the carrier of all history" (L 66); its intervention simply relocates the vantage point of the historian from the past to the present. Kracauer calls into question this fundamental premise of historicism and entertains the possibility that chronological time is "an empty, indifferent flow that takes along with it a conglomerate of unconnected events" (L 38). Elsewhere he describes chronological time as an "empty vessel" (L 201). Kracauer does not deny the existence of chronological time, but he draws attention to its ambiguity: it is "both the carrier and not the carrier of all significant historical forces and developments" (L 38).

In his effort to conceive of history beyond historicism, Kracauer develops a series of alternate temporalities that conceive of history in terms of metaphors of displacement. For example, he calls into question the unity of historical periods and the "belief that people actually 'belong' to their period" by pointing to examples of "chronological exterritoriality" (L 68). Exceptional thinkers and historians such as Vico and Burckhardt, Kracauer suggests, stand outside their historical contexts; their "complex and ambivalent physiognomy" cannot be derived "from the conditions under which [they] lived and worked" (L 68–69). Elsewhere Kracauer writes of the "temporal exterritoriality" of historical ideas, explaining that "they must be thought of as lying both within and outside flowing time" (L 199–200). Historical ideas are located on the continuum of chronological time, yet they are also dislocated and stand apart from their historical context. In accordance with the temporal exterritoriality of history, Kracauer understands the task of the historian through the lens of the experience of exile: "It is only in this state of self-effacement, or homelessness, that the historian can commune with the material of his concern.... A stranger to the world evoked by the sources, he is faced with the task—the exile's task—of penetrating its outward appearances, so that he may learn to understand the world from within" (L 84).[76] Kracauer's insight into the historian's situation of exile is remarkable for its spatialization of the historical continuum. The encounter with the past from the vantage point of the present becomes a journey—the operative metaphor of Kracauer's chapter on "The Historian's Journey"—of self-imposed exile.[77] Here Kracauer's critique of the present-interest theory of history reaches its pinnacle: he sides with Leo Strauss against Collingwood in claiming that "the historian does *not* retain his identity" in the process of undertaking a journey into the past and returning to the present (L 91). The historian finds stable footing neither in the past as a self-contained epoch nor in his present as the point of departure for his investigations. A sense of place emerges in history only in the mode of not-belonging. This is the state of alienation that Kracauer thinks can be produced both by photography and by history.

For Kracauer, history begins to take shape not in the flow of chronological time, but in the space of what he calls "historical ideas," which mark the final destination of the historian's journey (L 97). Citing Burckhardt's concept of the awakening individual in the Renaissance and Marx's theory of base and superstructure as examples, Kracauer notes that historical ideas are distinguished by the fact that they point beyond their material and introduce principles

---

76. Kracauer's insight into the historian's situation of exile succinctly expresses the key insight of David Lowenthal's *The Past Is a Foreign Country* (Cambridge: Cambridge University Press, 1985).

77. As Alan Itkin notes, "in using Orpheus' journey to the underworld as his guiding metaphor, Kracauer asks us to think of our engagement with the historical past as a journey through a space where we view figures representative of different times and different events from a number of different perspectives" ("Orpheus, Perseus, Ahasuerus: Reflection and Representation in Siegfried Kracauer's Underworlds of History," *Germanic Review* 87 [2012]: 175–202, at 178).

with which a range of historical relationships can be understood. Historical ideas "connect the particular with the general" such that the historical material begins to assume new shapes:

> Any such connection [between the particular and the general] being an uncertain venture, [historical ideas] resemble flashes illumining the night. This is why their emergence in the historian's mind has been termed a "historical sensation" and said "to communicate a shock to the entire system ... the shock ... of recognition." They are nodal points—points at which the concrete and the abstract really meet and become one. Whenever this happens, the flow of indeterminate historical events is suddenly arrested and all that is then exposed to view is seen in the light of an image or conception which takes it out of the transient flow to relate it to one or another of the momentous problems and questions that are forever staring at us. (*L* 101)[78]

The shock of recognition produced by a historical idea goes hand in hand with the disruption of the historical continuum and the emergence of a space of reflection. For this reason, Kracauer calls those moments in history that allow historical ideas to become visible "nodal points": points at which the concrete material of history no longer stands solely within the flow of chronological time. Chronological time is arrested when images of condensed historical recognition occupy a space of history that at once emanates from a given historical context and exceeds it. By engaging with historical material at the limit of its temporal exterritoriality, Kracauer pictures history as becoming manifest at the point of contact between the movement of time and the crystallization of historical ideas.

Drawing on the art historian George Kubler's book *The Shape of Time: Remarks on the History of Things* (1962),[79] Kracauer suggests that "instead of emphasizing matters of chronology ... the historian had better devote himself to the 'discovery of the mani-

---

78. Kracauer cites here Isaiah Berlin's "History and Theory: The Concept of Scientific History," but his remarks also resonate with Walter Benjamin's theory of the dialectical image. See Isaiah Berlin, "History and Theory: The Concept of Scientific History," *History and Theory* 1, no. 1 (1960): 1–31; and Benjamin, *Das Passagen-Werk*, 578.

79. George Kubler, *The Shape of Time: Remarks on the History of Things* (New Haven, CT: Yale University Press, 1962).

fold shapes of time'" (*L* 144). In Kubler's theory, shaped times result from the arrangement of works of art into sequences whose common element is that they "represent successive 'solutions' of problems originating with some need and touching off the whole series" (*L* 144). The construction of such series of art works places a given art object on a temporal trajectory that does not depend on its place in chronological time: "Each sequence evolves according to a time schedule all its own. Its time has a peculiar shape. This in turn implies that the time curves described by different sequences are likely to differ from each other" (*L* 144). The period into which a work of art falls, insofar as this period is defined by its chronological parameters, is far less significant than the relation it bears to other works of art that address similar problems, even if their chronological locations place them in vastly different periods. Shaped times, in other words, exist both alongside and beyond chronological time.

Kracauer extends Kubler's theory of shapes of time from its narrow application in the field of art history to a more general principle of historiography, one that offers an alternative to the historicist principles of chronological flow and temporal continuity. "At a given historical moment," Kracauer writes, "we are confronted with numbers of events which, because of their location in different areas, are simultaneous only in a formal sense.... The shaped times of diverse areas overshadow the uniform flow of time" (*L* 147). The formal simultaneity of events in chronological time, according to Kracauer, is secondary to the divergent shaped times that they occupy in a given historical moment. With this insight, Kracauer suggests a concept of history that recognizes the contemporaneity of noncontemporaneous historical figures:

> History consists of events whose chronology tells us but little about their relationships and meanings. Since simultaneous events are more often than not intrinsically asynchronous, it makes no sense indeed to conceive of the historical process as a homogeneous flow. The image of that flow only veils the divergent times in which substantial sequences of historical events materialize. In referring to history, one should speak of the march of times rather than the "March of Time." Far from marching, calendric time is an empty vessel. (*L* 149)

The "simultaneity of the asynchronous" is Kracauer's gloss and interpretation of the concept of the *Gleichzeitigkeit des Ungleichzeitigen*, a term whose range of meanings include "the simultaneity of the nonsimultaneous," "the synchronicity of the nonsynchronous," and "the contemporaneity of the noncontemporaneous." The concept played an important role in early twentieth-century writings on history by authors as diverse as Karl Barth and Ernst Bloch.[80] Recognizing the fine nuances that distinguish the concepts of contemporaneity, simultaneity, and synchronicity, each of which are present in the German *Gleichzeitigkeit*, Kracauer draws attention to the heterogeneity of events that take place in a given moment: they are simultaneous in the formal sense of taking place at the same moment in chronological time, but they are asynchronous insofar as they represent divergent shapes of time that do not fall into step with one another. The historicist concept of a chronological flow of time produces a distorted concept of history as "a process in homogeneous chronological time" that fails to do justice to the more complex shapes of time with which Kracauer is concerned.[81]

With its contrast between the empty vessel of calendric time and a diverse set of shaped times, Kracauer's critique of historicism invokes a set of spatial metaphors that recall the notions of "emptiness" (*Leere*) and "configuration" (*Gestaltung*), which played a prominent role in his Weimar essays. Similarly, his concept of the antinomy of time introduces images of "pockets" and "voids" to critique the idea of a homogeneous flow of historical time: "Because

---

80. In the wake of Kierkegaard, Barth invoked the contemporaneity of Christ at each moment in time, a contemporaneity that overcomes the noncontemporaneity of the present with past times. In *Erbschaft dieser Zeit* (*Heritage of Our Times*) (1935), Bloch argued in inverse terms that noncontemporaneous social strata coexist in a given present: "Not all people exist in the same Now" (*Heritage of Our Times*, trans. Neville and Stephen Plaice [Oxford: Polity, 1991], 97). Bloch's work builds on Herder's insight that "in actuality, every changing thing has the measure of its own time within itself.... No two worldly things have the same measure of time" (Johann Gottfried Herder, *Metakritik zur Kritik der reinen Vernunft* [Berlin: Aufbau, 1955], 68; cited in Reinhart Koselleck, *Futures Past: On the Semantics of Historical Time*, trans. Keith Tribe [Cambridge, MA: MIT Press, 1985], 247). Likewise, Kracauer refers, citing an article by W. von Leyden, to Herder's notion that "everything carries within itself . . . the measure of *its own time*" (*L* 146).

81. Kracauer, "Time and History," 381.

of the antinomy at its core, time not only conforms to the conventional image of a flow but must also be imagined as being not such a flow. We live in a cataract of times. And there are 'pockets' and voids amidst these temporal currents, vaguely reminiscent of interference phenomena" (L 199). The pockets and voids that Kracauer invokes here, which disrupt the smooth progression of history, can be compared to the hollow spaces (*Hohlräume*) of the city described in "Two Planes." Just as the empty void of the Bay bent straight lines into curves, so too do the pockets that punctuate the current of time disrupt the uniform and homogeneous path of history conceived as a straight line. For Kracauer, the pockets and voids of history suggest a curved shape of time comparable to surfaces of variable curvature such as valleys and saddles. Conceiving of history as a heterogeneous collection of conflicting currents, as a set of simultaneous yet asynchronous events, Kracauer suggests that pockets and voids appear when these currents interfere with one another, just as in the physics of wave interference waves can amplify or cancel each other out depending on whether they are in phase or out of phase. In light of these interferences, Kracauer pictures history as constantly changing shape, as being riven by temporal exterritorialities that interpenetrate the flow of time. Kracauer therefore understands historical ideas "as lying both inside and outside flowing time" (L 200).

The image of chronological time as an empty vessel resembles the classical understanding of space, going back to Newton and Kant, as a three-dimensional entity and formal unity pictured as a "container."[82] By contrast, Kracauer's concept of history as a set of shaped times resembles the topological notion of space defined by the relation of a set of elements or points to one another. These points define spaces with intrinsic shapes and curvatures that are independent of any set of coordinates in which they might be embedded, just as for Kracauer time curves are independent of the empty vessel of chronological time. Kracauer's picture of historical time as beset by pockets and voids, together with his idea of an antimony of time in which history stands both inside and outside the

---

82. See Günzel, "Raum—Topographie—Topologie," 16–17; and Lash, "Deforming the Figure," 261.

flow of time, can be profitably compared to Michel Serres's concept of a fold in time. In his conversations with Bruno Latour, Serres expounds on his understanding of the topology of time as follows:

> [Time is] not laminar [flowing smoothly]. The usual theory supposes time to be always and everywhere laminar. With geometrically rigid and measurable distances—at least constant.... No, time flows in a turbulent and chaotic manner; it percolates ... this time can be schematized by a kind of crumpling, a multiple, foldable diversity.... If you take a handkerchief and spread it out in order to iron it, you can see certain fixed distances and proximities. If you sketch a circle in one area, you can mark out nearby points and far-off distances. Then take the same handkerchief and crumple it by putting it in your pocket. Two distant points suddenly are close, even superimposed. If, further, you tear it in certain places, two points that were close can become very distant. The science of nearness and rifts is called topology, and the science of stable and well-defined distances is called metrical geometry.... Admittedly, we need the latter for measurements, but why extrapolate from it a general theory of time? People usually confuse time and the measurement of time which is a metrical reading on a straight line.[83]

The empty spaces, pockets, and voids that Kracauer detects in history are examples of what Serres calls the "foldable diversity" of time. They are instances where eddies and ripple effects deform the homogeneous flow of time, where the smooth surface of history is marked by tears and rifts, making it possible for shapes of time to emerge in which chronologically distant points in time become proximate. Topology provides a conceptual framework with which time can be conceived as a curved surface, as twisted and folded upon itself.

Finally, Kracauer's argument for the interpenetration of temporal exterritorialities within the flow of time suggests a nuanced understanding of the relation between history and eschatology. Kracauer famously refuses to tackle the last things directly in his book on history, attending instead to the "last before the last" in the hope of

---

83. Michael Serres and Bruno Latour, *Conversations on Science, Culture, and Time* (Ann Arbor: University of Michigan Press, 1995), 59–60. Cited in Mike Michael and Marsha Rosengarten, "HIV, Globalization and Topology: Of Prepositions and Propositions," *Theory, Culture & Society* 29, no. 4/5 (2012): 93–115, at 104.

generating "anteroom insight" (L 192).[84] He thus marks the space of eschatology as beyond history, yet he also defines history as a realm situated at the threshold of the last things. The advantage of anteroom insight is its concreteness: it helps us to "overcome our abstractness" and to "think *through* things, not above them" (L 192). At the same time, anteroom insight offers possibilities for the redemption of history. Both history and photography, Kracauer writes, "make it easier for us to incorporate the transient phenomena of the outer world, thereby redeeming them from oblivion" (L 192). Like the photographic archive, history preserves an archive of memory that can be reassembled and reconstituted. Despite its configuration as an "intermediary area," history represents, in its materiality, the shapes in which the last things may be given form (L 192).

Kracauer's interest in the redemption of history from oblivion, via its incorporation as historical memory, marks in some respects a departure from the pathos of his Weimar essays, which conceived of redemption in negative terms that required the destabilization of historical reality.[85] In "Die Hotelhalle" ("The Hotel Lobby") (1925), for example, Kracauer writes that "the relationship to the last things [*Beziehung zum Letzten*] demands the convulsion of the penultimate things [*Erschütterung der vorletzten Dinge*] without annihilating them" (O 163). In *History*, by contrast, the convulsion of historical reality is taken as a given, while a relationship to the last things offers a possibility of preserving history from oblivion.[86]

---

84. In a letter to Bloch written in 1965, Kracauer comments on his penchant for "anteroom insight" by noting that he finds himself inclined "im Verkehr mit den umliegenden Dingen und Verhältnissen zu verzögern und sie nicht gleich alle auf ein letztes Ende hin zu interpretieren." While this hesitation about the interpretation of things in relation to a "last end" might be read as symptomatic of an aversion to eschatological thought, Kracauer immediately clarifies: "Daher meine Überzeugung, daß einer, der nicht verstrickt ins Hier ist, niemals in ein Dort gelangen könne." These remarks make it clear that for Kracauer, dwelling in the anteroom area of history is the precondition for any approach to the eschatological space of the last things (Bloch, *Briefe, 1903–1975*, 1:400).

85. See O 49.

86. As Hansen notes, Kracauer's sense of redemption was significantly altered by the aporias of the history of the Holocaust: "The utopian motif of the last-minute rescue (*Rettung*)—the rescue of suffering individuals, of a people threatened with

Such a relationship to the last things becomes possible when history is grasped in its exterritoriality.

In conceiving of the last things before the last as an anteroom area, Kracauer, like Barth and Rosenzweig, detemporalizes the relation of history and eschatology and reconceives this relation in spatial terms.[87] The last things before the last are not only temporally prior; they also stand spatially *before* the last things.[88] As Kracauer notes, "the historian devotes himself to the last things before the last, settling in an area which has the character of an anteroom. (Yet it is this 'anteroom' in which we breathe, move, and live)" (*L* 195). As an anteroom area, history defines a liminal space that is both separated from the space of the last things and defined by its relationship to this space. It is a space of waiting, preliminary and provisional, yet also a space that can preserve in memory the concrete and material representations of what may lie in the space of eschatology proper.

---

annihilation—has given way to a more modest project of redemption (*Errettung*), a term that no longer promises critical-allegorical readings (as in Kracauer's Weimar texts) but rather seems to entail a mimetic adaptation to the world of things" ("Introduction," in Siegfried Kracauer, *Theory of Film: The Redemption of Physical Reality* [Princeton, NJ: Princeton University Press, 1997], vii–xlv, at xxiv).

87. Kracauer followed the emergence of Dialectical Theology closely and published reviews of the first issues of the journal *Zwischen den Zeiten* in the *Frankfurter Zeitung* ("Zwischen den Zeiten (1923)," in *Werke*, vol. 5, no. 1, ed. Inka Mülder-Bach [Berlin: Suhrkamp, 2011], 563–64, 634–36). Kracauer's claim that the last things can be approached only from an anteroom area echoes his interpretation of the work of Barth, Gogarten, and Thurneysen, whose position he summed up as follows: "Those who see a last thing [*ein Letztes*] that would give claim to redemption in cultural works or in the moral act block themselves off from God from within" ("Zur religiösen Lage in Deutschland (1924)," in *Werke*, vol. 5, no. 2, ed. Inka Mülder-Bach [Berlin: Suhrkamp, 2011], 155–59, at 158). Kracauer met Rosenzweig in the context of the circle around the Hungarian rabbi Nehemias Anton Nobel and exchanged letters with him, but only Rosenzweig's letters have been preserved (see Baumann, "Drei Briefe: Franz Rosenzweig an Siegfried Kracauer," 166).

88. Similarly, Kessler compares Kracauer's concept of the anteroom to the literal meaning of the profane as a realm that lies before the consecrated space of the sacred: "Die Problematik des Vorletzten . . . und das Stichwort vom Vorraum sind für Kracauers Geschichtsverständnis von höchster Bedeutung. Ersteres, der Bereich des Vorletzten, der ein Problemfeld ist, ist der Raum der Geschichte. Dieser aber ist Vorraum in dem prägnanten Sinne des *profanum*, des vor dem Fanum, dem geweihten, dem konsekrierten, dem heiligen Bezirk Liegenden" ("Entschleiern und Bewahren," 107).

These considerations, together with Kracauer's analysis of the antimony of time—its "inextricable dialectics between the flow of time and shaped times negating it"[89]—lead to a concept of history that can have no end, yet which stands at each moment in relation to the last things.[90] Kracauer's late work on history thus displays a unique combination of eschatological urgency and the persistent deferral of a temporal end.[91] These ideas come together in the figure of Ahasuerus, the Wandering Jew, whom Kracauer introduces as an emblem of the vexed relation between history and eschatology.[92] Ahasuerus embodies the antinomy of chronological time and shaped time. On the one hand, he has traveled continuously through time and its various periods; as Kracauer puts it, "He indeed would know firsthand about the developments and transitions, for he alone in all history has had the unsought opportunity to experience the process of becoming and decaying itself" (*L* 157). Ahasuerus thus functions as a medium for the passage of time that maintains the continuity of historical memory.

---

89. Kracauer, "Time and History," 388.

90. Kracauer first spelled out his criticism of an immanent realization of the messianic end in *Der Detektiv-Roman* (*The Detective Novel*) (1925), which refers to a "premature anticipation, a tremendous pressure" (*Vorgreifen, eine gewaltige Bedrängung*) that "begins at the end and can therefore already at the beginning advance through an illusory reality toward the end" (*Der Detektiv-Roman: Eine Deutung*, in *Werke*, vol. 1, ed. Inka Mülder-Bach [Frankfurt am Main: Suhrkamp, 2006], 103–209, at 208). Kracauer characterizes this mode of thought as "conclusive thinking" (*abschlußhafte[s] Denken*), a form of thought that is directed by the closure of an end. It posits an end or goal, whether it be the dawn of a classless society or permanent cultural progress, that can be realized only as an "illusion of reality" (*Scheinwirklichkeit*) and as a "distortion" (*Verzerrung*) of the messianic end (*Der Detektiv-Roman*, 208).

91. In a similar vein, Kessler notes that *Der Detektiv-Roman* is a text whose argumentation is at once "eschatologically charged" (*eschatologisch geladen*) and "critical of eschatology" (*eschatologiekritisch*) ("Entschleiern und Bewahren," 109). Criticizing an objectification of the last things, Kracauer makes it clear that the messianic realm represents a horizon that does not lie within the space of human reality: "The Messianic does not fall within human reality or only breaks in upon it [*nur einfällt*]. Surely it would disappear if reality were to dissipate on its account" (*Der Detektiv-Roman*, 206–7).

92. On the history of the Ahasuerus motif, see Baumann, *Im Vorraum der Geschichte*, 150–53. Baumann notes that the figure of Ahasuerus has been connected to antisemitic images of Jews as a rootless people, yet has also been reinterpreted by Jewish writers as a symbol of a modern concept of historical time.

On the other hand, Ahasuerus appears, in Kracauer's reading, as a surface on which the shapes of time, in their conflict with one another, are given form.[93] In a crucial parenthetical remark, Kracauer writes: "How unspeakably terrible he must look! To be sure, his face cannot have suffered from aging, but I imagine it to be many faces, each reflecting one of the periods which he traversed and all of them combining into ever new patterns, as he restlessly, and vainly, tries on his wanderings to reconstruct out of the times that shaped him the one time he is doomed to incarnate" (L 157).

The face of Ahasuerus is pictured here as a superimposition of historical periods that are endlessly recombined into different shapes. This picture implies a form of contemporaneity that differs from the chronological simultaneity that Kracauer claimed exists only in a formal sense. As we saw above, Kracauer sought to show that events that take place at the same moment in chronological time may in fact belong to different time curves, calling into question the significance of their chronological simultaneity. In the figure of Ahasuerus, by contrast, Kracauer is concerned with a metahistorical contemporaneity of time shapes that become condensed in the face of a single figure. The face of Ahasuerus, in short, appears as an inscription of the last things—that singular time that is reconstructed from the manifold times that shaped him.[94] This eschatological moment of superimposition opens up a perspective on an end in which noncontemporaneous figures are condensed into a single moment. How-

---

93. While the reading proposed here emphasizes how Ahasuerus embodies the antinomy of time, standing both within and outside chronological time, Itkin stresses the temporal exterritoriality of Ahasuerus to the exclusion of his embeddedness in historical time: "Ahasuerus, who does not grow or age, and who is not rooted in the world of everyday life . . . , is in a sense outside of the flow of time" ("Orpheus, Perseus, Ahasuerus," 196). Similarly, Johannes von Moltke pictures Ahasuerus as "standing above time," arguing that "both Ahasuerus and Benjamin's angel are messianic figures, engaged in a redemptive struggle against the inexorable flow of time and history" (*The Curious Humanist: Siegfried Kracauer in America* [Oakland: University of California Press, 2016], 194–95).

94. Itkin compares the images of Ahasuerus's face to the underworld of history that Orpheus was unable to hold before his eyes ("Orpheus, Perseus, Ahasuerus," 197). The implication of this interpretation—that the face of Ahasuerus stands at the limit of what can be represented—accords with the claim made here that the face embodies an eschatological superimposition of the last things.

ever, as much as Ahasuerus strives to give unity to the historical process, his efforts are in vain: the last things fail to coalesce because the movement of history permits no moment of stasis in its permutations and combinations.[95] The provisional character of historical reality precludes any resolution of the changing faces of Ahasuerus.

As a counterpoint to this concept of a history without an end, Kracauer considers the literary work of Marcel Proust, holding it up as an example of how the antinomy between the continuity of chronological time and the discontinuity of shaped time can be resolved in aesthetic terms. While the worlds of Proust's novel embody spaces of memory whose spontaneous emergence defies the continuity of chronological time, they are re-embedded in the chronological flow of time at the conclusion of the novel. As Kracauer puts it, Proust "establishes temporal continuity in retrospect. . . . Only now, after the fact, he recognizes that this way through time had a destination; that it served the single purpose of preparing him for his vocation as an artist" (*L* 162). This autobiographical conceit, in which the hero of the novel becomes one with its author and "sets out to write the novel he has written" (*L* 163), establishes an end or *telos* that structures a heterogeneous material from the standpoint of a destination that could not be foreseen in advance. Its retrospective gaze allows the beginning of the novel to anticipate, if only unconsciously, its subsequent trajectory, which for the reader becomes apparent only at its end. In Proust's novel, the end gives meaning to the narrative by serving as the temporal destination toward which it moves. The aesthetic redemption that Proust achieves is thus dependent on a vantage point from which the whole of a life can be surveyed.

---

95. Similarly, Mülder-Bach argues that the plurality of potential configurations of the last things means that for Kracauer there is always the possibility of an escape or exit, even in the end: "The confrontation with inescapability seems to be postponed until the very end—the end to which the title of the book ambiguously alludes: *History: The Last Things before the Last*. In German that would be: *Geschichte: Die letzten Dinge vor dem Letzten*, or *vor den letzten*, since grammatically 'the last' can be either a singular noun or a plural adjective. There is method in this ambiguity: if it were possible to multiply 'the last,' there might even be an exit in the end" ("History as Autobiography," 155–56).

Yet such an aesthetic reconciliation of the tension between chronological time and shaped time is not possible for history, as Kracauer notes: "Neither has history an end nor is it amenable to aesthetic redemption. The antinomy at the core of time is insoluble. Perhaps the truth is that it can be solved only at the end of Time. In a sense, Proust's personal solution foreshadows, or indeed signifies, this unthinkable end—the imaginary moment at which Ahasuerus, before disintegrating, may for the first time be able to look back on his wanderings through the periods" (L 163). What is most notable in this passage is the extreme variance attributed to different figures of the end, which oscillate between modes of possibility and impossibility, reality and imagination, existence and negation. An end to history is categorically ruled out, yet an end of time is posited, if only as a solution to the antinomy of time that is declared to be insoluble. The end of time is described as unthinkable, but it is also figured as an imaginary moment that entails a very specific scene in which Ahasuerus looks back on history. In the image of Ahasuerus's superimposed faces discussed above, with its meta-historical contemporaneity of periods, all present moments in history combine together in ever-changing forms. In the scene described in this passage, by contrast, Ahasuerus calls the past to mind in a retrospective gaze that requires a position of distance: it is only at the point of disintegration that the look back at history becomes possible. Recalling the dialectic of figuration and disintegration in "Figure and Disintegration," one might say that the ultimate shape of history, for which Ahasuerus stands, is given its contours only at the moment in which the legendary figure breaks apart and can no longer sustain any form.[96] For this reason, the end of time is unthinkable; it can only be imagined as a retrospective gaze from a moment in which the gaze itself is no longer possible.

The figure of Ahasuerus provides Kracauer with an embodiment of the relationship between history and eschatology. While denying

---

96. Koch contends with these paradoxes by reading Ahasuerus's terrible face as an assembly of the faces of the dead, such that the dissolution of Ahasuerus at the end of time coincides with the resurrection of the dead ("'Not Yet Accepted Anywhere,'" 107–8). The reading presented here emphasizes how Kracauer articulates these implicit theological motifs through a spatial metaphorics of figuration and disintegration.

that history can have an end or be redeemed, he recognizes that history is defined by its relation to the last things, which function as a threshold that cannot be crossed. Kracauer's theory of history, from the Weimar essays to *History*, uncovers an eschatological void in the fabric of history that becomes visible not at its temporal end but in its spatial configurations. In the place of an image of redemptive history, it is the lingering image of Ahasuerus's terrible face, with its superimposition of moments of suffering in history, that defines the liminal relation to the last things in a world in which a teleological end to history is impossible.[97] For Kracauer, history is comprised of a manifold of time curves that are constantly transforming and deforming. Just as mathematical topology attends to the consistency of these shapes under deformation, so does the historian trace the emergence of shapes of time from out of the flow of chronological time, putting the disintegrated elements of history into new configurations. For Kracauer, the shape of time, like the geometry of urban space, cannot be adequately described by the law of perspective or the coordinates of a Euclidean space in which it is embedded. Rather, history defines its own intrinsic forms of space in a kaleidoscope of changing modal relations. The ends of history are thus nonvisualizable in Kracauer's work because they are multidimensional. But by the same token, history stands at each moment in proximity to the last things, at all times fragmented by its pockets and voids, ever on the verge of dissolution.

---

97. As Pecora notes, Kracauer's rejects the "teleological imperative" according to which the "last things" are synonymous with the end or goal of history (*Secularization and Cultural Criticism*, 70).

5

IMAGES WITHOUT END

Robert Musil's Narrative Ruptures

In the literary and essayistic work of Robert Musil, the interplay of spatial form and the temporality of the eschatological moment is pursued in its literary and aesthetic dimensions. Like the other writers considered in this study, Musil works with a concept of an end that is decoupled from teleological development, and he emphatically critiques the principles of historicist historiography. Musil detaches the concept of history from the assumption of a narrative thread, breaking with the idea that history has the structure of a narrative with a beginning, middle, and end. Instead, he views the order of history as provisional and defined by contingency. Musil's concept of history is fundamentally open-ended and resembles a series of chance encounters rather than a teleological path toward a goal. In his magnum opus, *Der Mann ohne Eigenschaften* (*The Man without Qualities*) (1930–43), this concept of history is given expression in the literary form of the text, which consists of a series of spaces of reflection.

In the absence of a narrative trajectory, a new shape of time emerges in an experience that Musil calls the "other condition," the representation of which is one of the enduring concerns of his work, from his first novel *Die Verwirrungen des Zöglings Törleß* (*The Confusions of Young Törleß*) (1906) to his essays of the 1920s and his late work on *The Man without Qualities*. In his essays, Musil defines the other condition as a "a hypothetical limit case [*hypothetischer Grenzfall*] that one approaches in order to fall back again and again into the normal condition" (P 1154).[1] As such, the other condition "stands under the reign of exceptions to the norm [*Ausnahmen über die Regel*]" (P 1028) that emerge through "the explosion of the normal total experience" (*die Sprengung des normalen Totalerlebnisses*) (P 1145).[2] The sudden, epiphanic qualities of the other condition make it well suited to probing the possibilities of a "time of the now," though its rupture of the narrative continuum never stabilizes as a permanent state. Nevertheless, precisely because of its concentration in the moment, the other condition is in important respects without an end, and it resists the sense of narrative closure implied by teleological models of history. As a limit situation that can never be fully integrated into historical time, the other condition is closely aligned with the eschatological moment found in the modernist imagination.

Musil's account of the other condition takes important cues from key ideas and impulses in modernist mathematics. As a student of mathematics, Musil was well informed about the pressing mathematical questions of his day, and the major protagonists of his works, from Törleß to Ulrich, are students of mathematics or mathematicians themselves.[3] Spurred on by the discovery of non-Euclidean

---

1. See Robert Musil, "Ansätze zu neuer Ästhetik: Bemerkungen über eine Dramaturgie des Films" ("Toward a New Aesthetic: Observations on a Dramaturgy of Film") (1925), P 1137–54.
2. See Robert Musil, "Skizze der Erkenntnis des Dichter" ("Sketch of the Knowledge of the Poet") (1918), P 1025–30.
3. Justice Kraus notes that Musil wrote *The Confusions of Young Törleß* while studying philosophy, physics, and mathematics in Berlin ("Musil's *Die Verwirrungen des Zöglings Törleß*, Cantor's Structures of Infinity, and Brouwer's Mathematical Language," *Scientia Poetica: Jahrbuch für Geschichte der Literatur und der Wissenschaften* 14 [2010]: 72–103, at 72). In her analysis of Musil's essay

geometry and by Riemann's view that mathematics goes beyond objects that can be considered "abstractions from things in the 'real' world,"[4] the mathematician David Hilbert, in his groundbreaking *Grundlagen der Geometrie (Foundations of Geometry)* (1899), adopted an axiomatic approach that broke "with any kind of geometric intuition in defining geometric terms."[5] With Hilbert as its leading exponent, mathematical modernism shifted from the analysis of three-dimensional objects in empirical space, which present themselves to intuition (in the Kantian sense), to a study of spaces whose properties are defined solely by a set of axioms posited by the mathematician.[6] In short, mathematics found itself liberated to create spaces that were at times nonintuitive or even counterintuitive as a result of its "attenuated relationship with the real world."[7] Similarly, in *The Man without Qualities*, Ulrich experiences a destabilization of empirical space—a "peculiar spatial inversion"—that prompts him to ask: "Can one get out of one's space and into a second, hidden space?" (*Kann man denn aus seinem Raum hinaus, in einen verborgenen zweiten?*) (M 632).[8] Musil's second space exists in the realm of possibility and can be understood as a manifestation of the other condition. Like the constructed spaces of modernist mathematics, the

---

"Der mathematische Mensch" ("The Mathematical Human Being") (1913), Andrea Albrecht demonstrates Musil's deep familiarity with the problems of contemporary mathematics ("Mathematische und ästhetische Moderne: Zu Robert Musils Essay 'Der mathematische Mensch,'" *Scientia Poetica: Jahrbuch für Geschichte der Literatur und der Wissenschaften* 12 [2008]: 218–50).

4. Jeremy J. Gray, "Modernism in Mathematics," in *The Oxford Handbook of the History of Mathematics*, ed. Eleanor Robson and Jacqueline Stedall (Oxford: Oxford University Press, 2009), 663–83, at 671.

5. Gray, *Plato's Ghost*, 19.

6. See Gray, "Modernism in Mathematics," 675.

7. See Gray, *Plato's Ghost*, 184–85. For example, Hilbert's axiomatic method showed that Moulton's lines and planes are mathematically consistent yet cannot be embedded in any three-dimensional space.

8. Alexander Honold argues that Ulrich's inversion experience, and in particular the idea of a second space, offers an alternative exit to the labyrinth of *The Man without Qualities*, one that its unreached temporal end is unable to provide (*Die Stadt und der Krieg: Raum- und Zeitkonstruktion in Robert Musils Roman "Der Mann ohne Eigenschaften"* [Munich: Fink, 1995], 479). Yet while the second space offered by the other condition is a corollary of the book's deferred temporal end, it by no means resolves the narrative impasse at the center of *The Man without Qualities*.

space of the other condition is no longer embedded in the empirical space of the natural world. Instead, it posits itself as a state of exception.

In works such as *The Confusions of Young Törleß*, Musil and his protagonists are compelled to confront the groundlessness of the creative self-assertion of mathematics. In Törleß's uncertainty and anxiety about imaginary numbers, the concept of infinity, and intersecting parallel lines, Musil provides a literary exposition of the so-called foundational crisis of mathematics (*Grundlagenkrise der Mathematik*), which raised the question of whether the language, syntax, and order that mathematics creates are anything more than castles in the air in the absence of a foundation in intuition. These doubts contribute to the dizzying and disorienting qualities of the other condition, but they also reflect Musil's aesthetic commitment to prioritizing a sense of possibility over a sense of reality.[9] In this sense, Musil incorporates both the paradoxes and promises of mathematical modernism into his program of aesthetic modernism.[10] That is, he recognizes a deep affinity between the creative potential of modernist mathematics and the poetic possibilities of the imagination, commenting that "there is today no better possibility for fantastic feeling than that of the mathematician."[11]

In *The Confusions of Young Törleß*, the paradoxes of a mathematical sense of groundlessness are closely connected with thought processes cannot be brought to a conclusion. Similarly, the larger structure and organization of *The Man without Qualities* is defined by endless reflections in a narrative that lacks decisive turning points. Much has been made of the compositional difficulties that prevented Musil from completing his magnum opus. The impossibility of bringing the project to an end, however, is also symptomatic of the status of ends in modernist eschatology, which eschews teleology

---

9. Musil's discussion of the dialectic of reality and possibility can be found in chapter 4 of *The Man without Qualities*, which bears the title "Wenn es Wirklichkeitssinn gibt, muß es auch Möglichkeitssinn geben" ("If there is a sense of reality, there must also be a sense of possibility") (*M* 16–18).
10. On the conjunction of mathematical and aesthetic modernism in Musil's work, see Albrecht, "Mathematische und ästhetische Moderne."
11. Robert Musil, "Der mathematische Mensch," *P* 1004–8, at 1006.

in favor of proliferating spaces of reflection. The impossibility of an end in temporal terms is connected to the problem of the duration of the other condition, which is impossible to sustain but is broached again and again with images of transgression. Musil's work thus shows the influence of eschatological thought on narrative form in modernist literature.

In *The Sense of an Ending* (1966), Frank Kermode argues that in modern literature the theological problem of narrating the end of history is transposed from the realm of imminence to the sphere of immanence. According to Kermode, when the eschatological end could no longer be conceived as happening imminently (that is, within this lifetime), it was reconceived as something inner-historical, as a moment of crisis.[12] Kermode shows how the idea that the end is present in every moment, which has its origins in St. Paul and St. John, became a central concern in twentieth-century thought.[13] In literary terms, the immanence of the end in moments of crisis, as opposed to the deterministic sense of temporal ends, helps Kermode account for the rise of modern tragedy. With Shakespeare's *King Lear* as a cardinal example, he suggests that tragedy depends on "the notion of an endless world," such that "when the end comes it is not an end, and both suffering and the need for patience are perpetual."[14] Yet in contrast to modern tragedy, *The Man without Qualities* is devoid of moments of crisis. In place of the perpetual suffering of an endless world, Musil's experiment gives us an endless dispersion of reflective positions. Its horizon is therefore not an ever-present crisis but the other condition.[15] To be sure, one can find examples of what

---

12. Frank Kermode, *The Sense of an Ending: Studies in the Theory of Fiction* (Oxford: Oxford University Press, 1966), 5–6, 25–26.

13. Kermode, *The Sense of an Ending*, 25–26. Kermode refers to Rudolf Bultmann, Herbert Butterfield, and R. G. Collingwood, but the figures considered in this book show that the interest in a concept of the end located in the present was widespread in early twentieth-century thought.

14. Kermode, *The Sense of an Ending*, 82.

15. Kermode argues that Musil is a paradigmatic example of the modernist experiment, but that his experiment ultimately fails because he cannot reconcile the "non-narrative contingencies of modern reality" with fiction (*The Sense of an Ending*, 128). Kermode's verdict is based on the idea that a novel cannot avoid having beginnings and ends, even if the world is without beginning and end. As a question

Alexander Honold calls "small endings" in *The Man without Qualities*, but they are never sufficient to establish a teleological horizon for the book, or to resolve the problem of narrating an ending.[16] Musil's nonteleological narrative form can thus be considered a transposition of a specifically modernist eschatology—one that is tied not to a looming crisis but to images of a moment out of time.

For Musil, the shape of time is given in images that arise when time stands still. Like Barth, Rosenzweig, and Kracauer, he draws on the resources of spatial form to represent the aporias and antinomies of time. Yet Musil's use of spatial form is unique in its focus on the creation of images. In book 2 of *The Man without Qualities*, the other condition becomes manifest in spatialized images of time in a state of suspension. The dynamism of these images and their internal tensions allow Musil to develop alternatives to narrative structures of development but also to guard against the danger that the other condition itself might become static and lifeless. Musil captures interrupted time in images of unanchored space, but he also shows the dissolution of these images, which are ultimately groundless and unable to provide narrative ends. We are left with a series of images without end, a proliferation of moments of rupture without the promise of closure.

Bridges over an Abyss: Törleß and the Crisis of Mathematics

Musil's first literary foray into the realm of the other condition, his novel *The Confusions of Young Törleß*, is bound up with a set of mathematical problems and with a crisis of mathematical representation. Mathematical discourses on imaginary numbers, non-Euclidean space, and the foundations of mathematics underscore the radical divide between reality and the imagination that opens up in

---

of genre and literary form, however, it is debatable whether *The Man without Qualities* should even be considered a novel.

16. Alexander Honold, "Endings and Beginnings: Musil's Invention of Austrian History," in *Robert Musil's The Man without Qualities*, ed. Harold Bloom (Philadelphia: Chelsea House, 2005), 113–22, at 120.

the other condition. Mathematics provides a spatial imaginary for the complex temporality—both momentary and endless—of the other condition, as well as a paradigm for the creative freedom of modernist aesthetics. On its surface, *The Confusions of Young Törleß* is a novel about an adolescent searching for his place in the world while being educated at a Prussian boarding school. Törleß confronts the ethical dilemma of his friends' sadomasochistic humiliation of a schoolmate and must come to terms with his own homoerotic desires. Like much of Musil's early prose, *The Confusions of Young Törleß* probes the psychological dimensions of its protagonist's struggles, conveying the fine texture of his mental activity in the medium of literature.

The opening lines of the novel, ostensibly nothing more than an indication of the setting, already point to the mathematical dimensions of Törleß's existential crisis: "A small station on the route leading to Russia. Four parallel iron tracks [*vier parallele Eisenstränge*] run infinitely straight [*endlos gerade*] in both directions between the yellow gravel of the broad causeway; each track accompanied, like a dirty shadow, by a dark streak, burned into the ground by the exhaust steam" (*P* 7). The four "strands of iron" (*Eisenstränge*) that make up the train tracks are ascribed the geometrical properties of parallel lines that run "infinitely straight" in both directions. The term *Eisenstränge*, an unusual term in German, has an implicit narratological meaning that invokes the thread or line (*Strang*) of a plot. Already the opening lines of the novel signal that the clear-cut developmental line of the story will be shadowed by a darker second dimension, one defined by the "streak" (*Strich*) burned into the ground by the exhaust steam of the train. These two dimensions are represented in *The Confusions of Young Törleß* through the contrast between conscious and unconscious processes, between rational and irrational forms of thought, and between the daytime studies of the students and their nocturnal adventures and aberrations. Musil thus indicates from the outset that some form of the other condition may unsettle the pure geometrical unity of the infinitely straight parallel lines that make up the normal condition of the novel's development.

While critics have noted that early formulations of the other condition can be found in *The Confusions of Young Törleß*,[17] many scholars argue that its representation in the novel has a distinctly antireligious element, a claim based on Törleß's skeptical attitude toward religion.[18] Some critics have gone so far as to suggest that mathematics provides a substitute for religion in Musil's work.[19] This chapter shows that a more complex and nuanced relationship between mathematics and theological problems is at stake in the novel and in Musil's work more generally. The conflict early in the novel between Törleß and his aristocratic friend, referred to as "the Prince," does indeed turn on their incompatible views of religion, but it is also the staging of a conflict between "understanding" (*Verstand*) and "soul" (*Seele*), two mental faculties that Musil thought should be brought into productive relationship with one another.[20] In the early episode in *Törleß*, we read: "But one time they [Törleß and the Prince] had gotten into an argument about religious matters. And in this moment everything was at stake. For Törleß's reason, as if independent of him, struck out relentlessly against the delicate prince. He assailed him with the ridicule of the rational one, barbarically destroying the filigree edifice in which his soul was at home,

---

17. See Malcolm Spencer, "Violence and Love: The Search for the 'andere Zustand' in Robert Musil's *Der Mann ohne Eigenschaften*," in *Contested Passions: Sexuality, Eroticism, and Gender in Modern Austrian Literature and Culture*, ed. Clemens Ruthner and Raleigh Whitinger (New York: Peter Lang, 2011), 249–58, at 251.

18. See Allen Thiher, *Understanding Robert Musil* (Columbia: University of South Carolina Press, 2009), 79–80. While noting an affinity between Törleß's encounter with the infinity of the heavens and the work of Pascal, Thiher emphasizes that Törleß, in contrast to Pascal, refuses to embrace religion. Thiher structures his interpretation of Törleß's antireligious stance around a binary opposition of religion and the secular.

19. See Gwyneth E. Cliver, "Maddening Mathematics: The Kinship of the Rational and the Irrational in the Writing of Robert Musil," *Journal of Romance Studies* 7, no. 3 (2007): 75–85, at 78: "For Törless and Ulrich, speaking in mathematical metaphors provides an outlet through which to ponder such philosophical categories as morality, the soul and truth, without relying on theological foundations. In an increasingly secularized society, these modern protagonists require a new language for their speculation, and mathematics becomes this substitute in Musil's texts."

20. See Musil, "Ansätze zu neuer Ästhetik," P 1145–47.

and they parted ways in anger" (P 12). This episode has remarkably few consequences or ripple effects in the novel, but it serves as an early demarcation of boundaries. Törleß follows the path of reason as far as it will take him, even if he is not fully himself when he does so. Religious matters come to the surface in the novel at moments when reason is confronted with its own antinomies. Far from serving as a substitute for religion, Musil's mathematical metaphors provide a point of entry for reflection on theological problems.

In his conversations with Beineberg, Törleß draws a direct connection between religion and mathematics, suggesting that matters of religion can be treated in the same manner as mathematical calculations: "Religion? Oh well. That's something to look forward to.... I think when I really get going I could just as easily prove that twice two is five as that there can only be one God" (P 22). In another passage from early in the novel that anticipates the emergence of the other condition, Musil articulates the fragile relationship between Törleß's self and the world with reference to the mathematical figure of the asymptotic limit. Describing early indications of Törleß's talent for "astonishment" (*Staunen*), in which he feels a deep affinity with events, people, and things yet also senses their incomprehensibility, Musil writes:

> Between the events and his self, indeed between his own feelings and some innermost self that desired to understand them, there remained a dividing line, which, like a horizon, receded before his desire the closer he came to it. Indeed, the more precisely he comprehended his sensations with his thoughts, the more familiar they became to him, all the more foreign and incomprehensible they simultaneously seemed to become, so that it no longer appeared as if these sensations receded from him but rather as if he himself drew distant from them, and yet he could not shake off the impression of drawing closer. (P 25)

The image of the "dividing line" (*Scheidelinie*) is noteworthy for its double nature: it is an infinite limit separating not only Törleß's self from the world but also his self from his own feelings. In contrast to the concept of an "infinite approach" (*unendliche Annäherung*) that Nicholas of Cusa invoked in his mathematical formulation of the relationship to God, here the approach of the horizon does not provide for any sense of convergence, however liminal, but instead maintains a sense of distance even at the horizon.[21] The limit situation here involves an inability to distinguish between "drawing closer" (*sich nähern*) and "drawing away" (*sich entfernen*).

The scene of Törleß's encounter with the other condition arises, as Gwyneth E. Cliver notes, "as a kind of intermission, a break from the primary narrative."[22] It is framed by turbulences of the ongoing narrative concerning the humiliation of Basini, but it displaces the psychological and emotional dimensions of this relation onto another level. Lying on his back and gazing up at the sky, Törleß is struck by the sudden appearance of the other condition in a moment that has a clear epiphanic quality:

> And suddenly he realized—and it was as if it was happening for the first time—how high the sky really is. It was like a shock [*Erschrecken*]. Directly above him there shined a small, blue, unspeakably deep hole between the clouds. It seemed to him that with a long, long ladder one must be able to climb into it. But the further he pressed in and rose up with his eyes, the deeper the blue luminous ground withdrew.... It was as if his extremely tense vision were hurling glances like arrows between the clouds, and as if these glances, however farther they aimed, always came up just a little bit short.... "Indeed, there is no end," he said to himself, "it goes on and on, perpetually, into infinity." (P 62)

The terrifying nature of Törleß's epiphany bears comparison to the Kantian concept of the sublime, the force of which is magnified by the contrast between the tremendous magnitude of the cosmos and the smallness of the human observer.[23] Musil, however, emphasizes

---

21. On the role of mathematics in Nicholas of Cusa's work, see Martin Gessmann, *Montaigne und die Moderne: Zu den philosophischen Grundlagen einer Epochenwende* (Hamburg: Meiner, 1997), 131–32. See also Ruth Bendels, *Erzählen zwischen Hilbert und Einstein: Naturwissenschaft und Literatur in Hermann Brochs "Eine methodologische Novelle" und Robert Musils "Drei Frauen"* (Würzburg: Königshausen & Neumann, 2008), 208; and Genese Grill, "The 'Other' Musil: Robert Musil and Mysticism," in *A Companion to the Works of Robert Musil*, ed. Philip Payne, Graham Bartram, and Galin Tihanov (Rochester, NY: Camden House, 2007), 333–54, at 347.
22. Cliver, "Maddening Mathematics," 76.
23. The sense that the enormity of the heavens extends into infinity corresponds closely to Kant's definition of the sublime: "Nature is sublime in those of its

the negative qualities of this experience: it is not just the infinity of the heavens that strikes Törleß but the sense that there is an unspeakably deep hole in the sky, an image that suggests an infinite absence or lack. The dialectic of drawing close and drawing away discussed above is evident in this passage, in which the luminous ground in the sky withdraws ever more deeply as the eye's gaze attempts to penetrate it. Even as the goal appears within reach, the power of vision fails to reach it. The infinite limit expressed here is marked expressly as one of infinite distance rather than as an infinite approach. Whereas Barth worked with the mathematical figure of the infinitesimal limit, in which the approach to the limit takes place as a quantity is diminished to an infinitely small magnitude, Musil compares Törleß's experience of the other condition to a limit at infinity, in which a function tends toward infinity as its variable approaches an asymptote.

The irresolvable tension of Törleß's experience of the other condition is closely connected to his recognition that certain basic mathematical concepts have an uncertain foundation. Törleß immediately connects his experience of the infinity of the heavens to the concept of infinity used in his mathematics lessons, but his unsettling epiphany calls into question the mathematical principle that one can calculate with the infinite as though it were "something stable" (*etwas Feste[s]*) (P 63).[24] Törleß returns to these questions again following a mathematics lesson that introduces imaginary numbers, a concept he has trouble accepting. His friend Beineberg suggests that one can work with imaginary numbers as if they existed and points out that they are just one of many mathematical paradoxes: "Are the irrational numbers any different? A division that never comes to an end, a fraction whose value never emerges, no matter how long you calculate? And how can you conceive that parallel lines are supposed to

---

appearances whose intuition carries with the idea of its infinity" (*Kritik der Urteilskraft*, 255). In his discussion of the dynamic-sublime, Kant emphasizes that the terrible violence of the sublime highlights the smallness of our power to resist (261).

24. As Albrecht notes, Törleß's concerns about the concept of infinity are grounded in the history of mathematics, beginning with debates about the status of the infinitesimal as a quantity in the seventeenth and eighteenth centuries, and culminating in concerns about the antinomies of infinite sets discovered around 1900 in the wake of Cantor's set theory ("Mathematische und ästhetische Moderne," 227–28).

intersect in infinity? I think that if one were all too diligent, there would be no mathematics" (P 73). Together with the possibility of irrational numbers, the idea of parallel lines that intersect at infinity, which recalls the opening image of the novel with its endlessly straight parallel train tracks, is deeply troubling to Törleß: "I too find the whole thing *peculiar*. The idea of the irrational, the imaginary, of lines that are parallel and intersect in infinity—or at least somewhere—it gets me worked up. When I think about it, I'm stunned and dumbstruck" (P 81).

For Törleß, the experience of the other condition is shocking because it confronts him with a realm in which the elements of thought have no grounding in intuition. In non-Euclidean spaces with a positive degree of curvature, lines that appear to be parallel from the vantage point of flat Euclidean space will indeed intersect, and if the degree of positive curvature is exceedingly small, they will appear to do so at infinity. In the work of Riemann, Beltrami, Klein, and Hilbert, it was established that such spaces are mathematically consistent and can be derived from an axiom system in which Euclid's parallel postulate does not hold. But such spaces have no basis in a geometric intuition of the space of the natural, empirical world. Likewise, as Justice Kraus notes, for Hilbert an imaginary number such as $\sqrt{-1}$ has no foundation in reality or in intuition; rather, it "achieves meaning only within the context of mathematical language because it has no referential value in itself."[25] For Hilbert, the mathematical existence of imaginary numbers is secured because they function within a set of axioms and concepts that never lead to a contradiction.[26] By contrast, Törleß finds the existence of

---

25. Kraus, "Musil's *Die Verwirrungen des Zöglings Törleß*," 100.
26. As Hilbert puts it in his lecture "Mathematische Probleme" (1900), "Wenn man einem Begriffe Merkmale erteilt, die einander widersprechen, so sage ich: der Begriff existiert mathematisch nicht. So existiert z. B. mathematisch nicht eine reelle Zahl, deren Quadrat gleich −1 ist. Gelingt es jedoch zu beweisen, daß die dem Begriffe erteilten Merkmale bei Anwendung einer endlichen Anzahl von logischen Schlüssen niemals zu einem Widerspruche führen können, so sage ich, daß damit die mathematische Existenz des Begriffes ... bewiesen worden ist" (David Hilbert, "Mathematische Probleme," in *Die Hilbertschen Probleme*, ed. P. S. Alexandrov [Leipzig: Akademische Verlagsgesellschaft Geest & Portig, 1971], 22–80, at 38). As Kraus points out, Hilbert's position aligns with that of Törleß's mathematics teacher,

a second world untethered from the empirical reality to be profoundly paradoxical.

Törleß's doubts about these concepts are not just symptoms of an existential crisis; they are grounded in a set of contemporary problems in mathematics around 1900 known as the foundational crisis of mathematics. As Andrea Albrecht has argued, Musil was well aware of these problems, which arose from Georg Cantor's and Ernst Zermelo's discoveries of contradictions and antinomies in classical set theory.[27] In 1895 Cantor realized that for every set there exists a power set—the set of all of its subsets—that has a higher cardinality than the original set. As a result, a *set of all sets* would have to contain its own power set, with a higher cardinality, resulting in a contradiction of the principle that there is a hierarchy of infinities with different cardinalities.[28] These perplexities raised the question of how to give mathematics a foundation and provoked a dispute between Intuitionists such as Poincaré, who claimed that such antinomies are the result of a tautological logic and argued for a return to intuition, and Formalists such as Hilbert, who thought a secure foundation was possible with a fully consistent axiomatic approach.[29] Hilbert was eventually shown to be correct, but the mathematical uncertainties surrounding set theory were still unresolved as late as 1914, when the mathematician Felix Hausdorff noted: "Set theory is the foundation of all of mathematics. . . . But there is still no complete agreement on the foundation of this foundation; that is, about an indisputable foundation of set theory itself."[30]

In his essay "Der mathematische Mensch" ("The Mathematical Human Being") (1913), Musil gives an account of the *Grundlagen-*

---

who can be considered a mathematical Formalist ("Musil's *Die Verwirrungen des Zöglings Törleß*," 100).

27. See Albrecht, "Mathematische und ästhetische Moderne," 236–37. Kraus argues furthermore that Musil was familiar with Cantor's work on set theory and the theory of infinity when he composed *The Confusions of Young Törleß* ("Musil's *Die Verwirrungen des Zöglings Törleß*," 78–79, 92–94).

28. See Mehrtens, *Moderne—Sprache—Mathematik*, 152.

29. See Albrecht, "Mathematische und ästhetische Moderne," 237–38.

30. Felix Hausdorff, *Grundzüge der Mengenlehre* (Leipzig: Veit, 1914), 1. Cited in Mehrtens, *Moderne—Sprache—Mathematik*, 147.

krise der Mathematik that emphasizes the close kinship between mathematics and the creative possibilities of the imagination:

> And suddenly, after everything had been put in good order, the mathematicians—those who brood within themselves—realized that something in the foundations of the whole thing absolutely could not be put in order; indeed, they had a look at the very bottom and found that the whole building hovered in the air. But the machines were running! As a result, one has to imagine that our existence is a pale apparition; we live this life, but only on the basis of an error without which it never would have come into being. There is today no better possibility for fantastic feeling than that of the mathematician.[31]

In contrast to physics and engineering, which derive practical consequences and applications from mathematical principles, mathematics for Musil represents a domain of pure reflection where the boundary between the real and the imaginary breaks down. What is more, Musil recognizes the modernist stakes of mathematics in its engagement with its conditions of possibility and the premises of its intellectual activity. Like modernist literature and art, modernist mathematics is aware that the world it posits is fundamentally a construction. As Albrecht argues, Musil perceives a deep affinity between mathematics and aesthetic modernism: "Musil declares mathematics to be a realm of hypothetical thought experiments and an unparalleled experience of the fantastic, a realm of the not-yet-real, the unreal, the impossible, the imaginary, and the utopian."[32] For Musil, the foundational crisis of mathematics revealed the entwinement of conceptual thought and the imagination.

This attunement to the imaginative possibilities of mathematical antinomies motivated Musil's interest in non-Euclidean geometries as a vehicle for exploring alternative concepts of space. As Gray notes, the discovery of non-Euclidean geometry in the nineteenth century contributed to the foundational crisis of mathematics by displacing old certainties about the nature of mathematical objects, breaking with intuition in defining geometric terms, and raising the

---

31. Musil, "Der mathematische Mensch," *P* 1006.
32. Albrecht, "Mathematische und ästhetische Moderne," 242.

question of whether a true geometry could exist.[33] In an essay on Oswald Spengler titled "Geist und Erfahrung: Anmerkungen für Leser, welche dem Untergang des Abendlandes entronnen sind" ("Mind and Experience: Notes for Readers Who Have Escaped the Decline of the West") (1921), Musil uses non-Euclidean geometries as a way of imagining the possibility of multiple concepts of space. Mathematical spaces, of which the "empirical-metrical" space of Euclidean geometry is but one example, Musil argues, are in essence symbols that provide a conceptual framework for conceiving of the space of profane reality: "The space chosen for this presentation, just like the other mathematical symbols, is for the moment just a conceptual bridge for events in another space, that of profane reality" (P 1046). By defining mathematical space not as reality itself but as a "conceptual bridge" (*begriffliche Brücke*) between mathematical symbols and profane reality, Musil emphasizes the constructed nature of the space of mathematics, which can be mapped onto empirical reality but need not be derived from it.

This insight that space is constructed as a language of mathematical symbols accords with the key tenets of Hilbert's mathematical modernism. For Hilbert, mathematics is a "language of symbols": a set of signs and names that the mathematician produces as a "creation out of nothing."[34] Calling into question the idea that there is a single space, namely the "empirical-metrical" space of Euclidean geometry, Musil writes:

> It is nothing but the space of experience under the prevailing aspect of measurement, of which one can easily convince oneself by keeping in mind that, in a certain sense, there exist other seen, touched, and heard spaces besides the empirical-metrical space, in all gradations from the primary impression up to the completely conscious perception. These

---

33. Gray, *Plato's Ghost*, 18–19.
34. See Mehrtens, *Moderne—Sprache—Mathematik*, 123–24. The creative capacity of mathematics was already a key component of Hilbert's axiomatic approach in his *Grundlagen der Geometrie* (1899) and "Mathematische Probleme" (1900); the close kinship of this approach to the language of theological creation became apparent in Hilbert's "Neubegründung der Mathematik" (1922), in which he introduced his approach to number theory with the line "in the beginning was the sign" (cited in Gray, *Plato's Ghost*, 411).

spaces are definitely not Euclidean; for example, in the space of vision, parallel lines intersect, length is dependent on the relative position on a route, and the three dimensions are not equal. (P 1046)

The kinds of perceptual spaces described here as "seen, felt, and heard spaces" (*gesehene, getastete und gehörte Räume*) are very basic spatial configurations in which the principles of Euclidean geometry do not hold. They are not the only theoretically definable spaces in which this is the case. Musil's choice of these spaces of perception as examples of deviations from Euclidean geometries, however, serves a deeper purpose: it shows just how fragile the "conceptual bridge" between mathematical symbols and reality can be.

The existence of multiple and non-Euclidean spaces, together with the still unresolved paradoxes of set theory, called into question the status of mathematics, which could no longer be relied on to provide a stable ground. Indeed, the foundational crisis of mathematics consisted in the recognition of the absence of such a ground. As Musil notes in his essay "Skizze der Erkenntnis des Dichter" ("Sketch of the Knowledge of the Poet") (1918), the domain of rationality—or what Musil calls the "ratioid sphere"—is one in which we expect firm and stable certainties:

> One can say that the ratioid sphere is ruled by the concept of the solid for which a deviation does not come into consideration; by the concept of the solid as a *fictio cum fundamento in re*. At the very bottom the ground here shakes too, the deepest foundations of mathematics are logically unsecured, the laws of physics are valid only approximately, and the stars move in a coordinate system that has a place nowhere [*das nirgends einen Ort hat*]. (P 1027)[35]

No longer able to provide a solid foundation, mathematics cannot claim to operate solely in the realm of rationality. It is in this sense

---

35. Musil borrows the concept of a "fictio cum fundamento in re" from Franz Brentano. See Franz Brentano, *Über Aristotles: Nachgelassene Aufsätze*, ed. Rolf George (Hamburg: Meiner, 1986), 145: "Der Ausdruck fictio cum fundamento in re, den ich gebrauche, bedarf vielleicht eine Erläuterung. Ich meine dabei ein Verfahren, ähnlich wie es die Mathematiker einhalten, wenn sie negative Größen, echte Brüche, irrationale, imaginäre und unendlich kleine Zahlen einführen und mit ihnen ganz so, wie wenn sie ganze positive Zahlen wären, rechnen."

that Musil describes the mathematician as "not attentive to purpose, but uneconomical and passionate," a person defined by "total surrender and passion," or, as we saw above, a figure who provides access to a "fantastic feeling" (*phantastischen Gefühl*) (P 1005–6).[36] As Kraus notes, for Musil mathematics is "the epitome of rationality and simultaneously a structure without a base. It is systematic and anti-systematic at the same time."[37]

The borderline situation of mathematics between rationality and the imagination, which arises from the recognition that mathematics lacks a stable foundation, stands in the background of Törleß's experience of the other condition and is reflected in his doubts about imaginary numbers. The very possibility of moving back and forth between real and imaginary numbers gives him vertigo:

> But does not something entirely peculiar nevertheless adhere to the matter? How shall I put it? Just think about it this way for once: at the beginning of such a calculation there are entirely solid figures, which might represent meters or weights or something else tangible, and at any rate are real numbers. At the end there likewise stand such real numbers. But these two are connected together by something that does not exist. Is that not like a bridge for which only the beginning and ending piers exist [*eine Brücke, von der nur Anfangs- und Endpfeiler vorhanden sind*], yet which one nonetheless so securely crosses as if the bridge stood there in its entirely? For me such a calculation has something vertiginous [*etwas Schwindliges*] about it, as if a part of the path goes God knows where. But what is actually uncanny for me is the power embedded in such a calculation, a power that holds one so tight that one lands again properly after all. (P 74)

The image of a bridge consisting only of beginning and ending piers, with an imaginary span suspended over an abyss (articulated here as a metaphor for the mathematical operation involved in the use of imaginary numbers), is emblematic of Törleß's larger problem of finding a bridge between himself and the world. In this passage, the

---

36. Musil, "The Mathematical Man," in *Passion and Soul: Essays and Addresses*, ed. and trans. Burton Pike and David S. Luft (Chicago: University of Chicago Press, 1990), 41.
37. Kraus, "Musil's *Die Verwirrungen des Zöglings Törleß*," 89.

bridge is all the more vertigo-inducing because it appears to function. The passage through the realm of the imaginary induces a profound sense of disorientation in Törleß. Like the mathematicians who called into question the foundations of mathematics, Törleß is concerned not with the purpose or the result of the calculation but with the premises of its operation.

While the image of a bridge is typically a figure of mediation that establishes a connection between disparate realms of experience, Musil constructs images of bridges such that their capacity to mediate is undercut by the unfathomable abyss that they span. In Törleß's reflections on imaginary numbers, the bridge is anchored at its beginning and end points, but the passage across is consigned to the subjunctive mood of the "as if." While the piers of the bridge are anchored in the real, the crossing of the bridge requires a suspension in the realm of the imaginary. Törleß's experience of vertigo in this suspension is closely connected to his horror (*Erschrecken*) at the infinity of the heavens. Both experiences have the character of a shock: "A jolt goes through one's head, a dizziness, a shock" (*Es geht einem so ein Ruck durch den Kopf, ein Schwindel, ein Erschrecken*) (P 23). Ultimately, Törleß is unable to incorporate the realm of the imaginary into the real, and consequently he cannot grasp how the bridge is able to function.

The bridge is a pervasive metaphor in Musil's work, and it frequently serves as a spatial image for the temporality of the other condition. In the passage cited above, the eschatological dimensions of this temporality are apparent in the anchoring of the bridge in beginning and end points, which recall the eschatological connection of *Urgeschichte* and *Endgeschichte*. Significantly, the passage from beginning to end cannot be conceived in terms of a narrative trajectory but rather has the character of a leap, such that the movement from beginning to end is suspended over an abyss. In other iterations of the bridge metaphor, the failure to construct a bridge with a stable foundation is connected to the temporality of endlessness, which suggests the deferral and retreat of the end. An earlier scene in the novel, in which Törleß recognizes both his need for a bridge between himself and his experiences and his inability to complete the bridge, provides another set of images for a temporality that cannot reach an end:

> He had the desire to search tirelessly for a bridge, a connection, a comparison—between himself and that which stood silently before his mind. But every time he calmed himself down with a thought, an incomprehensible objection appeared again: you are lying. It was as if he had to perform an endless division [*unaufhörliche Division*] that again and again yielded a persistent remainder, or as if he were rubbing his feverish fingers sore in an attempt to untie an endless knot [*endlosen Knoten*]. (P 65)

The construction of a bridge, which is likened here to a comparison or "simile" (*Vergleich*), is troubled by a moment of displacement. The mediation sought between self and world is confronted with a displaced remainder that cannot be integrated, expressed here in the image of an interminable division. The image of the endless knot gives this temporality of the deferred end an even stronger spatial dimension. The futile effort to loosen and untie the endless knot suggests that the desire for mediation is thwarted by the very entangled complexity of the world to which a bridge is sought.

The other condition thus emerges as a form of perception or state of consciousness in which the subject is confronted with an infinite abyss that destabilizes the normal conditions of consciousness. Musil uses mathematical images to represent the nature of this confrontation, but mathematics, rather than providing a stable foundation, only intensifies the experience of shock implicit in the other condition. Significantly, however, the other condition cannot have an end in the sense of a goal or *telos* because of the endless nature of its approximation. The mathematical and spatial metaphors with which these problems are represented ("interminable division," "endless knot," "a bridge in which only beginning and end piers exist," "a deep hole...without end," "a shaking, receding ground," and so on) suggest that the other condition can never be conceived as the teleological endpoint of utopian yearnings because it can never be established as something permanent. In "Ansätze zu neuer Ästhetik: Bemerkungen über eine Dramaturgie des Films" ("Toward a New Aesthetic: Observations on a Dramaturgy of Film") (1925),

Musil clarifies these insights into the temporality of the other condition with his most radical iteration of the bridge metaphor:

> It is common knowledge that this condition [the other condition] is never permanent, except in a pathological form; it is a hypothetical limit case that one approaches in order to fall back again and again into the normal condition. And this is what distinguishes art from mysticism: art never loses the connection with the usual attitude; it appears, then, as a dependent condition, as a bridge that arches away from the solid ground as if it possessed an abutment in the imaginary [*eine Brücke, die vom festen Boden sich so wegwölbt, als besäß sie im Imaginären ein Widerlager*]. (P 1154)

In this passage, the bridge illustrates the precarity of the relationship between the other condition and the normal condition. In contrast to the image of the bridge suspended over an abyss in *The Confusions of Young Törleß*, where at least the beginning and end points of the bridge are secured, here only one side of the bridge is anchored; its abutment on the other side exists only in the hypothetical, subjunctive mood of the imaginary.

These figurations of the other condition suggest a concept of eschatology in which the figure of the end remains a hypothetical limit case that can alternatively be conceived as abyssal, as beyond representation, and as an imaginary or aesthetic construction, one that can be approximated and approached, but never established as a real form. Musil's images of bridges spanning an abyss articulate the conceptual aporias that surround the representation of the last things. These images join Barth's tangent to the circle, Rosenzweig's arc, and Kracauer's pockets and voids as part of an archive of modernist images that picture the liminal relationship of history and eschatology in spatial terms. Moreover, the centrality of mathematical paradoxes in Törleß's epiphanic moments shows the unique conjunction of the nonintuitive dimension of mathematics and a theological sense of groundlessness in the modernist imagination.[38]

---

38. While Thiher contrasts Pascal's turn to religion for consolation with Törleß's refusal to embrace religion (*Understanding Robert Musil*, 79–80), this chapter proposes that Pascal and Törleß provide two variations on a theological theme: in

Beyond the "Path of History": Non-Endings in *The Man without Qualities*

Whereas in *The Confusions of Young Törleß* the figure of endlessness becomes manifest in moments of the other condition induced by mathematical paradoxes, in *The Man without Qualities* Musil gives expression to this sense of endlessness in the literary form of the text, which is set up as a proliferation of spaces of reflection without a narrative trajectory toward an ending. Musil's nonteleological concept of eschatology—which bubbles to the surface in moments of the other condition—is thus closely connected to his effort to reconceive the literary form of the modernist novel. But it also has important implications for the concept of history, which Musil likewise views as contingent, decentered, and without a goal.

Although designated by its German editor Adolf Frisé as a "novel" (*Roman*), *The Man without Qualities* contains only the rudimentary elements of the genre. It begins with a chapter ironically titled "Woraus bemerkenswerter Weise nichts hervorgeht" ("Out of which, remarkably enough, nothing develops"), and throughout its more than one thousand pages it contains a minimum of narrative development. Instead, the book is replete with essayistic reflections, loosely organized around the scaffolding of monologues or dialogues, that constitute the core of the work. While *The Man without Qualities* constantly shifts back and forth between narration and essayistic modes of reflection, the interface between narration and reflection becomes most explicit when epiphanic moments enter the textual fabric of the modernist novel and disrupt its narrative trajectory.

Such moments have distinct beginnings in narrative ruptures, but their ends are much less clear. While epiphanic moments provide an intimation of fullness and completion, their literary representation cannot be brought to a resolution. The enigmatic expression of such moments is connected to the problem of the modernist novel, namely,

---

the former, the groundlessness of existence is stabilized through faith, while in the latter the intensification of this groundlessness reverberates in the space of the modernist imagination. The difference is not between a religious and a secular perspective, but between two interpretations of a theologically connoted sense of mathematical infinity.

the impossibility of its having an end. The epiphanic structure of the other condition thus changes the contours of the modernist novel and the way we read it, inducing a mode of reflexive reading that treats the novel as a complex spatial configuration of interconnected points, rather than a narrative, diegetic account of development toward a temporal end.[39] In this sense, the other condition responds to Musil's concern with the question of how to provide order and organization in a world that tends toward entropic dispersion.[40] Rather than following a clear-cut narrative trajectory, the states of reflection brought about by the other condition expose the difficulty of providing for such order and organization. The spaces of reflection opened up in *The Man without Qualities* are themselves subject to the entropic dispersion that for Musil is a key characteristic of the cultural situation of modernity.

In a chapter late in book 1 titled "Heimweg" ("The way home"), Musil sets up the novel as a spatial topography by attesting to the fundamental loss of a narrative order. The chapter narrates Ulrich's evening walk home through the streets of Vienna, but quickly passes over into a series of reflections on how perspectival foreshortening allows the mind to bring disparate visual impressions together into an ordered whole. The ring of trees surrounding the city, which serves as a metaphor for the continuity and wholeness of experience, is full of gaps, but perspectival foreshortening gives it the appearance of spatial unity. For Ulrich, however, this experience of wholeness and completeness is foreclosed. He is unable to synthesize a perspective that could give spatial unity to the world he perceives. Analogously, the narrative order that allows the self to construct a chronological series of events has come unraveled for Ulrich:

---

39. This mode of reading corresponds closely to Joseph Frank's concept of spatial form. Frank argues that to understand modern literature in terms of spatial forms "means that the reader is intended to apprehend their work spatially, in a moment of time, rather than as a sequence" ("Spatial Form in Modern Literature," 225).

40. On the centrality of the concept of entropy for *The Man without Qualities*, see Gerhard Meisel, "Verkehr und Entropie in Musils 'Kakanien,'" in *Medien und Maschinen: Literatur im technischen Zeitalter*, ed. Theo Elm and Hans Hiebel (Freiburg: Rombach, 1991), 304–32; and Christian Kassung, *Entropie-Geschichten: Robert Musils "Der Mann ohne Eigenschaften" im Diskurs der modernen Physik* (Munich: Fink, 2001).

> The law of this life, which, overburdened and dreaming of simplicity, one longs for, is none other than the law of narrative order [*erzählerischen Ordnung*]! The simple order that consists in being able to say: "When this took place, that occurred!" It is the simple sequence, the representation of the overwhelming manifold of life in a unidimensional order, as a mathematician would say, that reassures us; the stringing together of everything that has taken place in space and time on a thread, precisely the famous "narrative thread" [*Faden der Erzählung*] out of which the thread of life is also composed.... Most human beings relate to themselves fundamentally as storytellers [*Erzähler*].... [T]hey love the orderly succession of facts, because it has the look of necessity, and they somehow feel secure, in the midst of chaos, through the impression that their life has a "course." And Ulrich now realized that he had lost this basic epic mode of thought to which private life still clings, even as in public everything has already become non-narrative and no longer follows a "thread," but rather spreads out as an infinitely interwoven surface. (M 650)

Musil makes it clear in this passage that perspectival foreshortening is not only a function of spatial perception but also an essential ingredient in the creation of narrative order. Such narrative order allows the mind to find the thread that connects disparate moments and thus to find order in chaos. Ulrich, however, no longer has such narrative order at his disposal. Instead of narrative modes of thought, he is predisposed to essayistic spaces of reflection that spread out as an "infinitely interwoven surface" (*unendlich verwobenen Fläche*). There is, in other words, a persistent suspension of narrative trajectory in *The Man without Qualities* that opens up a space for the intrusion of the other condition into the narrative order of the novel.

The spatial metaphorics of an "infinitely interwoven surface" that spreads out in entropic dispersion stands in contrast to the temporal determination of the "thread of the story" or "narrative thread" (*Faden der Erzählung*), which, as Musil notes, depends on a concept of seriality or sequentiality. The basis of the narrative order, as the passage cited above makes clear, is a "simple sequence" (*die einfache Reihenfolge*) in a unidimensional order, the "stringing together" (*Aufreihung*) of everything that takes place in time and space on a single "thread" (*Faden*), such that events appear in "the sequence of their temporal course" (*in der Reihenfolge ihres zeitlichen Ablaufes*) (M 650). In Ulrich's conversation with Agathe in the chapter "Weiterer Verlauf des Ausflugs auf die Schwedenschanze; Die Moral des nächsten Schritts" ("Further course of the excursion to the Swedish Ramparts; the morality of the next step"), the semantics of the series (*Reihe*) lead to another set of reflections on the constitution of the narrative order. Here the infinite regress of steps, each of which is subordinate to a further step, is unable to provide order within chaos, and results instead in the inability to come to a decision:

> "That's nonsense," Ulrich replied emphatically. "What I said was that it does not come down to a false step but to the next step after that. But what does it come down to after the next step? Evidently to the one that follows after that? And after the nth step, the n-plus one step?! Such a human being would have to live without end and decision [*ohne Ende und Entscheidung*], indeed without reality. And yet it is the case that it always only comes down to the next step. The truth is that we have no method for dealing with this unending series." (M 735–36)

What is noteworthy in these reflections is the reduction of the series of steps to a mathematical series that extends into infinity. No single action, in this scenario, can emerge as the decisive step, because it will always be followed by an equally decisive step. Whereas calculus provides a method for resolving an integral as an infinite sum of rectangles of infinitesimal width, the temporal series of moral action cannot be grasped as such a function. It is an "unending series" (*ruhelose[] Reihe*) because each of its moments offers a decision tree of possible continuations, none of which is capable of marking the end of the series. Just as for Törleß an infinite series is unimaginable, for Ulrich a series of steps without end entails a life without a grounding in reality. Indeed, Ulrich's sense that he has "no method for dealing with this unending series" is characteristic of his status as a man without qualities, one who has prioritized the sense of possibility over the sense of reality. The unending series of steps with which Ulrich is confronted, and which he has no means of resolving through a process of decision making, results in a moral calculus that has no end. The resulting inability to carry out a decisive act is symptomatic of the collapse of the narrative order with which *The Man without Qualities* contends. If the "thread of narrative" gains its contours through its trajectory toward an end, the absence of such an end makes it impossible to sustain a narrative thread.

In a diary entry from the early 1920s, Musil provides a related gloss on the question of whether the order of life is directed by an end. This note is part of an effort to produce a concept of history that allows for contingency yet does not lead to complete dispersion. It is connected with Musil's larger concern of giving structure to the world without lapsing into ideology. He writes: "The complete task is: life without system and yet with order. Self-creative order. Generative order. An order not predetermined from a to z, but rather an order in the step from n to $n+1$. Perhaps direction instead of order, or rather orientation."[41] In contrast to the later formulation in *The Man without Qualities*, which expresses despair over the endlessness of the morality of the next step, Musil's diary entry detects a creative and generative possibility in the iterative step from n to $n+1$. The sense of provisional order in this concentration on the next step provides for the orientation of a direction without laying out in advance an order from beginning to end (from a to z). To avoid any sense of historical necessity, Musil searches for alternatives to the concept of order, suggesting "direction" (*Richtung*) and "orientation" (*Gerichtetheit*). Such orientation, however, remains provisional in that no single step can claim finality. The concept of narrative order, by contrast, depends on a stable temporal progression from beginning to end.

The collapse of the narrative order in *The Man without Qualities* is the consequence of a concept of history that is fundamentally open-ended. In this respect, Musil's concept of history has an affinity with Kracauer's insight into the provisional character of historical reality, as expressed in the image of the ever-changing faces of Ahasuerus. Yet even more than Kracauer, Musil emphasizes the indeterminate character and contingency of the direction that history takes. In the key chapter "Seinesgleichen geschieht oder warum erfindet man nicht Geschichte?" ("The like of it takes place, or why doesn't one invent history?"), for example, Musil satirizes the idea of a straightforward "path of history" (*Weg der Geschichte*). As in

---

41. Musil, *Tagebücher*, vol. 1, ed. Adolf Frisé (Reinbek bei Hamburg: Rowohlt, 1983), 653. Cited in Inka Mülder-Bach, "Der 'Weg der Geschichte' oder: Finden und Erfinden: Geschichtserzählung in Robert Musils Roman *Der Mann ohne Eigenschaften*," *Internationales Archiv für Sozialgeschichte der deutschen Literatur* 36, no. 1 (2011): 187–205, at 204.

the chapter "The way home," here too Ulrich's reflections are staged against the backdrop of an urban journey, such that the perambulations of Ulrich's thoughts are doubled in the spatial architecture of the city and its modes of transportation. As he is carried through the city by a trolley, Ulrich notes the peculiarity of history—"What a peculiar affair history is after all!" (M 359)—and reflects:

> It looks uncertain and matted, our history, when one observes it in close proximity, like a half trodden down morass, and yet strangely enough a path ultimately runs through it, precisely that "path of history" whose origin no one knows. This attitude that history serves as material outraged Ulrich. The luminous, rocking box [*leuchtende, schaukelnde Schachtel*] in which he rode seemed to him like a machine in which several hundred kilograms of human beings were shaken up in order to make futurity out of them. (M 360)

As Inka Mülder-Bach notes, Musil's metaphor of the "path of history" "is one of the oldest and most persistent chronotopoi of the epic and historiographical tradition."[42] The invocation of a "path of history" is consistent with the widespread use of spatial metaphors to think about temporal problems in the early twentieth century. However, Musil's ironic treatment of the concept undermines the sense of linearity that it implies. The "path of history" emerges in this passage from an uncertain point of origin, and it does not follow a regular course, but rather charts a path on the uncertain terrain of a swamplike morass. Adding a further dimension of contingency to the concept of history, Ulrich describes the trolley on which he is riding not as a means of traveling from a beginning to an end point, or as a network of connections—metaphors the system of transportation in a modern metropolis might suggest—but as a container that swings back and forth and shakes up its passengers in order to lend them futurity. Such a future, it is clear, does not follow from the causal determinations of the "course of history," but is radically defined by chance.

Musil reprises this theme of the radical contingency of history in a number of variations, offering alternative sets of images intended to dislodge the concept of history as defined by progressive development.

---

42. Mülder-Bach, "Der 'Weg der Geschichte,'" 192.

Entertaining the idea that world history might have a narrative structure "like all other stories," Ulrich raises the objection that an authorial vantage point cannot be attributed to history: "But for the most part history arises without authors. It does not arise from a center but from the periphery. From small causes" (*M* 360–61). By shifting the focus of history from the center to the periphery, Musil emphasizes the plurality of small causes and the multitude of perspectives that intersect in the constitution of history. History cannot be guided by teleology for the simple reason that there is no single guiding thread whose course history follows. The image of center and periphery recalls one of the fundamental thought-images of *The Man without Qualities*: a ring with an empty center.[43] The entropic dispersion of Ulrich's reflections, together with the expansion of the metropolis at its periphery, undermines the concept of history structured by teleology by making it impossible to unify the whole of history in the face of the multiplicity of its vectors and possible ends.

A further thought experiment in this series is directed against the idea of historical necessity. Musil not only proposes that history *could* have turned out differently, a proposition that emphasizes the possibility of other paths that history might take, but suggests that it *would* turn out differently if it took place again under the exact same conditions: "Thus if one were to relocate a generation of contemporary Europeans at the age of earliest childhood to the Egyptian year 5000 BCE and leave them there, world history would begin again in the year 5000, initially repeat itself for a while, and

---

43. The image of a ring with an empty center is established in the very first chapter of *The Man without Qualities*. In the opening scene, the victim of a traffic accident is embedded in "the depth of a hole" (*die Tiefe des Lochs*) constituted by the ring of spectators who gather around (*M* 10). Without a center point, there is nothing to hold the circle together, except for an attraction to the void of absence at its core. For more on the ethical dimensions of the concept of a void in *The Man without Qualities*, see Patrizia McBride, *The Void of Ethics: Robert Musil and the Experience of Modernity* (Evanston, IL: Northwestern University Press, 2006). McBride cites a key passage from book 2 of *The Man without Qualities* that likewise invokes the image of a circle without a center: "There's a whole circle of questions here, which has a large circumference and no center, and all these questions are: 'How should I live?'" (*M* 895, cited in *The Void of Ethics*, 4).

then, for reasons that no one can guess, gradually begin to deviate" (M 361). This hypothetical thought experiment targets the Christian idea of providence, according to which history proceeds according to a divine plan. The concept of providence is not as alien to Ulrich's reflections as it might seem at first glance. After all, Musil begins the chapter by noting that Clarisse had "aroused the peculiar desire in Ulrich to utter the word God" (M 357). Ulrich reflects thereupon that "God doesn't mean the world literally; it is an image, an analogy, a figure of speech to which he has to resort for some reason, and naturally always inadequately" (M 357). The providential concept of history appears to founder here because the world lacks the stability of a substance that could be adequately formed by God. By contrast, Musil's attribution of the linguistic qualities of an image, analogy, or figure of speech to God's relation to the world prioritizes the possibilities of the world over its reality.

Such a relation of God to the world is suggested toward the beginning of book 1 by the young Ulrich, who in his early years remarks: "that God too would probably prefer to speak of his world in the Conjunctivus potentialis (hic dixerit quispiam = here one could object), for God creates the world and in the course of doing so thinks that it could just as well be different [*es könnte ebensogut anders sein*]" (M 19). The concept of historical necessity is thus criticized by Musil both from within and outside theology. God himself, Musil suggests, has no stake in the providential necessity of the course of world history because he is enamored with the multiple possibilities of the world, which becomes an object of a linguistic play of signs. Needless to say, Musil's thought experiment about a repetition of history from its prehistorical beginnings that inevitably leads to a divergence also takes aim at secular corollaries of providence, such as the idea of natural law and the Enlightenment idea of progress.

In his most incisive critique of a concept of history structured by teleology and causality, Musil compares the "path of history" to a circuitous walk through an urban landscape that culminates in an unknown corner of the city. Criticizing the relation of cause and effect implicit in the path followed by a billiard ball, Musil reframes the course of history by using images of radical contingency that suggest an itinerary that is always going astray:

> The path of history is thus not that of a billiard ball, which, once struck, traverses a fixed course [*eine bestimmte Bahn*], but rather resembles the path of clouds, or the path of one who roams through the streets, who is distracted here by a shadow, there by a group of people or a strange blending of house facades, and finally arrives at a place that he neither knew nor wanted to reach. The course of world history is characterized by a certain getting lost [*Sich-Verlaufen*]. The present is always like the last house in a town that somehow no longer entirely belongs to the town houses. (M 361)

This passage expresses disorientation on several levels. Ulrich's reflections are set against his own wanderings through the city, such that he must figure out where he is when his thought process comes to an end. As Mülder-Bach notes, the similes Ulrich uses, from the billiard ball to the path of clouds and the figure who roams through the alleys, give the impression of getting lost in unconscious associations.[44] Emphasizing the way the discourse of physics enters Musil's literary work, Mülder-Bach points out that "the 'billiard ball' illustrates the concept of causality in classical physics, which is oriented toward mechanical-reversible processes, while the 'path of clouds' belongs to those irreversible processes with which thermodynamics are concerned."[45] Yet it is the contrast between these two images that is most pertinent here. The path of a billiard ball is the epitome of a predictable causal relation, while the path of clouds is the very image of unpredictability and contingency. In likening the path of history to that of clouds, Musil recalls the meteorological language of the barometric minimum with which *The Man without Qualities* begins. But the sense of "getting lost" (*Sich-Verlaufen*) that Ulrich considers to be characteristic of history is perhaps best expressed by the figure whose perambulations through the modern city, through a series of distractions, lead him to an unknown place that he never intended to reach. This image expresses not only the

---

44. Mülder-Bach, "Der 'Weg der Geschichte,'" 199.
45. Mülder-Bach, "Der 'Weg der Geschichte,'" 199.

contingency of history but also the error of assuming that its trajectory is defined by a teleological path toward a goal.[46]

In his essay "Das hilflose Europa oder Reise vom Hundertsten ins Tausendste" ("Helpless Europe, or a Journey from the Hundredth to the Thousandth") (1922), a critical reckoning with the contemporary political and cultural landscape, Musil extends his critique of historicism with an ironic account of the pretensions of historical objectivity. Invoking a commonplace of the theory of historiography, that those who are directly involved in world history cannot fully grasp its relations because they lack historical distance, Musil counters that participation in a historical constellation offers unique insights into history that are unavailable to subsequent observers: "We would know enough to form a judgment for ourselves about contemporary matters and those of the recent past; in any case we know more than future times will know" (P 1076). The possibilities opened up by historical distance—the ostensible precondition of historical objectivity—are predicated on the loss of historical facts, resulting in the freedom to structure what remains in an arbitrary fashion: "The famous historical distance [*historische Distanz*] consists in ninety-five out of a hundred facts getting lost, so that the remaining ones can be ordered as one pleases. But by viewing these five facts as a fashion from twenty years ago or as a lively conversation between people that one cannot hear, objectivity manifests itself" (P 1076). In this caricature of historical objectivity, Musil emphasizes that the overview afforded by historical distance provides for objectivity only in the sense that an arbitrary order or structure can be given to historical events. The construction of a historical narrative,

---

46. As Honold notes, Musil's sense of historical contingency is connected to the way he begins *The Man without Qualities*. By beginning with the movement of the clouds, Musil establishes a set of initial conditions that in no way determine what follows: "To respect contingency means to admit that history is not predictable, that it cannot be processed by sheer deduction. A beginning can thus give us initial conditions, but it cannot determine what follows, although everything that actually follows is determined. A beginning therefore could be described as a starting point, 'from which, remarkably enough, nothing develops' ('woraus bemerkenswerter Weise nichts hervorgeht'). So, why not start this beginning with the unpredictable *per se*, the movement of the clouds?" ("Endings and Beginnings," 118–19).

as Musil suggests in a manner reminiscent of Nietzsche, involves a necessary falsification of reality.[47]

For Musil, the construction of a unified narrative of history, one that proceeds from a center as if it had a single author, is problematic because it fails to acknowledge the role of chance in history. In a passage whose unmistakable political context is the experience of the First World War, Musil describes the contingency of history as follows:

> A very current feeling of chance [*Gefühl von Zufall*] accompanies everything that took place. The faith in the necessity of history would be considerably overextended if one were to discern an expression of uniform meaning in all the decisions that we have experienced. In retrospect it is easy to recognize a necessity in the failure of German diplomacy or military strategy: but everyone still knows that it just as well could have turned out differently, and that the result often hung on a hair's breadth. It almost looks as if the events were not necessary at all, but rather that necessity was first tolerated belatedly. (P 1077)

For Musil, this war, more than any other event in world history, confounded any attempt to recognize a "uniform meaning" in the many decisions that led up to it. Even if one dimension of the constellation of actions that led to the war—Musil mentions the failure of German diplomacy or military strategy—seems in retrospect to provide an example of the inevitability of the conflict, Musil is unable to shake the sense that "it could just as well have turned out differently" (*daß es ebensogut auch anders hätte kommen können*). The impression of necessity, he claims, is tolerated after the fact, but this by no means implies that the events themselves were necessary. The provisionality of history in the direction of futurity—in the step from n to $n+1$—is complemented here by an insight into the provisionality of the past: even in retrospect, Musil claims, history could have been otherwise.

Yet Musil does more than simply deny historical necessity: he shows how it is intertwined with chance and contingency. In this

---

47. In *Jenseits von Gut und Böse*, for example, Nietzsche famously argues for "untruth as a condition of life" (*Sämtliche Werke: Kritische Studienausgabe*, vol. 5, ed. Giorgio Colli and Mazzino Montinari [Munich: De Gruyter, 2002], 18).

sense, the essay "Helpless Europe, or a Journey from the Hundredth to the Thousandth" anticipates the famous dialectic of reality and possibility that Musil formulates in *The Man without Qualities*. Musil affirms the lawfulness of history, as shown in his reference to the proverbial case of the man who walks under a roof and is struck by a roof tile. Yet he insists that what we call historical necessity in fact contains both the lawfulness of physical reactions and the element of chance: "Laws may always be there..., but there is also always something there that is only there once, only this time" (P 1078). The paradoxical effort to negotiate between necessity and contingency, reality and possibility, defines the task that Musil set for himself in *The Man without Qualities*, a book whose unspoken but omnipresent horizon is the outbreak of the First World War. This fact is immediately apparent to the reader in the opening chapters, which locate the story in the year extending from August 1913 to August 1914, and it is further supported by Musil's unpublished fragment: "War. All lines end in the war [*Alle Linien münden in den Krieg*]" (M 1851). As Mülder-Bach argues, however, the book's orientation toward the outbreak of the war in no way implies

> that sufficient reasons and goals, causality and teleology are imputed belatedly to the historical process. What Musil expects of the narration is rather the entanglement [*Verschränkung*] of contingency and finalization. In one and the same process, he seeks to narrate how it could have turned out differently and how it nonetheless turned out such that out a plurality of possible realities, the real possibility of the war was selected and realized.[48]

Musil's opposition to the imputation of a teleological movement toward the outbreak of the war, which would imply its historical necessity, is coupled with his recognition of the reality of this outcome. His solution to the problem of how to adequately represent this state of affairs is to mark the outbreak of the war from the very beginning as the horizon of the book, but to defer the inception of this end, to the point of being unable to complete the work. Musil maintains the end in suspension by giving voice to the numerous

---

48. Mülder-Bach, "Der 'Weg der Geschichte,'" 191.

possibilities that by no means made the war inevitable, yet at the same time he is committed to showing how a certain constellation of possibilities, in which chance and contingency played a significant role, did indeed lead to the war.

Even as he accepts the reality (if not the necessity) of the outbreak of the war, Musil considers, as a thought experiment, what it would take to reverse the series of events that led to this outcome. His example of a hunter who shoots a deer in a forest leads him to the threshold of a concept of restitution that strikes an eschatological note. The imagination of the series of events that would be necessary for the historical process to be reversed has an almost cinematic quality, evoking the film technique of reverse motion.[49] With the precise attention to detail of an engineer, Musil describes this sequence as follows: "Return journey [*Rückfahrt*]: the deer stands up—but it would not be allowed to stand up, but would have to 'fall' upward, its antlers would first have to carry out the mirrored dance of the movements of striking the ground, and it would have to begin with the final velocity and end with the initial velocity" (P 1078). Musil appears to be well aware that such a reversal of beginning and end, which can be compared to the philosophical and theological figuration of inversion considered elsewhere in this book,[50] would require a radical restructuring of the world as we know it:

> In order to take back even a single step, it would not be sufficient to reverse the events, but one would also have to possess the most extensive powers of authority for the reconstruction [*Umbau*] of the entire world. Gravity would have to function upward, a vertical plane of earth would have to exist in the air, ballistics would have to change in an entirely unimaginable way, in short, if one plays a melody from the back to the front, it is no longer a melody, and one would have to unsettle time and space [*Zeit und Raum erschüttern*] itself for that to be different. (P 1078)

---

49. Musil may well have been aware of the technique of reverse motion, which was first discovered in 1896. See Andrew M. Tohline, "Towards a History and Aesthetics of Reverse Motion" (PhD diss., College of Fine Arts of Ohio University, 2015).

50. Consider, for example, Rosenzweig's concept of an "inversion of temporal sequence" (S 467), as discussed in chapter 3.

While the imagination of a multitude of possible worlds provides the resources to challenge the concepts of historical necessity and teleology, a reversal of the historical process would require a far more radical rupture of the spatial and temporal structure of the world. The convulsion of time and space and the authorization to reconstruct the entire world described in this passage have clear eschatological connotations, yet they are considered only in the realm of the subjunctive, as a limit of historical experience.

Within the frame of history itself, Musil concludes that "reversal" (*Umkehrung*) and "reinstatement" (*Widergutmachung*) are insufficient to put the deer back on its feet. With a strong sense of irony, Musil laconically writes that "something new" would have to take place for this to happen: "The truth is that in order even just to put a shot deer back on its feet, something entirely new must take place, not merely a reversal and reinstatement! The world is full of an unruly will for the new, full of the compulsive idea of doing things differently, of progress [*Zwangsidee des Andersmachens, des Fortschritts*]!" (P 1078). The ambiguity of the passage not only leaves open the question of whether the deer is returned to life or merely put back on its feet as a corpse; it also associates the compulsion to begin again or to "do things differently" with the idea of progress. In a cynical reading of this passage, one must conclude that Musil has little hope of the worldly restitution of wrongs. And if restitution is impossible, all that is left is a lingering in the possibilities of what might have been.

In *The Man without Qualities*, the preoccupation with the question of what might have been unfolds not from a retrospective viewpoint but in a present that stands forever on the threshold of an end that never comes to pass. The suspension of the end of the novel, which would entail not only the outbreak of the war but also the consummation of incest between Ulrich and his sister Agathe, has important implications, as we have seen, for the structure of the book, whose narrative trajectory flattens out radically after the introductory chapters and culminates, in book 2, in endless dialogues between Ulrich and Agathe interspersed with essayistic reflections. *The Man without Qualities* finds itself increasingly concentrated in the unfolding of moments, such that the rhetorical gesture of "at that very moment" is one of its signature expressions.[51] Moreover, based on what can be gleaned from the occasional temporal markers that track Ulrich's "one-year holiday from life," the passage of time slows significantly in the later sections of *The Man without Qualities*. The temporality of this suspension before the end is given ample expression in the rich world of images evoked in the book's essayistic sections. In the unpublished chapter draft "Ulrichs Tagebuch" ("Ulrich's diary"), one of a number of possible continuations of the Ulrich-Agathe narrative, Ulrich reflects in his diary on the character of his dialogues with Agathe:

> Every movement of the soul led to the discovery of a new, even more beautiful movement, in which they mutually helped one another, such that the impression arose of a never-ending intensification and of a discussion that lifted up without subsiding. The last word could never be spoken, for every end was a beginning and every final result was the first of a new opening, so that each second radiated like a rising sun, but at the same time carried with it the peaceful transience of the setting sun. (M 1417)

The images of a "never-ending intensification" (*nicht endenden Steigerung*) and of an exchange that "lifts up without subsiding" (*ohne Senkung hebend*) are emblematic of the structure of *The Man without Qualities* as a whole. Since the rising action of the plot never develops to a climax, there is no place for the falling action toward an end and denouement. Intensification takes place in Ulrich's and Agathe's dialogues without any trajectory toward an end, because any potential end is in fact a new beginning.

The image of each moment simultaneously embodying both the radiance of the rising sun and the ephemerality of the setting sun expresses a coincidence of beginning and end that is characteristic of modernist eschatology. The deferral of the end in a moment of

---

51. Compare Karl Heinz Bohrer, "Utopia of the Moment and Fictionality: The Subjectivization of Time in Modern Literature," in *Suddenness: On the Moment of Aesthetic Appearance*, trans. Ruth Crowley (New York: Columbia University Press, 1994), 197–245, at 221: "An unending chain of formulas for the moment of 'now' runs through Musil's *Man without Qualities*, such as 'only for a fleeting moment,' 'in that fraction of a second,' 'all at once.'"

intensified presence reconfigures the end in two key respects: it detaches the end from its position as the endpoint in a teleological movement, and by grasping the end as present in each moment, it enables an inversion of beginning and end. The theological dimensions of such considerations have clear consequences for the concept of history that Musil develops, as well as for the structure of *The Man without Qualities*. One consequence, as Alexander Honold notes, is that "all that we know about the novel's mysterious ending, we know by its beginning and we know it from the beginning."[52] A second consequence, as Burton Pike points out, is that *The Man without Qualities* has the structure of a fractal: the form of the whole is implicit in each of its parts, so that any given moment in book can serve as a microcosm of the book as a whole.[53]

The lack of an end to *The Man without Qualities* is thus not the result of a compositional problem of the sort that might befall a writer facing difficult personal circumstances or writer's block. Instead, the lack of closure in *The Man without Qualities* is a necessary result of Musil's concept of history, his opposition to teleology, and his preference for essayistic modes of reflection.[54] The endless deferral of the looming outbreak of the war, in a dizzying refusal of the historical *telos* surrounding the text, can only be understood in terms of Musil's commitment to a concept of history in which, as Michael André Bernstein puts it, "the future ramifications of the present are potentially unlimited," which serves to "guarantee human freedom in a way impossible with teleological readings of history."[55] Given Musil's struggle against linear time and his effort to reconceive of literature in essayistic terms, as a kaleidoscopic shaking and

---

52. Honold, "Endings and Beginnings," 116.
53. On the concept of the fractal as the form of *The Man without Qualities*, see Burton Pike, "Der Mann ohne Eigenschaften: Unfinished or without End?," in *A Companion to the Works of Robert Musil*, ed. Philip Payne, Graham Bartram, and Galin Tihanov (Rochester, NY: Camden House, 2007), 355–70, at 367–68.
54. Compare Pike, "Der Mann ohne Eigenschaften," 361: "The structure of *Der Mann ohne Eigenschaften* ensures the work's unfinishability. Musil constantly struggled to reconcile his commitment to presenting life as an open process with the novel as a form."
55. Michael André Bernstein, *Foregone Conclusions: Against Apocalyptic History* (Berkeley: University of California Press, 1994), 105.

reshaking of elements,[56] any search in Musil's notes for what must have been the likely continuation and conclusion of *The Man without Qualities* will ultimately be fruitless. Musil's sense that history has no *telos* or single meaning becomes the principle of his literary experiment in *The Man without Qualities*, and it results in a mode of literature that defers and resists any sense of narrative closure.

Images of the Standstill of Time

The eschatological subtext of *The Man without Qualities* becomes explicit in book 2, which bears the title "Ins Tausendjährige Reich" ("Into the Millennial Kingdom") and introduces a set of eschatological metaphors to describe the transgressive possibilities of the other condition.[57] The traditional eschatological metaphor of the Millennial Kingdom is appropriated to suggest a sense of narrative rupture, but its permanence and duration are foreclosed. This introduces a tension into the representation of the moment in the other condition: while it tends toward stasis, it is ultimately unable to withstand the larger process of entropic dispersion that is constitutive of *The Man without Qualities*. In the absence of narrative trajectory and temporal closure, the moment of the other condition emerges as a space of images. Indeed, Musil's appropriation of es-

---

56. See Pike, "Der Mann ohne Eigenschaften," 363, who cites a note by Musil: "Nicht in Zeitreihe erzählen. Sondern hintereinander, zum Beispiel: ein Menschen denkt 'a,' tut Wochen später das Gleiche, aber denkt 'b.' Oder sieht anders aus. Oder tut das Gleiche in einer anderen Umwelt. Oder denkt das Gleiche, aber es hat eine andere Bedeutung, und so weiter. Die Menschen sind Typen, ihre Gedanken, Gefühle sind Typen; nur das Kaleidoscop ändert sich" (*Der Mann ohne Eigenschaften*, ed. Adolf Frisé [Reinbek bei Hamburg: Rowohlt, 1952], 1636).

57. As Bohrer has pointed out, Musil's use of eschatological metaphors such as the "Millennial Kingdom" should be seen as "self-conscious quotations of an infinite conversation, that is, much like Benjamin's 'Theses on the Philosophy of History,' they can only be understood as allegories that 'make use of' a set of theological concepts" ("Utopia of the Moment and Fictionality," 224). In other words, far from implying an explicitly theological program, Musil's use of eschatological metaphors needs to be understood in terms of their aesthetic effects. As Bohrer argues, Musil "had to rely on eschatological metaphors, because of all the available images they most closely approached the boundary where one transgresses the cultural norm" (225).

chatological metaphors allows him to respond to the situation of narrative impasse by constructing images of the standstill of time. In the so-called garden chapters—part of a series of chapters intended as a continuation of book 2, some of which were submitted for publication in 1938 and whose galley proofs were withdrawn, others of which he worked on until his death in 1942 and left in fair copies—Musil provides spatial images of the other condition that express an underlying tension between the suspension of time and the reemergence of temporal movement. The garden chapters thus provide a compelling example of what Walter Fanta has called "Musil's intention to capture time—in the absence of teleology—in images."[58]

In the garden chapters, the eschatological motif of the Millennial Kingdom suggests a sense of endlessness, yet one in which the end is no longer set off against an unsustainable temporal or narrative trajectory. Instead, it is evoked in spatial terms, with the garden increasingly being treated as a space of signification outside the temporal arc of the year leading up to the outbreak of the war.[59] Standing at the threshold of the consummation of an incestuous union—a boundary that is never transgressed—Ulrich and Agathe conduct a series of "holy discourses" that circle around the other condition, but they reach a conceptual and linguistic limit in their attempt to grasp the other condition through dialogue.[60] Musil's

---

58. Walter Fanta, "The 'Finale' of Der Mann ohne Eigenschaften: Competing Editions and the 'Telos' of the Narrative," in *A Companion to the Works of Robert Musil*, ed. Philip Payne, Graham Bartram, and Galin Tihanov (Rochester, NY: Camden House, 2007), 371–94, at 392–93.

59. Similarly, Harmut Böhme notes the spatial dimension of Ulrich's and Agathe's interest in mysticism: "Ulrich and Agathe attempt to attain mystical experience through a spatialization of time [*Verräumlichung der Zeit*]" ("Die 'Zeit ohne Eigenschaften' und die 'Neue Unübersichtlichkeit': Robert Musil und die posthistoire," in *Kunst, Wissenschaft und Politik von Robert Musil bis Ingeborg Bachmann*, ed. Josef Strutz [Munich: Fink, 1986], 9–33, at 29).

60. In their discussion of a "mysticism bright as day" (*taghelle Mystik*) (M 1089–91), for example, Ulrich claims that mysticism consists in a state of being whose exceptional quality cannot be captured by language: "For the word has no traction in such a condition" (M 1088–89). Just as Ulrich approaches a state of nothingness, however, Agathe intervenes by reciting mystical texts from Ulrich's library, using the language of mysticism as a kind of lifeless, recitative convention. The

alternative to this impasse consists in tracing the spatial configurations of the other condition as they become manifest in a space of images. Agathe's attempt to imagine what it would be like to be released from all feelings of life, for example, is pictured as a reconfiguration of the space of the garden scene:

> Even space itself, this empty cube that always remains the same, had now changed, she thought. When she held her eyes closed for a while and then opened them again, so that the garden entered her view untouched, as if it had just been created, she perceived as clearly and immaterially as a vision how the line connecting her with her brother was distinguished from all the others: the garden "stood still" around this line, and without anything about the trees, paths, and other parts of the actual surroundings having changed, of which she could easily convince herself, everything was put in relation to this connection as its axis and was thereby in a visible way invisibly changed. (M 1093–94)

Agathe searches here for a perspective on the space of the garden that is unencumbered by a habitual perceptual schema. Not the objects in space but space itself is transformed in this experiment, which consists of an attempt to apprehend the garden in such a way that it is "untouched" by any conscious disposition. In particular, the notion of space as an empty three-dimensional container—a set of unchanging Euclidean coordinates in which objects are embedded—is displaced by a vision of space that is shaped by the presence of its occupants. The resulting view of the garden "as if it had just been created" recalls Ulrich's prior encounter with a "retracted and formless creation" (M 1090). It does not signal the return of an original state of paradise, despite the persistent pull of the garden metaphor in this direction. Rather, the almost imperceptible change in the garden consists in a reorientation of the space of the garden around the affective and erotic axis of Agathe's relationship to Ulrich. This subtle transformation induces a moment of stasis, such that "the garden 'stood still' around this line ['*stand*' *um diese Linie*]."

This freezing of the space of garden around the axis of the siblings' affective bond, however, is not pure motionlessness, but rather main-

---

approach to the other condition through dialogue thus becomes mired in an ossified and conventional form of language.

tains a tension between movement and rest. This tension is expressed in Agathe's observation "that all of the surrounding figures stood there abandoned in the uncanniest way, but were also delightfully animated in the uncanniest way" (M 1094). Agathe's experience of the spatial reconfiguration of the garden gives rise to a unique temporality that oscillates between a *nunc stans* and a temporal flow:

> And something similar to this impression of space had moreover come to pass with the sense of time: this flowing belt [*fließende Band*], this rolling staircase [*rollende Treppe*] with its uncanny affair with death, appeared in some moments to stand still [*stillzustehn*], and in others it flowed away without connection. In the course of a single external moment, time could disappear internally, without a trace of whether an hour or a minute had transpired. (M 1094)

In a characteristic blending of philosophical and religious traditions with modern technology, Musil adapts the image of time as a flowing current to the factory conveyor belt and the department store escalator. This ironic note adds an element of humor to Musil's characterization of the unsteady movement of time, alternating between standstill and flow. The effect of these images, much like Kracauer's image of pockets and voids in the flow of time, is to call into question the notion that historical time is a uniform and homogeneous continuum.

At the same time, the oscillation of flowing and arrested time suggests a liminal encounter of time and eternity, history and eschatology. Ulrich's and Agathe's experiment in the garden chapters, with its intimation of the other condition, approaches again and again a moment of *nunc stans*, but is never able to leave behind the flow of time. Confronted with both the collapse of narrative trajectory and the persistence of temporal movement, *The Man without Qualities* gives expression to these contradictions in the medium of the image. Often in conjunction with the resources of the simile (*Gleichnis*), Musil's images approach the other condition by maintaining opposite extremes in a state of tension.[61] An epiphanic moment at the

---

61. As Rebekka Schnell notes, Musil's *Gleichnis* "always refers to an Other," but "this element of the Other, the disjunctive is never entirely subsumed" ("'Die plötzliche

conclusion of the chapter "Mondstrahlen bei Tage" ("Moonbeams in the daytime") provides a fitting illustration of this point. As the siblings' dialogue recedes into silence and Agathe entertains doubts about whether the Millennial Kingdom is really at hand, her experience of the "enchanted garden" (*verzauberte Garten*) is condensed in the following images: "But only a hammock came to mind, held taut between two enormous fingers and rocked with infinite patience; then a silent sense of being towered over, as by tall trees between which one feels oneself lifted up and disappearing; and finally a nothing [*ein Nichts*] that in an inexplicable way had a tangible content" (*M* 1095). These images comment on the garden scene as a liminal space in several ways. First, the figure of the hammock suggests the phenomenon of suspension, much like Musil's images of bridges. The "infinite patience" with which the hammock is rocked points to the limit at which movement becomes rest, implying an approach of stasis. Together, these ideas evoke a sense of peacefulness, yet this peacefulness is undercut by the lingering terror resulting from the enormity of the fingers holding the hammock taut.

A similar tension marks the figure of tall trees: the subject experiences both a sense of "being towered over" (*Überragtwerden*) by these trees and being "lifted up" and "disappearing" (*sich emporgehoben und verschwunden*) between them. The lateral rocking and horizontal suspension of the hammock is transposed onto the vertical plane with its implied upward movement. The disappearance of the subject in its elevation into the heights of the towering trees anticipates the third figure of a "nothingness" (*ein Nichts*). The dizzying heights evoke a feeling of ecstasy so strong that the subject is dissolved the moment these heights are reached. All of these tensions and contradictions are concentrated in the image of "a nothingness that in an inexplicable way had a tangible content."[62] The contradictions built into these images intensify to the point that they

---

enthüllte Zärtlichkeit der Welt . . .': Liebe als ästhetische und religiöse Utopie in Robert Musils 'Der Mann ohne Eigenschaften,'" in *Emotionale Grenzgänge*, ed. Lisanne Ebert [Würzburg: Königshausen & Neumann, 2011], 91–112, at 105, 109).

62. In addition to the tension between "nothingness" and "content," the reader is also struck here by the juxtaposition of *greifbar* ("tangible") and *unbegreiflich* ("inexplicable").

can no longer be imagined. The standstill of time—as an intimation of an eschatologically connoted Millennial Kingdom—is thus visualized as an image space that borders on the nonintuitive.

In the chapter "Atemzüge eines Sommertags" ("Breaths of a summer day"), arguably the centerpiece of the garden chapters,[63] Musil presents his most complex and fully developed articulation of the standstill of time in an image—a crystallization of the moment in an image of the garden suspended between life and death.[64] The chapter begins with the garden scene: Ulrich and Agathe lie on the meadow in the middle of the summer day; their conversation has ebbed to a standstill, and they have not noticed that the sun has risen higher. In this context they become spectators of an image that condenses the garden scene into a significant moment:

> A soundless current of a lusterless blossom snow, coming from a group of fading trees, hovered in the sunshine; and the breath that carried it was so soft that no leaf stirred. No shadow of these blossoms fell on the green grass, but the lawn appeared to darken from within like an eye. The trees and bushes, tenderly and extravagantly in leaf from the young summer, which stood to one side or formed a background, gave the impression of speechless spectators, who, astonished and spellbound in their cheerful attire, took part in this funeral procession and festival of nature. Spring and autumn, the language and silence of nature, and the magic of life and death were blended together in this image; the hearts

---

63. Musil's *Nachlass* contains no fewer than four versions of this chapter, which he began drafting in 1938 and continued to work on until his death in 1942 (M 1232–39, 1240–49, 1306–24, 1324–37).

64. "Atemzüge eines Sommertags" rightly enjoys a privileged place in the Musil scholarship, but Fanta's characterization of the chapter as "Das letzte Kapitel dieser letzten Serie von Annäherungen an die letzten Dinge" misjudges the reasons for its importance ("Statt Religion Literatur, statt Literaturwissenschaft Theologie: Zum Gottesbegriff bei Robert Musil," in *Die Gottesfrage in der europäischen Philosophie und Literatur des 20. Jahrhunderts*, ed. Rudolf Langthaler [Vienna: Böhlau, 2007], 187–205, 199). Despite the fact that the chapter, like the rest of *The Man without Qualities*, stands before the threshold of the outbreak of the war, its composition history, especially the fact that it was the last chapter Musil worked on before his death, leads Fanta to attribute it the status of the end of a series that contains the ultimate and most decisive expression of Ulrich's and Agathe's interest in the eschatological moment. Fanta thus implicitly grants the chapter a teleological sense that is at odds with Musil's concept of history and the literary form of *The Man without Qualities*.

appeared to stand still [*stillzustehen*], to have been torn from their breasts, to join the silent procession through the air. (*M* 1232)

This "moment of the procession of blossoms" (*Augenblick des Blütenzugs*) (*M* 1233) uses a range of techniques to visualize the standstill of time as an image: it depicts a coincidence of opposite extremes, approaches a boundary at which life and death are indistinguishable, and above all sets up a tension between drive and stasis that points to the lack of closure in this representation of the moment. Among the numerous examples of the unification of opposites in this passage, one can observe a tension between the "snow of blossoms" (*Blütenschnee*) and the setting of the scene on a "summer day" (*Sommertag*); between the "current" (*Strom*) of these blossoms and the way that they "hover" (*schweben*) in the air; and between the characterization of the scene as a "funeral procession" (*Begräbniszug*) and as a "festival of nature" (*Naturfest*). What is remarkable about this scene is that all of these contradictions are "blended together in the image" (*mischten sich in dem Bild*). The hovering blossoms do not simply represent a moment outside time but rather a moment in which silence, rest, and stillness—all the privative attributes associated with death and loss—are inextricably entwined with the libidinal drive and erotic charge of life and generation.[65] Musil's image holds "spring and autumn, the language and silence of nature, and the magic of life and death" in a state of suspension.[66]

In Agathe's response to this scene of hovering blossoms, she places it under the interpretive lens of mysticism, grasping the standing still

---

[65]. It is important to emphasize the combination of drive and stasis in the image of the blossoms, especially because many interpretations focus exclusively on the ossification of nature in the image. Honold, for example, writes: "Wenn in den 'Atemzügen eines Sommertags' sich der Einstand der Zeit in der Unberührbarkeit der zum Bilde erstarrten Natur offenbart, so ist dies ein Wink jenes *Kosmogonischen Eros*, wie er von Ludwig Klages, dem Kopf der Münchner Kosmiker-Runde, propagiert wurde" (*Die Stadt und der Krieg*, 423–24).

[66]. Comparing "Atemzüge eines Sommertags" to Musil's draft of the chapter "Die Reise ins Paradies" (1924/25), Honold notes that while in "Die Reise ins Paradies" the siblings descend into "incest, boredom and despair," in "Atemzüge eines Sommertags" they "persist in a state of limbo that keeps all possibilities open" (*Die Stadt und der Krieg*, 452–53).

of time as an intimation of the Millennial Kingdom. In this spatialization of the other condition, the motif of timelessness is transposed into a paradoxical sense of spacelessness:

> Time stood still [*die Zeit stand still*], a thousand years weighed as lightly as the opening and closing of an eye, she had arrived at the Millennial Kingdom, perhaps even God let himself be felt. And while she experienced all of this one thing *after* another, although time was supposed to have ceased to exist; and while her brother, so that she would not suffer from anxiety during this dream, lay *next* to her, although there appeared to be no more space: the world appeared, despite these contradictions, to be fulfilled by transfiguration in all respects. (*M* 1233)

This passage comes the closest of any in *The Man without Qualities* to making a claim for the reality of the Millennial Kingdom, yet even here this claim is relativized and clearly marked as speculative: the observation that "she had arrived at the Millennial Kingdom [*ans Tausendjährige Reich gelangt*]" suggests the approach of a threshold but not quite its transgression, while the claim "perhaps [*vielleicht*] even God let himself be felt" undercuts with its "perhaps" any certainty in the divine revelation.[67] Agathe experiences the contradiction between the suspension of time and space and the persistence of temporal succession and spatial juxtaposition, but she subsumes this tension within an all-encompassing transfiguration of the world.

For Agathe, the Millennial Kingdom represents a state of "ecstasy" (*Entzückung*) that entails a thoroughgoing stasis of the self and the elimination of all desire. Claiming that "one has to free oneself from reality and from the desire to devote oneself to it," she adopts a posture of asceticism that aims for a "highest selflessness" (*M* 1234). But she fails to recognize the extent to which, in adopting such a disposition, she has become an utterly passive spectator of the garden scene and its central image of the procession of blossoms. Agathe's reflections on the image of the procession of blossoms reveal the danger that, with the collapse of the tension between desire and fulfillment,

---

67. Compare Fanta, "Statt Religion Literatur," 200: "Ein 'vielleicht' ist für die volle Erfüllung zuwenig."

the Millennial Kingdom will become an ossified form.[68] The important achievement of Musil's chapter is that he does not allow this stasis to solidify.[69] Agathe's vision of transfiguration resulting in complete selflessness and the abolition of desire is followed by a sudden resurgence of passion and drive. Agathe's effort to assert self-control, which, as Musil remarks with a tinge of irony, takes place "as if she wanted to feign death" (*als wollte sie sich totstellen*), comes to nothing, and she realizes that it is an "impossible task to bring thoughts, sense impressions, and declarations of the will to a complete halt [*ganz stillzustellen*]" (M 1234). Drawn back into the world of feeling and desire, Agathe returns in thought to her troubled relationships with Hagauer and Lindner and to her associated feelings of hatred and melancholy.

Musil sums up this resurgence of feeling as a "moment of drastic change in life, of the flight of passions and conditions, of the strange current of feeling [*wunderlichen Strom des Gefühls*]" (M 1235). The vitality, energy, and movement in this characterization of Agathe's passions echoes similar motifs, especially that of the current (*Strom*), that play an important role in the image of the procession of blossoms. Agathe finds herself torn between such moments of passionate feeling and a return to the "motionless dreaminess" (M 1235) of the Millennial Kingdom. In other words, she experiences an oscillation between drive and stasis as two distinct mental states, whose alternation is comparable to a dialectic of waking and dreaming states. Whereas the motifs of life and death, speech and silence, movement and rest coexist

---

68. Schnell comes to a similar conclusion with regard to Musil's early draft "Die Reise ins Paradies": "Der vollzogene Inzest führt darin letztlich zum Scheitern der Geschwisterutopie, die Erfüllung der Begierde mündet in Stillstand, Überdruss und Leere" ("Liebe als ästhetische und religiöse Utopie," 109).

69. This is an important consequence of Musil's conviction that in aesthetic terms the other condition never loses its connection with the normal condition (P 1154). Similarly, Bohrer notes that "the mystical moments of the 'garden' and the 'summer day' remain integrated" in the siblings' dialogue. While acknowledging the openness of these moments to eschatological metaphors, he concludes: "Such an eschatological moment has always been in opposition to the normal state. From Musil's theoretical compulsion to wrest a 'dual aspect' from every state, we can conclude concerning the reduction to the utopian moment that the reality of the 'whole' never disappears" ("Utopia of the Moment and Fictionality," 224).

simultaneously in the image of the procession of blossoms, Agathe moves back and forth between them in temporal succession.

The garden chapters are thus structured by a fundamental tension between the reduction of space and time to an image and the collapse of the image with the resurgence of the passions and drives. Musil strikes a delicate balance in these chapters, as he does throughout *The Man without Qualities*, between keeping the representation of the moment open to the transgressive possibilities of eschatological metaphors and guarding against their becoming ossified as ideological positions.[70] By keeping the representation of the utopian moment open, Musil rules out the possibility of narrative closure. The emergence of images in the garden chapters is thus bound up with the problem of the end, and specifically with the impossibility of representing the end. *The Man without Qualities* works with images that interrupt the narrative continuum and induce a narrative rupture. But no image proves capable of standing in as a last image, as we will see, both because the image is like an abyss that has no ground and hence no end and because the production of images itself has a serial quality that is not structured by the narrative convention of beginning, middle, and end.

The draft of a chapter for book 2 of *The Man without Qualities* composed in the late 1920s, titled "... Seit dem Traum ..." ("... Since the dream ..."), illustrates these aporias of the image. This chapter is part of complex of chapters that explores the possibilities of Ulrich's and Agathe's experimental union, and it anticipates many of the ideas later found in the garden chapters. In the following passage, Musil describes a moment detached from time and space as an image that leads to an abyss:

> Sometimes already in the morning there lay between the house they occupied and the street a nothing [*ein Nichts*] through which neither Ulrich

---

70. As Bohrer notes, while the other condition shakes up the established order of perception, it runs the risk of "hardening into a mythology" of its own. Bohrer concludes that the "reflective intellectual attitude" in Ulrich's and Agathe's dialogues prevents the emergence of a "new eschatology" in the form of a myth, but the point remains that the citation of eschatological metaphors such as the Millennial Kingdom must guard against such ossification ("Utopia of the Moment and Fictionality," 224).

nor Agathe could break through; the stimuli of life lost the power to provoke the trivial, small decisions necessary to put on a hat or insert a key, these small strokes with which one propels oneself forward. But the space in the rooms was as if polished, and everything was full of a faint music that only lapsed into silence when one listened closely to hear it more clearly. And therefore an affectionate anxiety was present; the silence behind the sound of a word could often suddenly detach a moment from the series of other moments, throw off the chains of temporal and spatial relations, and send the moment out into an infinite depth over which it rested motionlessly. Life then stood still [*Das Leben stand dann still*]. In sweet agony, the eye could not retreat from the image. It sank into existence as if into a wall of flowers. It sank ever deeper and ever more slowly. It arrived at no ground; it could not turn back! (M 1504–5)

The perceptual transformation of the space of the house in this scene has all the hallmarks of the other condition. Of particular significance is the suddenness with which "a moment detaches itself from the series of other moments" (*einen Augenblick aus der Reihe der übrigen loslös[t]*).[71] The epiphanic quality of such a moment involves a suspension of spatial and temporal continuity, which Musil glosses here as a "throwing off" (*abstreifen*) of "the chains of temporal and spatial relations" (*die Ketten der zeitlichen und räumlichen Beziehungen*). In language that evokes the scene of the procession of blossoms, this moment is said to "rest motionlessly" over an "infinite depth." Here too the standing still of life is attributed the qualities of image, one that provides a scene for a spectator. But what is distinctive about this representation of the epiphanic moment as an image is the dynamic that develops between the spectating eye and the image, for the eye finds it impossible to detach itself from the image. Sinking ever deeper into the image, the eye "arrives at no ground; it cannot turn back!" (*es kam auf keinen Grund; es konnte nicht umkehren!*). The image that arises in the standstill of time, in other words, is bottomless. This not only results in the petrification of the spectator, who is transfixed by the image; it also indicates that the

---

71. For more on the momentary, sudden quality of the other condition as an "intensive moment," see David Wachter, *Konstruktionen im Übergang: Krise und Utopie bei Musil, Kracauer und Benn* (Freiburg im Breisgau: Rombach, 2013), 136.

image is unable to provide a narrative end. The abyssal qualities of the image are such that its end is beyond representation.

To this endlessness of *the* image, Musil appends a postscript on the endlessness of the plurality of *images* because of their serial proliferation. The flip side of Musil's concept of history, the pendant to the suspension of time in the other condition, is that each moment has a provisional status in accordance with the contingency of history. Or, as Musil puts it at the end of the chapter ". . . Since the dream . . . ," "each thing is an ossified particular case [*erstarrter Einzelfall*] of its possibilities" (M 1509). Noting the significant differences between contemporary reality and past worlds, Ulrich asks: "But are there not many realities?" He concludes: "So much [in the past] was as it is today and so much was different, as if it wanted to be a language in images [*eine Sprache in Bildern*] of which none is the last [*keines das letzte ist*]" (M 1508). Musil conceives history here as a fundamentally open-ended series of images, and while any given image has the ability to bring time momentarily to a halt, no image can claim to provide a vantage point on history as a whole.[72] This is a consequence of the commitment to a plurality of possible realities. Narrative rupture and the impossibility of a narrative end to history are the twin pillars of Musil's interest in the other condition, and they are maintained in a characteristic tension between stasis and drive. Insofar as everything can become an image—this is what Musil, in earlier drafts of the chapter "Breaths of a summer day," called *Bildwerdung* (M 1310, 1330)—the procession of images is endless. Yet insofar as each image involves a suspension of narrative trajectory, it exists outside the logic of beginnings and endings. Images such as the blossom scene can thus be considered eschatological images that are radically decoupled from teleology. They represent openings in history and possibilities for the transgression of boundaries, but they do not allow for narrative closure.

---

72. Compare Schnell, who reads this infinite series of images as a product of the "imaginative activity of the subject," whose "unfulfilled longing" produces "images and possibilities" in an "interminable process" ("Liebe als ästhetische und religiöse Utopie," 105).

Musil's Aesthetics of the End

As Musil's work illustrates, the encounter with eschatological thought in the early twentieth century had ceased to be tied to its secularization in modern philosophies of history. Whereas such philosophies of history, in the wake of Hegel, posit a goal or *telos* that determines a history of development from the standpoint of its end, Musil works with a concept of history that is decentered and structured by contingency. When eschatological motifs enter *The Man without Qualities*, they do not provide a teleological structure for the book's narrative. On the contrary, Musil's eschatological figures deepen the sense of narrative rupture in his work, suggest possibilities for the transgression of boundaries, provide openings for liminal experiences such as the other condition, and are concentrated in the moment as a space of images. Accordingly, Musil's utopia of the moment, as it is articulated in the other condition, has, as Bohrer notes, "become aesthetic."[73] No longer tied to the concrete expectation of a future state, Musil's utopian moment replaces "eschatological history by language" through a "strategy of discharge for the no longer anticipated but not excluded real case."[74] This kind of reorientation of eschatological thought—from the *telos* of history to the present moment, from the temporality of the end to spatial representations—is a common feature of the modernist imagination of eschatology, one that we have observed in different variations in Barth, Rosenzweig, and Kracauer. Musil's approach to this reorientation shares these general features and lends them a uniquely aesthetic and literary dimension. In Musil's work, eschatological history is discharged into language as a set of images and metaphors. These images, however, are not representations of the end but manifestations of narrative rupture that are never able to realize the utopian moment.

Bearing in mind the theological insight into the incommensurability of time and eternity, history and eschatology, it should come as little surprise that the aesthetic realm of images proves fruitful

---

73. Bohrer, "Utopia of the Moment and Fictionality," 232–33.
74. Bohrer, "Utopia of the Moment and Fictionality," 232–33.

for the use of eschatological motifs as hypothetical possibilities that never quite manage to crystalize as realities. As Bohrer notes, "the utopia of the moment shares the criterion of incommensurability with the phenomenon of the aesthetic itself."[75] Ulrich's and Agathe's experiments with the other condition remain liminal experiences that neither provide *The Man without Qualities* with narrative closure nor resolve the problem of the collapse of the narrative order. And indeed, the space of images that Musil develops in order to capture the utopian moment of the other condition remains doubly precarious. It is not just that this space of images is riven with internal tensions, such as the tension between drive and stasis. The status of such images is also complicated by their uncertain relationship to the normal condition to which they are juxtaposed. This normal condition is characterized by discord, strife, and violence, and the siblings' reclusive garden is never entirely able to shut out these realities. One must therefore entertain doubts about Honold's claim that an "aesthetic process" (*ästhetischer Vorgang*) that Musil calls "becoming an image" (*Bildwerdung*) is able to "balance out moments of precipitous violence that flare up."[76]

The problem that Musil confronts time and again is that the experiment of the other condition, even when it is directed toward the union of an erotic-affective bond, can also be conceived in terms of the liminal experiences of war, violence, and madness. In Harmut Böhme's view, Musil's novel fails for precisely this reason: "The failure of the novel, it seems, is essentially the result of the fact that Musil's attempt to extract an aesthetic figure [*ästhetische Gestalt*] from the 'other condition' was shattered by its liaison with the excesses of war, violence (Moosbrugger), and madness (Clarisse)."[77] The anarchic quality of the other condition, which resists domestication through a system of order, proves difficult if not impossible to control. Yet what could have been a failure instead testifies to the enduring relevance of Musil's project. The fact that Musil's eschatological metaphors never result in a stable aesthetic figure is

---

75. Bohrer, "Utopia of the Moment and Fictionality," 233.
76. Honold, *Die Stadt und der Krieg*, 454.
77. Böhme, "Die 'Zeit ohne Eigenschaften,'" 31.

precisely what saves *The Man without Qualities* from the descent into an aesthetic ideology.[78]

This aesthetic instability of the image, in turn, is indicative of the unstable boundary in Musil's work between religion and the secular and between theology and the profane. As in his description of the other condition as a form of a "profane religiosity" (*profane Religiosität*), here too these seemingly oppositional categories are maintained in a state of suspension.[79] To say that the eschatological moment has "become aesthetic"—insofar as it is expressed in the space of images—is of course to think through how eschatological motifs can be activated outside theology, but it also makes a claim about the afterlife of theological thought in modernity. To characterize this afterlife as a form of secularization would be misleading, because such a definition fails to recognize that aesthetic forms and theological motifs are engaged in a productive interaction in the modernist imagination of eschatology.[80] Thus, while Bohrer rightly points out that Musil dramatizes a "cultural-religious image for a 'now' that . . . cannot exist as a total presence" but that is capable of challenging "the compulsion of cultural norms," it is worth revisiting his claim that for Musil "this messianic-eschatological set of metaphors does not have a theological character but is a symbolic seman-

---

78. For Terry Eagleton and Paul de Man, aesthetic ideology consists in the reduction of the contingent and aporetic dimensions of the aesthetic sphere to a single, natural meaning. See Terry Eagleton, *The Ideology of the Aesthetic* (Oxford: Basil Blackwell, 1990), 10.

79. Musil, "Der deutsche Mensch als Symptom," *P* 1353–1400, at 1398.

80. The proximity of the other condition to religious experience has led critics to characterize it as a "secularized form of mystical experience" (*säkularisierte Form mystischen Erlebens*) (Irmgard Scharold, *Epiphanie, Tierbild, Metamorphose, Passion und Eucharistie: Zur Kodierung des "Anderen" in den Werken von Robert Musil, Clarice Lispector und J.M.G. Le Clézio* [Heidelberg: Universitätsverlag C. Winter, 2000], 88; see also Barbara F. Hyams, "Was ist 'säkularisierte Mystik' bei Musil?," in *Robert Musil: Untersuchungen*, ed. Uwe Baur and Elisabeth Castex [Königstein im Taunus: Athenäum, 1980], 85–98). Apart from the fact that such claims often fail to adequately define the concept of secularization that they invoke (see Hyams, "Was ist 'säkularisierte Mystik' bei Musil?," 86), the substitutional logic of such a secularization thesis, which implies that an originally religious substance is replaced by a secular substitute, cannot be reconciled with Ulrich's observation that the other condition is "more original than religions" (*M* 766).

tics for a utopian horizon that evades any reduction to a concept and that uses eschatological concepts merely as an allegorical reference."[81] While there is no doubt that Musil has no attachment to a theological system, any more than Kracauer does, to say that eschatological metaphors are used "*merely* as an allegorical reference" ignores the transformation that eschatological thought undergoes in the early twentieth century. No longer is it tied to any concrete expectation of the apocalyptic end or to the teleology of a philosophy of history, whether religious or secular. Instead, it has become a set of images and metaphors for thinking through the discontinuities of historical time and the incommensurability of the present moment with the utopian possibilities of the "now." Eschatological motifs are not just allegorical references useful for challenging cultural norms; they are dynamic images of the complexities of the moment and representations of the rupture of the historical continuum.

Musil's modernist reimagination of eschatology is not limited to explicit eschatological metaphors such as the Millennial Kingdom. It is also at work in images that have only an implicit eschatological subtext, especially those that represent figures of the end in nonteleological terms and transform the temporal end into a liminal space at the boundary between life and death. Whenever Musil provides images that contend with the collapse of the narrative order and detemporalize the end, we find ourselves in the space of the modernist imagination of eschatology. By way of conclusion, let us now examine two final images that demonstrate in nuce just how decisively Musil breaks with figures of the end that imply narrative closure.

First, and perhaps surprisingly, we have the inverted temporality of the telegram informing Ulrich of the death of his father, an event that could very well be considered a narrative ending par excellence. However, the telegram, which Ulrich's father himself composes, blurs in a peculiar way the line between life and death: in the moment of its composition, the father is alive and must anticipate an end in death that has yet to transpire, while in the moment of its communication, the subject who communicates his own death can no longer speak. Musil carefully constructs these paradoxes in the

---

81. Bohrer, "Utopia of the Moment and Fictionality," 226–27.

final chapter of book 1, where we read: "The telegram now reported to him [Ulrich], in a detailed manner and with a curious mixture of half-suppressed reproaches and a complete celebration of death [*Todesfeierlichkeit*], which his father had apparently himself meticulously arranged and drawn up, the demise of its progenitor [*das Ableben seines Erzeugers*]" (M 655). While the temporality of the telegram anticipates an end in death, and in this way attempts to give form and structure to the end in advance, in a manner not unlike Musil's collection of essays and prose images entitled *Nachlass zu Lebzeiten* (*Posthumous Papers of a Living Author*) (1935), the proper end of the father's death in the execution of his will is subjected to an indefinite deferral.[82] Indeed, Ulrich denies his father's inheritance in two ways. Already during his father's life Ulrich had refused to take advantage of his father's social position and definite qualities by staking out a position as a man "without qualities." Similarly, Ulrich's and Agathe's experiment with the other condition in book 2 is predicated on a postponement of the execution of the father's will, which Agathe is prepared to falsify if necessary.[83] Musil undercuts the testament's characteristic "proleptic access to death"[84] by deferring its execution. This scene, which mediates between book 1 and book 2, expresses in miniature the larger structural paradox of *The Man without Qualities*, namely that the end is already evident in the beginning, but is never reached in narrative terms.

---

82. In the "Vorbemerkung" to his *Nachlass zu Lebzeiten*, Musil comments on the inverted temporality of the project. Aiming to prevent the publication of posthumous papers before it is too late to do so, and thereby to claim sovereignty over death before it has taken place, the *Nachlass zu Lebzeiten* paradoxically treats the writings of its author as literary remains while the author is still living: "Was immer sich von der Frage vermuten ließe, wann ein Nachlaß von Wert sei, und wann bloß einer vom Werte: ich habe jedenfalls beschlossen, die Herausgabe des meinen zu verhindern, ehe es soweit kommt, daß ich das nicht mehr tun kann. Und das verläßlichste Mittel dazu ist es, daß man ihn selbst bei Lebzeiten herausgibt" (P 473).

83. For an insightful reading of the problem of the testament in *The Man without Qualities*, see Lars Friedrich, "Favor testamenti: Letztwillige Verfügungen in Musils 'Mann ohne Eigenschaften,'" in *Urteilen/Entscheiden*, ed. Cornelia Visman and Thomas Weitin (Munich: Fink, 2006), 72–90, esp. 86.

84. Friedrich, "Favor testamenti," 78: "So offenbart der proleptische Zugriff auf den Tod die für jedes Testament konstitutive Heterogenität mit sich selbst und der Zeit."

Second, we have Musil's short prose piece "Das Fliegenpapier" ("The Flypaper"), first published under the title "Römischer Sommer" ("Roman Summer") in 1913 and included as the first of a series of "images" (*Bilder*) in Musil's *Posthumous Papers of a Living Author*. "The Flypaper" recounts in minute detail the various stages of resistance of a fly that becomes trapped in flypaper. The prose piece presents a series of images, momentary snapshots of the fly's gradual descent deeper and deeper into the flypaper, with variations of the constellations of body parts that can become entrapped. In terms of its temporal and narrative trajectory, the death of the fly appears inevitable: "A nothing, an 'it' draws it in [*Ein Nichts, ein Es zieht sie hinein*]. So slowly that one is hardly capable of following, and usually with a sudden acceleration at the end, when it is overcome by a final inner collapse" (P 477). This flattening out of narrative movement, followed by a sudden moment of collapse, bears comparison with the slowing of time in the garden chapters interspersed with momentary outbreaks of the other condition. Here the image of the fly caught in the flypaper suggests an irreversible descent into death, which, like the death of Ulrich's father, appears to be a classic case of a narrative ending.

Yet Musil manages, however subtly, to subvert the sense of closure in this ending. He does so first through the sheer multiplicity of images of trapped flies, whose various configurations of resistance and submission prove aesthetically more productive than any representation of their death would be: "They lie there like this. Like crashed planes with a wing protruding in the air. Or like dead horses. Or with endless gestures of desperation. Or like sleepers" (P 477). But even beyond these variations, the final lines of "The Flypaper" defer the end in death indefinitely: "And only on the side of the body, in the vicinity of the base of the leg, they [the flies] have some very small organ that lives for a long time [*das lebt noch lange*]. It opens and closes, one cannot identify it without a magnifying glass, it looks like a tiny human eye that opens and closes endlessly [*das sich unaufhörlich öffnet und schließt*]" (P 477). The boundary between life and death, between movement and its cessation, considered in the previous passage in terms of the temporal dynamic of slow decline and sudden collapse, is expressed here in spatial terms. The infinite approach to the limit in death is transposed onto the infinitesimal space of a tiny organ of perception. The incessant opening and closing of this organ, comparable to the blinking of a human eye, is unconscious and involuntary, yet it throws into doubt the demarcation between life and death, drive and stasis, much like the image of the procession of blossoms. The image of the endless opening and closing of the eye, precisely in the context of what appears to be an inevitable end in death, is emblematic of Musil's commitment to openings in history that forego the finality of narrative closure. In this way, Musil's work explores in an unparalleled fashion the literary and aesthetic possibilities for conceiving of the end in spatial rather than temporal terms.

Epilogue

The Ends of Modernism

In keeping with the fundamental plurality of spaces in modernist mathematics—not one, but rather many geometries—the investigation of the imagination of time in early twentieth-century thought has revealed a multiplicity of shapes of time. Combining religious thought and key principles of non-Euclidean geometry, Barth, Rosenzweig, Kracauer, and Musil provided a range of new images of time that expanded and complicated the repertoire of concepts of history handed down by historicism. They showed that history can be illuminated in multiple aspects and that chronological time exists side by side with shaped times. For these writers, history is defined not only by the continuous flow of time toward the future, in stages of development, but also by eddies and voids that cannot be incorporated into the march of time. In the modernist imagination of history, the linearity of time's arrow is bent into the shape of a curve, the uniformity of the temporal continuum no longer an empty vessel for history but rather given shape by a manifold of historical figures.

The attribution of a larger trajectory of a universal history gives way to a sense of history defined not by a single meaning or *telos*, but by a plurality of voices that arise from the periphery.

A comparison of the multiple shapes of time given expression in the work of Barth, Rosenzweig, Kracauer, and Musil reveals striking affinities but also important differences, suggesting that modernism has a multitude of possible ends. The cross-pollination of time and space, and of mathematics and religious thought, offers a range of images for picturing the end of history and the shape of time. Nevertheless, no image is capable of standing in as the last. For Barth and Rosenzweig, geometrical figures provide images of history in its relation to eschatology, but a gulf separates Barth's image of the liminal relation of the tangent line to the circle from Rosenzweig's picture of an arc of history suspended between an eternal beginning and end. Yet they each manage to push their images, with the help of non-Euclideanism, to the limit of visualizability. For Barth, the liminal point of contact between tangent and circle—in which a gap always persists—marks the circle as a nonintuitive and inaccessible space. For Rosenzweig, each point on the arc is equally proximate to beginning and end, undermining the idea of a trajectory toward the end.

While Barth and Rosenzweig each explore models of history and eschatology in which beginning and end, *Urgeschichte* and *Endgeschichte*, become interchangeable, their responses to historicism are quite different. Barth is much more resolutely antihistoricist and views the space of history as utterly incommensurable with that of eternity. Rosenzweig, in contrast, both carries forward and revises the historicist paradigm. In light of his training with Meinecke—a key heir of the historicist approach—and his commitment to a narrative philosophy, his dual-aspect theory of history does not problematize history so much as overlay it with spatial figures of eternity. These divergent responses to historicism are epitomized in Barth's and Rosenzweig's contrasting mobilizations of images of parallel lines. For Barth, the failure of parallel lines to meet underscores the incommensurability of humanity and God and of history and eschatology. For Rosenzweig, the meeting of parallel lines in infinity provides an image for the two aspects of history—it is both a chain of temporal moments and the crystallization of figures of eternity.

Barth's interest in spatial forms as representations of a moment outside narrative sequence, meanwhile, has a strong affinity with Musil's exploration of images of the standstill of time. They each describe the encounter with infinity as a shock that is profoundly disorienting, and they construct the eschatological now as a realm of possibility. Both Barth and Musil picture the relationship of eschatology and history in terms of images of rupture, yet they remain grounded in their concrete historical situations. While Musil's work is unique in that his nonnarrative approach is tied to a new understanding of literary form, he shares with Barth an interest in rethinking the end through an encounter at the boundary of life and death. For Barth this threshold offers an opportunity for a diagonal cut through history in the figure of resurrection. In Musil's images of flies caught in flypaper, by contrast, the persistence of life at the boundary of death suggests a sense of endlessness even at the end.

This sense of endlessness, together with an insight into the provisionality of history, is shared by Kracauer, who, like Musil, thought there could never be an ultimate shape of history. Kracauer's image of the ever-changing faces of Ahasuerus, who wanders endlessly through history, finds a counterpart in Musil's reflection that every step is followed by a subsequent step—not as a path from a to z, but in an infinite series in which every step n is followed by $n+1$. But in terms of methodology, Kracauer's thinking about the shape of time is most closely connected to that of Rosenzweig. Both Kracauer and Rosenzweig approach the shape of time in terms of figures or *Gestalten*. For Rosenzweig, the geometrical imagination of history involves the production of spatial figures as *Gestalten*—most famously the figure of the star of redemption. Such figures provide an interface for reimagining the shape of time and the temporality of the moment. Similarly, Kracauer understands history as a set of time curves—as figures that are given shape as time is folded on itself. Kracauer thus shares Rosenzweig's commitment to a dual-aspect theory of history in which time is always both chronological time and shaped time.

Yet whereas Rosenzweig explores the construction of figures between the points of humanity, God, and world—and their corollaries creation, revelation, and redemption—Kracauer emphasizes the disintegration and deformation of figures. In topological terms,

Kracauer explores the modulation of shapes when figures are deformed. For him, the shape of time emerges not through the production of figures, but by tracing the patterns with which the elements of history crumble and decay. In this respect, Kracauer's approach has an affinity with Barth's negative theology: both writers explore the space of eschatology in terms of hollow spaces, rifts, and fractures in the surface of history.

The diverse shapes of time put forward by Barth, Rosenzweig, Musil, and Kracauer reflect the different branches of mathematics that animate their thinking. While an encounter with non-Euclideanism and the constructivist principles of modernist mathematics is common to all, there are subtle differences in the way each writer uses mathematics to imagine history in new spatial terms. Barth's use of the infinitesimal limit from modern calculus and the geometry of points, lines, and planes is perhaps the most conventional of these approaches, yet he discovers new possibilities in these images—for example, by describing grace and sin as points in different spaces or questioning the point of contact of a tangent line. Similarly, while Rosenzweig's comparison of the quantitative infinity of rational numbers to the qualitative infinity of the uncountable but spatially intuitable irrational number—a metaphor for the contrasting temporalities of Christianity and Judaism—is consistent with the geometry of Euclid, his paradoxical claims about equidistant midpoints rely on a projection from Euclidean to non-Euclidean space. The projective geometry of Beltrami, Klein, and Poincaré thus provides a metaphor for mapping historical moments onto the space of the end of history. In translating ideas from mathematics into images of spatialized time, Barth and Rosenzweig often operate at the boundary of what Rosenzweig calls supra-mathematical shapes.

In Kracauer's texts on urban space, photography, and history, we can observe a non-Euclidean approach that is turned in the direction of mathematical topology—one that emphasizes the potential deformation of shapes and the capacity of objects to define their own spaces. The bending of straight lines into curves in the geometry of urban space provides a prelude to an approach to history in which the shape of time is characterized by its shapelessness and the looming disintegration of its form. For Kracauer, the modal relation of spatial figures in topology provides a model for thinking about history not as a fixed and homogeneous temporal continuum, but as a mobile set of time curves. Our inability to picture the face of Ahasuerus because it is constantly undergoing shape deformations has a counterpart in Musil's concept of history as entropic dispersion. But for Musil, the motor of a new concept of time can be found in the uncertain foundations of mathematics, as exemplified in Törleß's doubts about imaginary numbers, the concept of infinity, and the intersection of parallel lines. As a realm of pure possibility that requires no ground in intuition, modernist mathematics provides a template for Musil's imagination of a new shape of time in the other condition, an experience in which the moment is cut loose from its anchor in chronological time.

The upshot of these transpositions of various mathematical models—whose common denominator can be found in the non-Euclidean revolution and its aftermath—is a new language for the nonintuitive character of history in its relation to eschatology. For Barth the accent lies on the utterly nonintuitive character of the divine, and via its liminal contact with history, of the nonintuitive double determination of each now. Similarly, for Rosenzweig the temporality of the "today" becomes nonintuitive when perceived from the standpoint of redemption, requiring an alternative geometry to understand its proximity to beginning and end. In Kracauer's work, by contrast, the end of history is nonvisualizable because it is multidimensional—an emblem for which is given in the superimposition of Ahasuerus's multiple faces. Likewise, Musil's images of the standstill of time are groundless, leading their spectator into an abyss whose end is beyond representation.

These nonintuitive shapes of time offered alternatives to the historicist understanding of the teleology of history. In modernism, it became possible to picture time not only as a unidirectional arrow, accelerating and progressing into the future in stages of development, but also as a time in suspension and exile, a time on the verge of disintegration, and a time that does not follow the thread of a narrative. Rather than driving toward a *telos*, history was given shape in the moment—constructed, assembled, and configured in images at the eschatological limit. By imagining history in the absence of teleology, Barth, Rosenzweig, Kracauer, and Musil advanced new ways of thinking about ends and endings that would prove to be productive far beyond the domains of mathematics and religious thought. To be sure, there are tensions in the ways modernism understands the end. For Kracauer and Musil, there can be no end of history because the modulation of history's shape and its entropic dispersion is truly endless. Barth and Rosenzweig, by contrast, reconceive the end in nontemporal terms, such that the end is always present, even from the beginning. Nevertheless, their approaches have in common a reimagination of the end in spatial terms as a limit phenomenon, recalling the etymology of the Greek *eschaton*.[1]

A productive afterlife of this spatial concept of the end and modernism's nonteleological model of history can be found in the work of Theodor W. Adorno and Hans Blumenberg, who explored the paradoxes of endings in terms that resonate strongly with modernism's shapes of time. The following vignettes from their work provide two key examples of the enduring legacy of the modernist imagination of history and eschatology. First, let us consider the final text in Adorno's *Minima Moralia: Reflexionen aus dem beschädigten Leben* (*Minima Moralia: Reflections from Damaged Life*) (1951). Although it bears the title "Zum Ende" ("Finale"), the text offers a conclusion to the collection of aphorisms in name only, for the sequence of the texts is highly contingent. While the three parts of the book are organized according to the years of their composition, the structure of the book lacks any discernible narrative arc. In his preface, Adorno emphasizes the open-endedness of its concluding aphorisms: "The concluding aphorisms of each part lead on thematically also to philosophy, without ever pretending to be complete or definitive [*abgeschlossen und definitiv*]: they are all intended to mark out points of attack or to furnish models for a future exertion

---

1. In their *Greek-English Lexicon*, Liddell and Scott give as the first definition of *eschatos* the spatial meaning of "*the furthest, uttermost, extreme,*" followed by "*uttermost, highest*" (of degree), "*lowest, meanest*" (of persons), and finally "*last*" (of time). See Henry George Liddell and Robert Scott, *A Greek-English Lexicon*, 8th ed. (New York: Harper & Brothers, 1897), 588–89.

of thought."² The title "Zum Ende," which, as Gerhard Richter has noted, is only inadequately rendered by the English "Finale,"³ bears out the inconclusiveness of its reflections: the text gestures "toward" an end without claiming to embody the end or even have access to it.

What might be signified by the end is indicated in the aphorism's first sentence: "The only philosophy which can be responsibly practiced in the face of despair is the attempt to contemplate all things as they would present themselves from the standpoint of redemption [*wie sie vom Standpunkt der Erlösung aus sich darstellen*]."⁴ While at first glance the standpoint of redemption might be taken as the end toward which *Minima Moralia* is oriented, a closer examination suggests that for Adorno this standpoint does not lie at the end of the horizontal line of history, but is instead a standpoint capable of shedding light on our present. While availing himself of a metaphorics of light typical of Enlightenment thought— "Knowledge has no light but that shed on the world by redemption" (*Erkenntnis hat kein Licht, als das von der Erlösung her auf die Welt scheint*)⁵—the state of the world to be illuminated by the standpoint of redemption is anything but an enlightened one. Indeed, invoking spatial metaphors reminiscent of Barth and Kracauer, Adorno writes: "Perspectives must be fashioned that displace and estrange the world, reveal it to be, with its rifts and crevices [*Risse und Schründe*], as indigent and distorted [*bedürftig und entstellt*] as it will appear one day in the messianic light."⁶ To contemplate the world from the standpoint of redemption, in other words, entails an apprehension, not of the world as redeemed, but of the world in its fallenness as eminently in need of redemption. The "tears" (*Risse*) and "cracks" (*Schründe*) of an estranged and displaced world call to mind Barth's "hollow spaces" (*Hohlräume*)

---

2. Theodor Adorno, *Minima Moralia: Reflections from Damaged Life*, trans. E.F.N. Jephcott (London: NLB, 1974), 18; Theodor W. Adorno, *Minima Moralia: Reflexionen aus dem beschädigten Leben* (Frankfurt am Main: Suhrkamp, 1969), 11.
3. Gerhard Richter, *Afterness: Figures of Following in Modern Thought and Aesthetics* (New York: Columbia University Press, 2011), 62–63.
4. Adorno, *Minima Moralia* (English), 247; *Minima Moralia* (German), 333.
5. Adorno, *Minima Moralia* (English), 247; *Minima Moralia* (German), 333.
6. Adorno, *Minima Moralia* (English), 247; *Minima Moralia* (German), 334.

and "fractures" (*Bruchstellen*) and Kracauer's "holes and fractures" (*Löcher und Risse*).

By forging perspectives in which the world appears as fractured and alienated from itself, Adorno rejects the historicist ideology of progress according to which history tends toward the realization of the ideal, and replaces it with a recognition of the negativity and brokenness of existence. He focuses his gaze not on the future constitution of the world in a messianic era but on the world as it exists in the present. Indeed, the standpoint of redemption that Adorno seeks as an antidote to despair proves elusive. Much like the concept of the end as a threshold or limit phenomenon in modernism, Adorno claims that the attainment of a standpoint of redemption verges on the impossible: "But it is also the utterly impossible thing, because it presupposes a standpoint [*Standort*] removed [*entrückt*], even though by a hair's breadth, from the scope of existence, whereas we well know that any possible knowledge must not only be first wrested from what is, if it shall hold good, but is also marked, for this very reason, by the same distortion and indigence [*Entstelltheit und Bedürftigkeit*] which it seeks to escape."[7] Adorno unearths the paradoxical nonpresence of a standpoint of redemption in two key respects. First, it remains a liminal possibility because it is removed from the conditions of existence and therefore unable to provide knowledge about existence, however small its deviation from the world may be. Yet the standpoint of redemption is also impossible because the contact that it retains with worldly existence subjects its knowledge to the same limitations and distortions as those already present in the world in need of redemption. The paradoxes of Adorno's standpoint of redemption can thus be compared to those of his friend Paul Tillich, who claimed that the standpoint of eschatology is one upon which it is impossible to stand, a standpoint that unsettles and sublates all possible standpoints.[8]

The end of *Minima Moralia* is ultimately defined not by the standpoint of redemption as such, but by the dual recognition of the im-

---

7. Adorno, *Minima Moralia* (English), 247; *Minima Moralia* (German), 334.
8. Tillich, *Kairos*, 2–3. For more on Tillich's concept of the eschatological standpoint, see chapter 2.

possibility of this standpoint and of the necessity of contemplating the world, in the face of despair, from this perspective. The "reality or unreality of redemption,"⁹ Adorno concludes, is of little importance compared to the philosophical imperative to formulate a standpoint of redemption despite the recognition of its impossibility: "Even its own impossibility [*seine eigene Unmöglichkeit*] it must at last comprehend for the sake of the possible [*um der Möglichkeit willen*]."¹⁰ Here Adorno echoes Barth's concept of an "impossible possibility" (*unmögliche Möglichkeit*) (R 72–73). The end of *Minima Moralia*, much like the end of modernist eschatology, is defined not by teleology, completion, and closure, but by an open-ended and aporetic space of reflection in the present.

Similarly, Hans Blumenberg concludes his *Arbeit am Mythos* (*Work on Myth*) (1979) with a set of reflections on what it might mean to bring a myth to an end. Taking as a point of departure Franz Kafka's "Prometheus," a text that perhaps comes closer than any other to consigning the myth to oblivion, Blumenberg is nevertheless convinced that there is no end to myth: "There is no end of myth, although aesthetic feats of strength aimed at bringing it to an end occur again and again" (*Es gibt kein Ende des Mythos, obwohl es die ästhetischen Kraftakte des Zuendebringens immer wieder gibt*).¹¹ Here the aesthetic experimentation with an eschatology of myth is confronted with the endlessness of its variations. The conviction that there will always still be something left to be said is definitive of Blumenberg's search for a model of thought that resists closure. Yet Blumenberg's reading of Kafka's text also suggests a model of an ending that resonates with the modernist concept of history.

Kafka's "Prometheus" consists of a series of four legends, which, as Blumenberg notes, are "not arbitrarily interchangeable but rather a sequence that exhibits the process, in its form, all the way to the

---

9. Adorno, *Minima Moralia* (English), 247; *Minima Moralia* (German), 334.
10. Adorno, *Minima Moralia* (English), 247; *Minima Moralia* (German), 334.
11. Hans Blumenberg, *Work on Myth*, trans. Robert M. Wallace (Cambridge, MA: MIT Press, 1985), 633; Hans Blumenberg, *Arbeit am Mythos* (Frankfurt am Main: Suhrkamp, 1979), 685.

end [*zum Ende hin*]."¹² In the first variation, which bears the closest resemblance to the mythic tradition, Prometheus is chained to the Caucasus because he betrayed the gods to men, and eagles are sent by the gods to feed on his liver. With each subsequent variation, Prometheus descends further and further into oblivion:

> According to the second, Prometheus pressed himself deeper and deeper into the rock to escape the pain of the hacking beaks, until he became one with the rock. According to the third, in the course of the millennia his betrayal was forgotten, the gods forgot it, the eagles, he himself forgot. According to the fourth, everyone became weary of what had become groundless. The gods grew weary, the eagles grew weary. The wound closed wearily. There remained the inexplicable mountain of rock [*Blieb das unerklärliche Felsgebirge*].¹³

As Blumenberg notes, Kafka's "amendments" of the Prometheus myth "belong to its eschatology" (*gehören in dessen Eschatologie*).¹⁴ The end of these variations of the myth is the disappearance of Prometheus into the rock to which he is chained. Prometheus not only becomes one with the rock, but is forgotten and his wound healed. All that remains is the inexplicable rock. For Blumenberg, this retreat of myth into the inorganic substance of rock amounts to an attempt to "embed history in the non-historical" (*die Geschichte in das Nicht-Geschichtliche einzubetten*) because "the inorganic alone outlasts history."¹⁵

In a reversal of the traditional schema of eschatological judgment, the end of history in Kafka's "Prometheus" is marked not by the alternatives of catastrophe or salvation, nor by the realization of utopian happiness, but by the mute persistence of the non-historical substrate of existence, condensed here in the image of the "inexplicable

---

12. Blumenberg, *Work on Myth*, trans. Wallace, 633–34, translation modified; Blumenberg, *Arbeit am Mythos*, 686.
13. Franz Kafka, "Prometheus," in *Kafka's Selected Stories*, trans. Stanley Corngold (New York: W. W. Norton, 2007), 129, translation modified; Franz Kafka, "Prometheus," in *Nachgelassene Schriften und Fragmente*, vol. 2, ed. Jost Schillemeit (Frankfurt am Main: Fischer, 1992), 69–70.
14. Blumenberg, *Arbeit am Mythos*, 685.
15. Blumenberg, *Arbeit am Mythos*, 687–88.

mountain of rock." It is an end that bears the hallmarks of a disappearance, a fading away, and a barely recognized passage into nothingness. As Kafka notes in his preface to "Prometheus": "Legend attempts to explain the inexplicable; because it arises from a ground of truth, it must end again in the inexplicable."[16] For Kafka the end of myth is a return to the situation of its origin, the reinstitution of the inexplicable for which myth seeks to provide an explanation.

Kafka's paradoxical end of the Prometheus myth echoes the tension in modernist eschatology between end and beginning, *Endgeschichte* and *Urgeschichte*: the origin anticipates the end, while the end repeats its origin. As Blumenberg argues, while Kafka's text seeks to eliminate the freedom of variation inherent in myth, it in fact results in a form of stasis that is itself generative of new variations and interpretations:

> The evidence of a density without any room to maneuver—the density that the rock possesses—is manufactured. Only a temporal reversal would still be thinkable: Prometheus steps forth from the rock again and presents himself afresh to his tormentors. The eschatological melancholy that lies over the whole forbids one to give way, even for a moment, to this imaginative license. Why should the world have to continue in existence if there is nothing more to say? But what if there were still something to say, after all? [*Wie aber, wenn doch noch etwas zu sagen wäre?*][17]

While the mountain of rock stands for a "density without any room to maneuver" (*spielraumlose[] Dichte*), Blumenberg turns to the imagination to create new spaces of movement. In a temporal reversal reminiscent of Rosenzweig's reversal of beginning and end, Blumenberg imagines Prometheus reemerging to new life from out of the rock. The eschatological melancholy according to which the world will cease to exist when there is nothing left to say is overcome by a resurgence of the work on myth. Blumenberg's reflections on the end of myth thus recall the dialectic of drive and stasis in Musil's representation of the other condition. As much as Kafka

---

16. Kafka, "Prometheus" (English), 129; Kafka, "Prometheus" (German), 69–70.
17. Blumenberg, *Work on Myth*, trans. Wallace, 636, translation modified; Blumenberg, *Arbeit am Mythos*, 688–89.

strives to bring the Prometheus myth to an end by exhausting its form, he is unable to repress what Blumenberg calls the "permanent productivity" of myth that can be reactivated at any moment. As an inexhaustible form, the prospect of a "last myth" is necessarily a fiction.[18]

Kafka's mountain of rock is emblematic of the key elements of history and eschatology in the modernist imagination. In place of a temporal concept of the end of time, it substitutes a spatial representation of what lies beyond history. In contrast to the teleological trajectory of history toward a goal, it posits a form of dissolution that leaves the end open to new variations. Here we can observe the legacy of the concepts of history and eschatology developed by Barth, Rosenzweig, Kracauer, and Musil, which probe the possibilities of imagining shapes of time beyond its representation in historicism. Their reconfiguration of the end in proximity to the present moment—as a liminal and threshold space, in ways that stretch its intuitive representation—ultimately had an impact not only on the concept of history in modernism, but also, as we have seen in Adorno and Blumenberg, on a mode of philosophical reflection for which the opening of thought to paradoxes, aporias, and antinomies takes priority over the semblance of completion and closure. In the wake of the modernist imagination of history and eschatology, the end of history gives way to an endlessness of reflection.

---

18. See Blumenberg, *Arbeit am Mythos*, 300, 295.

Bibliography

Abbott, Edwin A. *Flatland: A Romance of Many Dimensions*. Edited by Rosemary Jahn. Oxford: Oxford University Press, 2006.
Adorno, Theodor W. *Minima Moralia: Reflections from Damaged Life*. Translated by E.F.N. Jephcott. London: NLB, 1974.
———. *Minima Moralia: Reflexionen aus dem beschädigten Leben*. Frankfurt am Main: Suhrkamp, 1969.
Agamben, Giorgio. *The Time That Remains: A Commentary on the Letter to the Romans*. Stanford, CA: Stanford University Press, 2005.
Albrecht, Andrea. "Mathematische und ästhetische Moderne: Zu Robert Musils Essay 'Der mathematische Mensch.'" *Scientia Poetica: Jahrbuch für Geschichte der Literatur und der Wissenschaften* 12 (2008): 218–50.
Altmann, Alexander. "Rosenzweig on History." In *The Philosophy of Franz Rosenzweig*, edited by Paul Mendes-Flohr, 124–37. Hanover, NH: University Press of New England, 1988.
Ash, Mitchell G. *Gestalt Psychology in German Culture, 1890–1967: Holism and the Quest for Objectivity*. Cambridge: Cambridge University Press, 1995.
Assmann, Aleida. *Is Time out of Joint? On the Rise and Fall of the Modern Time Regime*. Translated by Sarah Clift. Ithaca, NY: Cornell University Press, 2020.

Auerbach, Erich. "Figura (1939)." In *Gesammelte Aufsätze zur romanischen Philologie*, edited by Fritz Schalk, 55–92. Bern: Francke, 1967.
Balthasar, Hans Urs von. *The Theology of Karl Barth*. Translated by Edward T. Oakes. San Francisco: Ignatius Press, 1992.
Barile, Margherita, and Eric W. Weisstein. "Arc." *MathWorld—A Wolfram Web Resource*. https://mathworld.wolfram.com/Arc.html.
Barnouw, Dagmar. *Critical Realism: History, Photography, and the Work of Siegfried Kracauer*. Baltimore: Johns Hopkins University Press, 1994.
Barth, Karl. *Die Auferstehung der Toten: Eine akademische Vorlesung über 1. Kor. 15*. Munich: Kaiser, 1924.
———. *The Epistle to the Romans*. Translated by Edwyn C. Hoskyns. London: Oxford University Press, 1933.
———. *Ethik II: Vorlesung Münster Wintersemester 1928/29, wiederholt in Bonn, Wintersemester 1930/31*. In *Gesamtausgabe*, vol. 2, edited by Dietrich Braun. Zurich: Theologischer Verlag Zürich, 1978.
———. "Not und Verheißung der christlichen Verkündigung." *Zwischen den Zeiten* 1 (1923): 3–25.
———. "Das Problem der Ethik in der Gegenwart." *Zwischen den Zeiten* 2 (1923): 30–57.
———. *Der Römerbrief (zweite Fassung), 1922*. 17th ed. Zurich: Theologischer Verlag Zürich, 2005.
Barth, Karl, and Eduard Thurneysen. *Zur inneren Lage des Christentums: Eine Buchanzeige und eine Predigt*. Munich: Kaiser, 1920.
Baumann, Stephanie. "Drei Briefe: Franz Rosenzweig an Siegfried Kracauer." *Zeitschrift für Religions- und Geistesgeschichte* 63, no. 2 (2011): 166–76.
———. *Im Vorraum der Geschichte: Siegfried Kracauers "History. The Last Things before the Last."* Konstanz: Konstanz University Press, 2014.
Beckett, Samuel. "Watt." In *The Grove Centenary Edition*, vol. 1, edited by Paul Auster, 169–379. New York: Grove Press, 2006.
Beintker, Michael. *Die Dialektik in der "dialektischen Theologie" Karl Barths: Studien zur Entwicklung der Barthschen Theologie und zur Vorgeschichte der "Kirchlichen Dogmatik."* Munich: Kaiser, 1987.
Beiser, Frederick C. *The German Historicist Tradition*. Oxford: Oxford University Press, 2011.
Bell, John L. *The Continuous, the Discrete and the Infinitesimal in Philosophy and Mathematics*. Cham, Switzerland: Springer, 2019.
Beltrami, Eugenio. "Essay on the Interpretation of Non-Euclidean Geometry." In *Sources of Hyperbolic Geometry*, edited by John Stillwell, 7–34. Providence: American Mathematical Society, 1996.
Bendels, Ruth. *Erzählen zwischen Hilbert und Einstein: Naturwissenschaft und Literatur in Hermann Brochs "Eine methodologische Novelle" und Robert Musils "Drei Frauen."* Würzburg: Königshausen & Neumann, 2008.
Benjamin, Walter. *The Arcades Project*. Translated by Howard Eiland and Kevin McLaughlin. Cambridge, MA: Belknap Press of Harvard University Press, 1999.

———. *Das Passagen-Werk*. In *Gesammelte Schriften*, vol. 5, edited by Rolf Tiedemann and Hermann Schweppenhäuser. Frankfurt am Main: Suhrkamp, 1980.

———. "Über den Begriff der Geschichte." In *Gesammelte Schriften*, vol. 1, edited by Rolf Tiedemann and Hermann Schweppenhäuser, 691–704. Frankfurt am Main: Suhrkamp, 1980.

Benn, Gottfried. "Brief an Friedrich Wilhelm Oelze, 2. Oktober 1936." In *Die Zeit*, July 1, 1977. https://www.zeit.de/1977/27/erkenne-die-lage-rechne-mit-deinen-defekten.

Berlin, Isaiah. "History and Theory: The Concept of Scientific History." *History and Theory* 1, no. 1 (1960): 1–31.

Bernstein, Michael André. *Foregone Conclusions: Against Apocalyptic History*. Berkeley: University of California Press, 1994.

Bieberich, Ulrich. *Wenn die Geschichte göttlich wäre: Rosenzweigs Auseinandersetzung mit Hegel*. St. Ottilien: EOS, 1990.

Bloch, Ernst. *Briefe, 1903–1975*, vol. 1. Edited by Karola Bloch, Jan Robert Bloch, Anne Fromman, Hanna Gekle, Inge Jens, Martin Korol, Inka Mülder-Bach, Arno Münster, Uwe Opolka, and Burghart Schmidt. Frankfurt am Main: Suhrkamp, 1985.

———. *Erbschaft dieser Zeit*. Zurich: Oprecht & Helbling, 1935.

———. *Heritage of Our Times*. Translated by Neville and Stephen Plaice. Oxford: Polity, 1991.

Blumenberg, Hans. *Arbeit am Mythos*. Frankfurt am Main: Suhrkamp, 1979.

———. *The Legitimacy of the Modern Age*. Translated by Robert M. Wallace. Cambridge, MA: MIT Press, 1983.

———. *Die Legitimität der Neuzeit*. 3rd ed. Frankfurt am Main: Suhrkamp, 1997.

———. *Work on Myth*. Translated by Robert M. Wallace. Cambridge, MA: MIT Press, 1985.

Bodenstein, Walter. *Neige des Historismus: Ernst Troeltschs Entwicklungsgang*. Gütersloh: Gütersloher Verlagshaus, 1959.

Böhme, Hartmut. "Die 'Zeit ohne Eigenschaften' und die 'Neue Unübersichtlichkeit': Robert Musil und die posthistoire." In *Kunst, Wissenschaft und Politik von Robert Musil bis Ingeborg Bachmann*, edited by Josef Strutz, 9–33. Munich: Fink, 1986.

Bohrer, Karl Heinz. "Utopia of the Moment and Fictionality: The Subjectivization of Time in Modern Literature." In *Suddenness: On the Moment of Aesthetic Appearance*, translated by Ruth Crowley, 197–245. New York: Columbia University Press, 1994.

Borges, Jorge Luis. "The Fearful Sphere of Pascal." In *Labyrinths, Selected Stories and Other Writings*, 189–92. New York: New Directions, 1964.

Boyd, Ian R. *Dogmatics among the Ruins: German Expressionism and the Enlightenment as Contexts for Karl Barth's Theological Development*. Oxford: Peter Lang, 2004.

Braiterman, Zachary. *The Shape of Revelation: Aesthetics and Modern Jewish Thought*. Stanford, CA: Stanford University Press, 2007.

Brentano, Franz. *Über Aristotles: Nachgelassene Aufsätze*. Edited by Rolf George. Hamburg: Meiner, 1986.

Bressoud, David M. *Calculus Reordered: A History of the Big Ideas*. Princeton, NJ: Princeton University Press, 2019.

Brunner, Emil. *Erlebnis, Erkenntnis und Glaube*. Tübingen: Mohr Siebeck, 1923.

———. *Die Mystik und das Wort: Der Gegensatz zwischen moderner Religionsauffassung und christlichem Glauben dargestellt an der Theologie Schleiermachers*. 2nd ed. Tübingen: Mohr Siebeck, 1928.

Bud, Robert, and Morag Shiach. "Being Modern: Introduction." In *Being Modern: The Cultural Impact of Science in the Early Twentieth Century*, edited by Robert Bud, Paul Greenhalgh, Frank James, and Morag Shiach, 1–19. London: University College of London Press, 2018.

Butler, Judith, Jürgen Habermas, Charles Taylor, and Cornell West. *The Power of Religion in the Public Sphere*. Edited by Eduardo Mendieta and Jonathan Van Antwerpen. New York: Columbia University Press, 2011.

Cassirer, Ernst. *The Myth of the State*. New Haven, CT: Yale University Press, 1946.

Clark, T. J. *Farewell to an Idea: Episodes from a History of Modernism*. New Haven, CT: Yale University Press, 1999.

Cliver, Gwyneth E. "Maddening Mathematics: The Kinship of the Rational and the Irrational in the Writing of Robert Musil." *Journal of Romance Studies* 7, no. 3 (2007): 75–85.

Cohen, Hermann. "Die Messiasidee." In *Hermann Cohens Jüdische Schriften*, vol. 1, edited by Bruno Strauß, 105–24. Berlin: C. A. Schwetschke & Sohn, 1924.

———. *Die Religion der Vernunft aus den Quellen des Judentums*. Leipzig: Gustav Fock, 1919.

Coolidge, J. L. "The Story of Tangents." *American Mathematical Monthly* 58, no. 7 (1951): 449–62.

Corngold, Stanley. "Kafka (with Nietzsche) as Neo-Gnostic Thinkers." In *Franz Kafka: The Ghosts in the Machine*, edited by Stanley Corngold and Benno Wagner, 151–78. Evanston, IL: Northwestern University Press, 2011.

Craver, Harry T. *Reluctant Skeptic: Siegfried Kracauer and the Crises of Weimar Culture*. New York: Berghahn, 2017.

Cremer, Douglas. "Protestant Theology in Early Weimar Germany: Barth, Tillich, and Bultmann." *Journal of the History of Ideas* 56, no. 2 (1995): 289–307.

Cremona, Luigi. *Elemente des graphischen Calculs*. Translated by Maximilian Curtze. Leipzig: Quandt & Händel, 1875.

De Vries, Hent. "Inverse versus Dialectical Theology: The Two Faces of Negativity and the Miracle of Faith." In *Paul and the Philosophers*, edited by Ward Blanton and Hent de Vries, 466–511. New York: Fordham University Press, 2013.

———. *Philosophy and the Turn to Religion*. Baltimore: Johns Hopkins University Press, 1999.

De Vries, Hent, and Lawrence E. Sullivan, eds. *Political Theologies: Public Religions in a Post-Secular World*. New York: Fordham University Press, 2006.
Dickinson, Colby, and Stéphane Symons, eds. *Walter Benjamin and Theology*. New York: Fordham University Press, 2016.
Dilthey, Wilhelm. *Der Aufbau der geschichtlichen Welt in den Geisteswissenschaften*. Berlin: Verlag der Königlichen Akademie der Wissenschaften, 1910.
Döring, Jörg, and Tristan Thielmann. "Einleitung: Was lesen wir im Raume? Der Spatial Turn und das geheime Wissen der Geographen." In *Spatial Turn: Das Raumparadigma in den Kultur- und Sozialwissenschaften*, edited by Jörg Döring and Tristan Thielmann, 7–45. Bielefeld: Transcript, 2008.
Dostoevsky, Fyodor. *The Brothers Karamazov*. Translated by Constance Garnett. New York: Macmillan, 1922.
Dostojewski, Fjodor Michailowitsch. *Die Brüder Karamasoff*. Translated by E. K. Rahsin. Munich: Piper, 1914.
Droysen, Johann Gustav. *Grundriss der Historik*. Leipzig: Veit, 1868.
———. *Texte zur Geschichtstheorie*. Edited by Günther Birtsch and Jörn Rüsen. Göttingen: Vandenhoeck & Ruprecht, 1972.
———. "Theologie der Geschichte: Vorwort zur Geschichte des Hellenismus II (1843)." In *Historik: Vorlesungen über Enzyklopädie und Methodologie der Geschichte*, edited by Rudolf Hübner, 369–85. Darmstadt: Wissenschaftliche Buchgesellschaft, 1960.
Drügh, Heinz. *Anders-Rede: Zur Struktur und historischen Systematik des Allegorischen*. Freiburg im Breisgau: Rombach, 2000.
Eagleton, Terry. *The Ideology of the Aesthetic*. Oxford: Basil Blackwell, 1990.
Edgerton, Samuel Y. *The Mirror, the Window, and the Telescope: How Renaissance Linear Perspective Changed Our Vision of the Universe*. Ithaca, NY: Cornell University Press, 2009.
Edwards, C. H. *The Historical Development of the Calculus*. New York: Springer, 1979.
Engel, Manfred, and Ritchie Robertson, eds. *Kafka, Religion, and Modernity*. Würzburg: Königshausen & Neumann, 2014.
Engelhardt, Nina. *Modernism, Fiction and Mathematics*. Edinburgh: Edinburgh University Press, 2018.
Fanta, Walter. "The 'Finale' of Der Mann ohne Eigenschaften: Competing Editions and the 'Telos' of the Narrative." In *A Companion to the Works of Robert Musil*, edited by Philip Payne, Graham Bartram, and Galin Tihanov, 371–94. Rochester, NY: Camden House, 2007.
———. "Statt Religion Literatur, statt Literaturwissenschaft Theologie: Zum Gottesbegriff bei Robert Musil." In *Die Gottesfrage in der europäischen Philosophie und Literatur des 20. Jahrhunderts*, edited by Rudolf Langthaler and Wolfgang Treitler, 187–205. Vienna: Böhlau, 2007.
Fenves, Peter. *The Messianic Reduction: Walter Benjamin and the Shape of Time*. Stanford, CA: Stanford University Press, 2011.
Frank, Joseph. "Spatial Form in Modern Literature: An Essay in Two Parts." *Sewanee Review* 53, no. 2 (1945): 221–40.

Friedrich, Lars. "Favor testamenti: Letztwillige Verfügungen in Musils 'Mann ohne Eigenschaften.'" In *Urteilen/Entscheiden*, edited by Cornelia Visman and Thomas Weitin, 72–90. Munich: Fink, 2006.

Frisch, Ralf. *Theologie im Augenblick ihres Sturzes: Theodor W. Adorno und Karl Barth; zwei Gestalten einer kritischen Theorie der Moderne*. Vienna: Passagen, 1999.

Fritsch, Matthias. *The Promise of Memory: History and Politics in Marx, Benjamin, and Derrida*. Albany: SUNY Press, 2005.

Fuchs, Anne. *Precarious Times: Temporality and History in Modern German Culture*. Ithaca, NY: Cornell University Press, 2019.

Gessmann, Martin. *Montaigne und die Moderne: Zu den philosophischen Grundlagen einer Epochenwende*. Hamburg: Meiner, 1997.

Gibbs, Robert. "Lines, Circles, Points: Messianic Epistemology in Cohen, Rosenzweig and Benjamin." In *Toward the Millennium: Messianic Expectations from the Bible to Waco*, edited by Peter Schäfer and Mark Cohen, 363–82. Leiden: Brill, 1998.

Giles, Steve. "Making Visible, Making Strange: Photography and Representation in Kracauer, Brecht and Benjamin." *New Formations: A Journal of Culture/Theory/Politics* 61 (2007): 64–75.

Glazova, Anna, and Paul North, eds. *Messianic Thought outside Theology*. New York: Fordham University Press, 2014.

Gogarten, Friedrich. "Zwischen den Zeiten (1920)." In *Anfänge der dialektischen Theologie; Teil II: Rudolf Bultmann, Friedrich Gogarten, Eduard Thurneysen*, edited by Jürgen Moltmann, 95–101. Munich: Kaiser, 1967.

Gordon, Peter Eli. *Rosenzweig and Heidegger: Between Judaism and German Philosophy*. Berkeley: University of California Press, 2003.

———. "Science, Finitude, and Infinity: Neo-Kantianism and the Birth of Existentialism." *Jewish Social Studies* 6, no. 1 (1999): 30–53.

Gray, Jeremy J. *János Bolyai, Non-Euclidean Geometry, and the Nature of Space*. Cambridge: Burndy Library, 2004.

———. "Modernism in Mathematics." In *The Oxford Handbook of the History of Mathematics*, edited by Eleanor Robson and Jacqueline Stedall, 663–83. Oxford: Oxford University Press, 2009.

———. *Plato's Ghost: The Modernist Transformation of Mathematics*. Princeton, NJ: Princeton University Press, 2008.

———. *Worlds out of Nothing: A Course in the History of Geometry in the 19th Century*. London: Springer, 2011.

Grill, Genese. "The 'Other' Musil: Robert Musil and Mysticism." In *A Companion to the Works of Robert Musil*, edited by Philip Payne, Graham Bartram, and Galin Tihanov, 333–54. Rochester, NY: Camden House, 2007.

Gropius, Walter. *Die neue Architektur und das Bauhaus: Grundzüge und Entwicklung einer Konzeption*. Edited by Hans M. Wingler. Mainz: Florian Kupferberg, 1965.

Grossheim, Michael. "Erkennen oder Entscheiden: Der Begriff der 'Situation' zwischen theoretischer und praktischer Philosophie." In *Internationales Jahr-*

buch für Hermeneutik, vol. 1, edited by Günter Figal, 279–300. Tübingen: Mohr Siebeck, 2002.

Gualtieri, Elena. "The Territory of Photography: Between Modernity and Utopia in Kracauer's Thought." *New Formations: A Journal of Culture/Theory/Politics* 61 (2007): 76–89.

Günzel, Stephan. "Physik und Metaphysik des Raums: Einleitung." In *Raumtheorie: Grundlagentexte aus Philosophie und Kulturwissenschaften*, edited by Jörg Dünne and Stephan Günzel, 19–43. Frankfurt am Main: Suhrkamp, 2006.

——. "Raum—Topographie—Topologie." In *Topologie: Zur Raumbeschreibung in den Kultur- und Medienwissenschaften*, edited by Stephan Günzel, 13–29. Bielefeld: Transcript, 2007.

Hahn, Hans. "Die Krise der Anschauung." In *Krise und Neuaufbau in den exakten Wissenschaften: Fünf Wiener Vorträge*, 41–64. Leipzig and Vienna: Franz Deuticke, 1933.

Handelman, Matthew. *The Mathematical Imagination: On the Origins and Promise of Critical Theory*. New York: Fordham University Press, 2019.

Hansen, Miriam Bratu. *Cinema and Experience: Siegfried Kracauer, Walter Benjamin, and Theodor W. Adorno*. Berkeley: University of California Press, 2012.

——. "Decentric Perspectives: Kracauer's Early Writings on Film and Mass Culture." *New German Critique* 54 (1991): 47–76.

——. "Introduction." In Siegfried Kracauer, *Theory of Film: The Redemption of Physical Reality*, vii–xlv. Princeton, NJ: Princeton University Press, 1997.

——. "Kracauer's Photography Essay: Dot Matrix—General (An-)Archive—Film." In *Culture in the Anteroom: The Legacies of Siegfried Kracauer*, edited by Gerd Gemünden and Johannes von Moltke, 93–110. Ann Arbor: University of Michigan Press, 2012.

Harries, Karsten. "The Infinite Sphere: Comments on the History of a Metaphor." *Journal of the History of Philosophy* 13, no. 1 (1975): 5–15.

Hart-Nibbrig, Christian. "'Die Weltgeschichte ist das Weltgericht': Zur Aktualität von Schillers ästhetischer Geschichtsdeutung." *Jahrbuch der deutschen Schillergesellschaft* 20 (1976): 255–77.

Harvey, David. *The Condition of Postmodernity: An Enquiry into the Origins of Cultural Change*. Oxford: Basil Blackwell, 1989.

Hausdorff, Felix. *Grundzüge der Mengenlehre*. Leipzig: Veit, 1914.

Hegel, Georg Wilhelm Friedrich. *Enzyklopädie der philosophischen Wissenschaften im Grundrisse (1830)*. Edited by Friedhelm Nicolin and Otto Pöggeler. Hamburg: Meiner, 1991.

——. *Grundlinien der Philosophie des Rechts*. In *Gesammelte Werke*, vol. 14, no. 1, edited by Klaus Grotsch and Elisabeth Weisser-Lohmann. Hamburg: Meiner, 2009.

——. *Vorlesungen über die Philosophie der Geschichte*. In *Werke*, vol. 12, edited by Eva Moldenhauer and Karl Markus Michel. Frankfurt am Main: Suhrkamp, 1970.

Heidegger, Martin. *Phänomenologie des religiösen Lebens (1920/1921)*. In *Gesamtausgaube*, vol. 60. Frankfurt am Main: Vittorio Klostermann, 1995.
——. *Sein und Zeit*. 11th ed. Tübingen: Niemeyer, 1967.
Heine, Heinrich. "Lutezia II." In *Historisch-kritische Gesamtausgabe der Werke*, edited by Manfred Windfuhr, 9–98. Hamburg: Hoffmann und Campe, 1990.
Heinzmann, Gerhard. "Hypotheses and Conventions in Poincaré." In *The Significance of the Hypothetical in the Natural Sciences*, edited by Michael Heidelberger and Gregor Schiemann, 169–92. Berlin: De Gruyter, 2009.
Helmholtz, Hermann von. "Über den Ursprung und die Bedeutung der geometrischen Axiome: Vortrag gehalten im Docentenverein zu Heidelberg 1870." In *Vorträge und Reden*, vol. 2, 1–31. Braunschweig: Vieweg, 1883.
Henderson, Linda D. *The Fourth Dimension and Non-Euclidean Geometry in Modern Art*. Cambridge, MA: MIT Press, 2013.
Herder, Johann Gottfried. "Auch eine Philosophie der Geschichte zur Bildung der Menschheit (1777)." In *Werke*, vol. 4, edited by Jürgen Brummack and Martin Bollacher, 9–107. Frankfurt am Main: Deutscher Klassiker Verlag, 1994.
——. *Metakritik zur Kritik der reinen Vernunft*. Berlin: Aufbau, 1955.
Heussi, Karl. *Die Krisis des Historismus*. Tübingen: Mohr Siebeck, 1932.
Hilbert, David. "Mathematische Probleme." In *Die Hilbertschen Probleme*, edited by P. S. Alexandrov, 22–80. Leipzig: Akademische Verlagsgesellschaft Geest & Portig, 1971.
Hinton, Charles H. "What Is the Fourth Dimension? (1884)." In *Speculations on the Fourth Dimension: Selected Writings by Charles H. Hinton*, edited by Rudolf v. B. Rucker, 1–22. New York: Dover, 1980.
Hollander, Dana. "On the Significance of the Messianic Idea in Rosenzweig." *Cross Currents* 53, no. 4 (2004): 555–65.
Honold, Alexander. "Endings and Beginnings: Musil's Invention of Austrian History." In *Robert Musil's The Man without Qualities*, edited by Harold Bloom, 113–22. Philadelphia: Chelsea House, 2005.
——. *Die Stadt und der Krieg: Raum- und Zeitkonstruktion in Robert Musils Roman "Der Mann ohne Eigenschaften."* Munich: Fink, 1995.
Huyssen, Andreas. *Miniature Metropolis: Literature in an Age of Photography and Film*. Cambridge, MA: Harvard University Press, 2015.
——. "Modernist Miniatures: Literary Snapshots of Urban Spaces." *PMLA* 122, no. 1 (2007): 27–42.
——. "The Urban Miniature and the Feuilleton in Kracauer and Benjamin." In *Culture in the Anteroom: The Legacies of Siegfried Kracauer*, edited by Gerd Gemünden and Johannes von Moltke, 213–25. Ann Arbor: University of Michigan Press, 2012.
Hyams, Barbara F. "Was ist 'säkularisierte Mystik' bei Musil?" In *Robert Musil: Untersuchungen*, edited by Uwe Baur and Elisabeth Castex, 85–98. Königstein im Taunus: Athenäum, 1980.
Innes, Keith. "Towards an Ecological Eschatology: Continuity and Discontinuity." *Evangelical Quarterly* 81, no. 2 (2009): 126–44.

Itkin, Alan. "Orpheus, Perseus, Ahasuerus: Reflection and Representation in Siegfried Kracauer's Underworlds of History." *Germanic Review* 87 (2012): 175–202.
Jaspers, Karl. *Die geistige Situation der Zeit*. Berlin: De Gruyter, 1931.
Jennings, Michael W. "Towards Eschatology: The Development of Walter Benjamin's Theological Politics in the Early 1920s." In *Walter Benjamins anthropologisches Denken*, edited by Carolin Duttlinger, Ben Morgan, and Anthony Phelan, 41–57. Freiburg: Rombach, 2012.
Jüngel, Eberhard. *Ganz werden: Theologische Erörterungen V*. Tübingen: Mohr Siebeck, 2003.
Kafka, Franz. "Prometheus." In *Kafka's Selected Stories*, translated by Stanley Corngold, 129. New York: W. W. Norton, 2007.
———. "Prometheus." In *Nachgelassene Schriften und Fragmente*, vol. 2, edited by Jost Schillemeit, 69–70. Frankfurt am Main: Fischer, 1992.
Kant, Immanuel. *Critique of Pure Reason*. Translated by Werner S. Pluhar. Indianapolis: Hackett, 1996.
———. "Das Ende aller Dinge." In *Was ist Aufklärung? Ausgewählte kleine Schriften*, edited by Horst D. Brandt, 62–76. Hamburg: Meiner, 1999.
———. "Idee zu einer allgemeinen Geschichte in weltbürgerlicher Absicht." In *Was ist Aufklärung? Ausgewählte kleine Schriften*, edited by Horst D. Brandt, 3–19. Hamburg: Meiner, 1999.
———. *Kritik der reinen Vernunft*. Edited by Jens Timmermann. Hamburg: Meiner, 1998.
———. *Kritik der Urteilskraft*. In *Gesammelte Schriften*, vol. 5, edited by Königlich Preußische Akademie der Wissenschaften. Berlin: Reimer, 1908.
Kantorowicz, Ernst H. *The King's Two Bodies: A Study in Mediaeval Political Theology*. Princeton, NJ: Princeton University Press, 1957.
Kaplan, Gregory. "In the End Shall Christians Become Jews and Jews, Christians? On Franz Rosenzweig's Apocalyptic Eschatology." *Crosscurrents* 53, no. 4 (2004): 511–29.
Kaplan, Leonard V., and Rudy Koshar, eds. *The Weimar Moment: Liberalism, Political Theology, and Law*. Lanham, MD: Lexington Books, 2012.
Kassung, Christian. *Entropie-Geschichten: Robert Musils "Der Mann ohne Eigenschaften" im Diskurs der modernen Physik*. Munich: Fink, 2001.
Kermode, Frank. *The Sense of an Ending: Studies in the Theory of Fiction*. Oxford: Oxford University Press, 1966.
Kessler, Michael. "Entschleiern und Bewahren: Siegfried Kracauers Ansätze für eine Philosophie und Theologie der Geschichte." In *Siegfried Kracauer: Neue Interpretationen*, edited by Michael Kessler and Thomas Y. Levin, 105–28. Tübingen: Stauffenburg, 1990.
Kierkegaard, Søren. "Philosophische Brocken (1844)." In *Gesammelte Werke*, vol. 10, translated by Emanuel Hirsch. Düsseldorf: Eugen Diederichs, 1952.
Klein, Felix. *Vorlesungen über Nicht-Euklidische Geometrie*. Berlin: Springer, 1928.

Klén, Riku, and Matti Vuorinen. "Apollonian Circles and Hyperbolic Geometry." *Journal of Analysis* 19 (2011): 41–60.

Koch, Gertrud. "'Not Yet Accepted Anywhere': Exile, Memory, and Image in Kracauer's Conception of History." *New German Critique* 54 (1991): 95–109.

———. *Siegfried Kracauer: An Introduction*. Translated by Jeremy Gaines. Princeton, NJ: Princeton University Press, 2000.

Koselleck, Reinhart. "Geschichte." In *Geschichtliche Grundbegriffe: Historisches Lexikon zur politisch-sozialen Sprache in Deutschland*, vol. 2, 593–718. Stuttgart: Ernst Klett, 1975.

———. *Futures Past: On the Semantics of Historical Time*. Translated by Keith Tribe. Cambridge, MA: MIT Press, 1985.

———. *Zeitschichten: Studien zur Historik*. Frankfurt am Main: Suhrkamp, 2000.

Koshar, Rudy. "Where Is Karl Barth in Modern European History?" *Modern Intellectual History* 5, no. 2 (2008): 333–62.

Kracauer, Siegfried. "Aus dem Fenster gesehen." In *Straßen in Berlin und anderswo*, 53–55. Frankfurt am Main: Suhrkamp, 2009.

———. *Der Detektiv-Roman: Eine Deutung*. In *Werke*, vol. 1, edited by Inka Mülder-Bach, 103–209. Frankfurt am Main: Suhrkamp, 2006.

———. "Gestalt und Zerfall." In *Werke*, vol. 5, no. 2, edited by Inka Mülder Bach, 283–88. Frankfurt am Main: Suhrkamp, 2004.

———. *History: The Last Things before the Last*. New York: Oxford University Press, 1969.

———. *The Mass Ornament: Weimar Essays*. Translated by Thomas Y. Levin. Cambridge, MA: Harvard University Press, 1995.

———. *Das Ornament der Masse: Essays*. Frankfurt am Main: Suhrkamp, 1963.

———. "Zur religiösen Lage in Deutschland (1924)." In *Werke*, vol. 5, no. 2, edited by Inke Mülder-Bach, 155–59. Berlin: Suhrkamp, 2011.

———. *Soziologie als Wissenschaft*. In *Werke*, vol. 1, edited by Inka Mülder-Bach, 9–101. Frankfurt am Main: Suhrkamp, 2006.

———. *Theory of Film: The Redemption of Physical Reality*. Princeton, NJ: Princeton University Press, 1997.

———. "Time and History." In *Werke*, vol. 4, edited by Inka Mülder-Bach, 377–93. Frankfurt am Main: Suhrkamp, 2009.

———. "Two Planes." In *The Mass Ornament: Weimar Essays*, translated by Thomas Y. Levin, 37–39. Cambridge, MA: Harvard University Press, 1995.

———. "Zwei Arten der Mitteilung." In *Werke*, vol. 5, no. 3, edited by Inka Mülder-Bach, 180–87. Berlin: Suhrkamp, 2011.

———. "Zwei Flächen." In *Das Ornament der Masse: Essays*, 11–13. Frankfurt am Main: Suhrkamp, 1963.

———. "Zwischen den Zeiten (1923)." In *Werke*, vol. 5, no. 1, edited by Inka Mülder-Bach, 563–64, 634–36. Berlin: Suhrkamp, 2011.

Kraus, Justice. "Musil's *Die Verwirrungen des Zöglings Törleß*, Cantor's Structures of Infinity, and Brouwer's Mathematical Language." *Scientia Poetica:*

Jahrbuch für Geschichte der Literatur und der Wissenschaften 14 (2010): 72–103.
Kreck, Walter. *Die Zukunft des Gekommenen: Grundprobleme der Eschatologie*. Munich: Kaiser, 1961.
Kreienbrock, Jörg. "Erkenne die Lage! Medien des Konkreten im Nachkrieg." Presentation at the Institut für Medienwissenschaft, Ruhr-Universität Bochum, April 19, 2016. https://ifmlog.blogs.ruhr-uni-bochum.de/forschung/mediendenken/kreienbrock-erkenne-die-lage/.
Kristeller, Paul Oskar. "Foreward." In Siegfried Kracauer, *History: The Last Things before the Last*, v–x. New York: Oxford University Press, 1969.
Kubler, George. *The Shape of Time: Remarks on the History of Things*. New Haven, CT: Yale University Press, 1962.
Lash, Scott. "Deforming the Figure: Topology and the Social Imaginary." *Theory, Culture & Society* 29, no. 4/5 (2012): 261–87.
Laube, Reinhard. *Karl Mannheim und die Krise des Historismus: Historismus als wissenssoziologischer Perspektivismus*. Göttingen: Vandenhoeck & Ruprecht, 2004.
———. "Zwischen Budapester und Berliner Historismus: Eine Pathologie der 'Krise des Historismus' aus der Sicht eines ungarischen Emigranten." In *Krise des Historismus, Krise der Wirklichkeit: Wissenschaft, Kunst und Literatur, 1880–1932*, edited by Otto Gerhard Oexle, 207–46. Göttingen: Vandenhoeck & Ruprecht, 2007.
Lazier, Benjamin. *God Interrupted: Heresy and the European Imagination between the World Wars*. Princeton, NJ: Princeton University Press, 2008.
Leslie, Esther. "Kracauer's Weimar Geometry and Geomancy." *New Formations: A Journal of Culture/Theory/Politics* 61 (2007): 34–48.
Liang, Hong. *Leben vor den letzten Dingen: Die Dostojewski-Rezeption im frühen Werk von Karl Barth und Eduard Thurneysen (1915–1923)*. Neukirchen-Vluyn: Neukirchener Theologie, 2016.
Liddell, Henry George, and Robert Scott. *A Greek-English Lexicon*. 8th ed. New York: Harper & Brothers, 1897.
Lissitzky, El. "Kunst und Pangeometrie (1925)." In *Rußland: Architektur für eine Weltrevolution*, 122–29. Braunschweig: Vieweg & Sohn, 1989.
———. "Proun." *De Stijl* 5, no. 6 (1922): 81–85.
Lowenthal, David. *The Past Is a Foreign Country*. Cambridge: Cambridge University Press, 1985.
Löwith, Karl. *Meaning in History*. Chicago: University of Chicago Press, 1949.
———. *Weltgeschichte und Heilsgeschehen: Die theologischen Voraussetzungen der Geschichtsphilosophie*. Stuttgart: Kohlhammer, 1953.
Lucas, Hans-Christian. "Die Weltgeschichte als das Weltgericht: Zur Modifikation von Hegels Geschichtsbegriff in Heidelberg." *Hegel-Jahrbuch* (1981/82): 82–96.
Lury, Celia, Luciana Parisi, and Tiziana Terranova. "Introduction: The Becoming Topological of Culture." *Theory, Culture & Society* 29, no. 4/5 (2012): 3–35.

Luther, Martin. *Die gantze heilige Schrift*, vol. 2, edited by Hans Volz. Bonn: Lempertz, 2004.

Mach, Ernst. *Space and Geometry in the Light of Physiological, Psychological and Physical Inquiry*. Translated by Thomas J. McCormack. La Salle, IL: Open Court, 1906.

Magee, Glenn. "Hegel's Philosophy of History and Kabbalist Eschatology." In *Hegel and History*, edited by Will Dudley, 231–46. Albany: SUNY Press, 2009.

Mannheim, Karl. "Historismus." *Archiv für Sozialwissenschaft und Sozialpolitik* 52, no. 1 (1924): 1–60.

Marquardt, Friedrich-Wilhelm. *Theologie und Sozialismus: Das Beispiel Karl Barths*. Munich: Kaiser, 1972.

Mattson, Brian G. "Bavinck's 'Revelation and the Future': A Centennial Retrospective." *Kuyper Center Review* 2 (2011): 126–56.

McBride, Patrizia. *The Void of Ethics: Robert Musil and the Experience of Modernity*. Evanston, IL: Northwestern University Press, 2006.

McCarthy, John A. "Disciplining History: Schiller als Historiograph." *Goethe Yearbook* 12 (2004): 209–25.

McCormack, Bruce L. *Karl Barth's Critically Realistic Dialectical Theology*. Oxford: Oxford University Press, 1995.

McDowell, John C. *Hope in Barth's Eschatology: Interrogations and Transformations beyond Tragedy*. Aldershot: Ashgate, 2000.

McGillen, Michael. "Erich Auerbach and the Seriality of the Figure." *New German Critique* 45, no. 1 (2018): 111–54.

———. "Eschatology and the Reinvention of History: Theological Interventions in German Modernism, 1920–1938." PhD diss., Princeton University, 2012.

———. "Theology's Weimar Moment: History before the Eschatological Limit." In *The Weimar Moment: Liberalism, Political Theology, and Law*, edited by Leonard Kaplan and Rudy Koshar, 267–87. Lanham, MD: Lexington Books, 2012.

Mehrtens, Herbert. *Moderne—Sprache—Mathematik: Eine Geschichte des Streits um die Grundlagen der Disziplin und des Subjekts formaler Systeme*. Frankfurt am Main: Suhrkamp, 1990.

Meier, Hans, Richard Newald, and Edgar Wind, eds. *Kulturwissenschaftliche Bibliographie zum Nachleben der Antike*. Leipzig: B. G. Teubner, 1934.

Meisel, Gerhard. "Verkehr und Entropie in Musils 'Kakanien.'" In *Medien und Maschinen: Literatur im technischen Zeitalter*, edited by Theo Elm and Hans Hiebel, 304–42. Freiburg: Rombach, 1991.

Mendes-Flohr, Paul. "Franz Rosenzweig and the Crisis of Historicism." In *The Philosophy of Franz Rosenzweig*, edited by Paul Mendes-Flohr, 138–61. Hanover, NH: University Press of New England, 1988.

Meyers Großes Konversations-Lexikon, vol. 11. Leipzig: Bibliographisches Institut, 1907.

Michael, Mike, and Marsha Rosengarten. "HIV, Globalization and Topology: Of Prepositions and Propositions." *Theory, Culture & Society* 29, no. 4/5 (2012): 93–115.

Mosès, Stéphane. *The Angel of History: Rosenzweig, Benjamin, Scholem*. Translated by Barbara Harshav. Stanford, CA: Stanford University Press, 2009.
———. "Hegel beim Wort genommen: Geschichtskritik bei Franz Rosenzweig." In *Zeitgewinn: Messianisches Denken nach Franz Rosenzweig*, edited by Gotthard Fuchs and Hans Hermann Henrix, 67–89. Frankfurt am Main: Josef Knecht, 1987.
Moyn, Samuel. "Is Revelation in the World?" *Jewish Quarterly Review* 96, no. 3 (2006): 396–403.
Mülder-Bach, Inka. "History as Autobiography: The Last Things before the Last." *New German Critique* 54 (1991): 139–57.
———. "Schlupflöcher: Die Diskontinuität des Kontinuierlichen im Werk Siegfried Kracauers." In *Siegfried Kracauer: Neue Interpretationen*, edited by Michael Kessler and Thomas Y. Levin, 249–66. Tübingen: Stauffenburg, 1990.
———. *Siegfried Kracauer—Grenzgänger zwischen Theorie und Literatur: Seine frühen Schriften, 1913–1933*. Stuttgart: J. B. Metzlersche Verlagsbuchhandlung, 1985.
———. "Der 'Weg der Geschichte' oder: Finden und Erfinden: Geschichtserzählung in Robert Musils Roman *Der Mann ohne Eigenschaften*." *Internationales Archiv für Sozialgeschichte der deutschen Literatur* 36, no. 1 (2011): 187–205.
Musil, Robert. "Ansätze zu neuer Ästhetik: Bemerkungen über eine Dramaturgie des Film." In *Prosa und Stücke, Kleine Prosa, Aphorismen, Autobiographisches, Essays und Reden, Kritik*. In *Gesammelte Werke*, vol. 2, edited by Adolf Frisé, 1137–54. Reinbek bei Hamburg: Rowohlt, 1978.
———. "Der deutsche Mensch als Symptom." In *Prosa und Stücke, Kleine Prosa, Aphorismen, Autobiographisches, Essays und Reden, Kritik*. In *Gesammelte Werke*, vol. 2, edited by Adolf Frisé, 1353–1400. Reinbek bei Hamburg: Rowohlt, 1978.
———. *The Man without Qualities*. Translated by Sophie Wilkins. New York: Alfred A. Knopf, 1995.
———. *Der Mann ohne Eigenschaften*. In *Gesammelte Werke*, vol. 1, edited by Adolf Frisé. Reinbek bei Hamburg: Rowohlt, 1978.
———. *Der Mann ohne Eigenschaften*, edited by Adolf Frisé. Reinbek bei Hamburg, Rowohlt, 1952.
———. "The Mathematical Man." In *Passion and Soul: Essays and Addresses*, edited by Burton Pike and David S. Luft, 39–42. Chicago: University of Chicago Press, 1990.
———. "Der mathematische Mensch." In *Prosa und Stücke, Kleine Prosa, Aphorismen, Autobiographisches, Essays und Reden, Kritik*. In *Gesammelte Werke*, vol. 2, edited by Adolf Frisé, 1004–8. Reinbek bei Hamburg: Rowohlt, 1978.
———. *Nachlass zu Lebzeiten*. In *Prosa und Stücke, Kleine Prosa, Aphorismen, Autobiographisches, Essays und Reden, Kritik*. In *Gesammelte Werke*, vol. 2, edited by Adolf Frisé, 471–562. Reinbek bei Hamburg: Rowohlt, 1978.
———. *Prosa und Stücke, Kleine Prosa, Aphorismen, Autobiographisches, Essays und Reden, Kritik*. In *Gesammelte Werke*, vol. 2, edited by Adolf Frisé. Reinbek bei Hamburg: Rowohlt, 1978.

———. "Skizze der Erkenntnis des Dichters." In *Prosa und Stücke, Kleine Prosa, Aphorismen, Autobiographisches, Essays und Reden, Kritik*. In *Gesammelte Werke*, vol. 2, edited by Adolf Frisé, 1025–30. Reinbek bei Hamburg: Rowohlt, 1978.

———. *Tagebücher*, vol. 1, edited by Adolf Frisé. Reinbek bei Hamburg: Rowohlt, 1983.

Myers, David N. *Resisting History: Historicism and Its Discontents in German-Jewish Thought*. Princeton, NJ: Princeton University Press, 2003.

Newman, Jane O. "Auerbach's Dante: Poetical Theology as a Point of Departure for a Philology of World Literature." In *Approaches to World Literature*, edited by Joachim Küpper, 39–58. Berlin: Akademie, 2013.

Nietzsche, Friedrich. *Jenseits von Gut und Böse*. In *Sämtliche Werke: Kritische Studienausgabe*, vol. 5, edited by Giorgio Colli and Mazzino Montinari. Munich: De Gruyter, 2002.

Nitsche, Jessica. "Dem Tod ins Auge (ge)sehen: Protagonistinnen der Fotografietheorie bei Döblin, Kracauer, Barthes und Benjamin." In *Blick.Spiel.Feld*, edited by Malda Denana, Jule Hillgärtner, Eva Holling, Anneka Mezger, Matthias Naumann, Jessica Nitsche, Lars Schmid, and Silke Schuck, 93–109. Würzburg: Königshausen & Neumann, 2008.

Ottmann, Henning. "Die Weltgeschichte (§§ 341–360)." In *G.W.F. Hegel, Grundlinien der Philosophie des Rechts*, 3rd ed., edited by Ludwig Siep, 267–86. Berlin: Akademie, 2014.

Ouspensky, P. D. *Tertium Organum: The Third Canon of Thought; a Key to the Enigmas of the World*. Translated by Nicholas Bessaraboff and Claude Bragdon. New York: Alfred A. Knopf, 1922.

Overbeck, Franz. *Christentum und Kultur: Gedanken und Anmerkungen zur modernen Theologie von Franz Overbeck*. Edited by Carl Albrecht Bernoulli. Basel: Benno Schwabe, 1919.

Pecora, Vincent P. *Secularization and Cultural Criticism: Religion, Nation, and Modernity*. Chicago: University of Chicago Press, 2006.

Pike, Burton. "Der Mann ohne Eigenschaften: Unfinished or without End?" In *A Companion to the Works of Robert Musil*, edited by Philip Payne, Graham Bartram, and Galin Tihanov, 355–70. Rochester, NY: Camden House, 2007.

Poincaré, Henri. "Space and Geometry." In *Science and Hypothesis*, 51–71. London: Walter Scott, 1905.

Ranke, Leopold von. "Geschichten der romanischen und germanischen Völker." In *Sämtliche Werke*, vol. 33. Leipzig: Duncker und Humblot, 1874.

———. "Idee der Universalhistorie (1831/32)." In *Aus Werk und Nachlass*, vol. 4, edited by Volker Dotterweich and Walther Peter Fuchs, 72–89. Munich: Oldenbourg, 1975.

———. *Über die Epochen der neueren Geschichte*. Leipzig: Duncker und Humblot, 1899.

Rendtorff, Trutz. "Karl Barth und die Neuzeit: Fragen zur Barth-Forschung." *Evangelische Theologie* 46 (1986): 298–314.

Richter, Gerhard. *Afterness: Figures of Following in Modern Thought and Aesthetics*. New York: Columbia University Press, 2011.
Richter, Silvia. "Language and Eschatology in the Work of Emmanuel Levinas." *Shofar: An Interdisciplinary Journal of Jewish Studies* 26, no. 4 (2008): 54–73.
Riedel, Wolfgang. "'Weltgeschichte ein erhabenes Object': Zur Modernität von Schillers Geschichtsdenken." In *Prägnanter Moment: Studien zur deutschen Literatur der Aufklärung und Klassik*, edited by Peter-André Alt, Alexander Kosenina, and Hartmut Reinhardt, 193–214. Würzburg: Königshausen & Neumann, 2002.
Rosenzweig, Franz. *Briefe*. Edited by Ernst Simon and Edith Rosenzweig. Berlin: Schocken, 1935.
———. "Globus: Studien zur weltgeschichtlichen Raumlehre." In *Der Mensch und sein Werk: Gesammelte Schriften*, vol. 3, edited by Reinhold Mayer and Annemarie Mayer, 313–68. Dordrecht: Nijhoff, 1984.
———. *Die "Gritli"-Briefe: Briefe an Margrit Rosenstock-Huessy*. Edited by Inke Rühle and Reinhold Mayer. Tübingen: Bilam, 2002.
———. "Das neue Denken: Einige nachträgliche Bemerkungen zum 'Stern der Erlösung.'" In *Der Mensch und sein Werk: Gesammelte Schriften*, vol. 3, edited by Reinhold Mayer and Annemarie Mayer, 139–61. Dordrecht: Nijhoff, 1984.
———. "'The New Thinking': A Few Supplementary Remarks to *The Star of Redemption*." In *Franz Rosenzweig's "The New Thinking,"* edited and translated by Alan Udoff and Barbara E. Galli, 67–102. Syracuse, NY: Syracuse University Press, 1999.
———. *The Star of Redemption*. Translated by Barbara E. Galli. Madison: University of Wisconsin Press, 2005.
———. *Der Stern der Erlösung*. Frankfurt am Main: Suhrkamp, 1988.
Rosenzweig, Franz, and Eugen Rosenstock. "Franz Rosenzweig und Eugen Rosenstock: Judentum und Christentum." In Franz Rosenzweig, *Briefe*, edited by Ernst Simon and Edith Rosenzweig, 637–720. Berlin: Schocken, 1935.
———. *Judaism despite Christianity: The 1916 Wartime Correspondence between Eugen Rosenstock and Franz Rosenzweig*. Translated by Dorothy Emmet. Chicago: University of Chicago Press, 2011.
Rothacker, Erich. *Geschichtsphilosophie*. Berlin: Oldenbourg, 1934.
Scharold, Irmgard. *Epiphanie, Tierbild, Metamorphose, Passion und Eucharistie: Zur Kodierung des "Anderen" in den Werken von Robert Musil, Clarice Lispector und J.M.G. Le Clézio*. Heidelberg: Universitätsverlag C. Winter, 2000.
Schelling, Friedrich Wilhelm Joseph von. *Die Weltalter Fragmente: In den Urfassungen von 1811 und 1813*. Edited by Manfred Schröter. Munich: C. H. Beck'sche Verlagsbuchhandlung, 1993.
Schellong, Dieter. "Jacob Taubes zu Karl Barth." In *Abendländische Eschatologie: Ad Jacob Taubes*, edited by Richard Faber, Eveline Goodman-Thau, and Thomas H. Macho, 385–405. Würzburg: Königshausen & Neumann, 2001.

Schiller, Friedrich. "Resignation." In *Sämtliche Werke*, vol. 1, edited by Albert Meier, 129–33. Munich: Hanser, 1962.

———. "Was heißt und zu welchem Ende studiert man Universalgeschichte?" In *Schillers Werke: Nationalausgabe*, vol. 17, no. 1, edited by Karl-Heinz Hahn, 359–76. Weimar: Böhlau, 1970.

Schmied-Kowarzik, Wolfdietrich. "Cohen und Rosenzweig zu Vernunft und Offenbarung." In *Die Gottesfrage in der europäischen Philosophie und Literatur des 20. Jahrhunderts*, edited by Rudolf Langthaler and Wolfgang Treitler, 47–66. Vienna: Böhlau, 2007.

Schnell, Rebekka. "'. . . die plötzliche enthüllte Zärtlichkeit der Welt . . .': Liebe als ästhetische und religiöse Utopie in Robert Musils 'Der Mann ohne Eigenschaften.'" In *Emotionale Grenzgänge*, edited by Lisanne Ebert, 91–112. Würzburg: Königshausen & Neumann, 2011.

Schröter, Michael. "Weltzerfall und Rekonstruktion: Zur Physiognomik Siegfried Kracauers." *Text + Kritik* 68 (1980): 18–40.

Schweitzer, Albert. *Von Reimarus zu Wrede: Eine Geschichte der Leben-Jesu-Forschung*. Tübingen: Mohr Siebeck, 1906.

Serres, Michael, and Bruno Latour. *Conversations on Science, Culture, and Time*. Ann Arbor: University of Michigan Press, 1995.

Shields, Rob. "Cultural Topology: The Seven Bridges of Königsburg, 1736." *Theory, Culture & Society* 29, no. 4/5 (2012): 43–57.

Simonis, Annette. "'Gestalt' als ästhetische Kategorie: Transformationen eines Konzeptes vom 18. bis 20. Jahrhundert." In *Morphologie und Moderne: Goethes "anschauliches Denken" in den Geistes- und Kulturwissenschaften seit 1800*, edited by Jonas Maatsch, 245–66. Berlin: De Gruyter, 2014.

Simons, Oliver. *Raumgeschichten: Topographien der Moderne in Philosophie, Wissenschaft und Literatur*. Munich: Fink, 2007.

Smith, John H. "The Infinitesimal as Theological Principle: Representing the Paradoxes of God and Nothing in Cohen, Rosenzweig, Scholem, and Barth." *MLN* 127 (2012): 562–88.

Snyder, John P. *Flattening the Earth: Two Thousand Years of Map Projections*. Chicago: University of Chicago Press, 1993.

Spencer, Malcolm. "Violence and Love: The Search for the 'andere Zustand' in Robert Musil's *Der Mann ohne Eigenschaften*." In *Contested Passions: Sexuality, Eroticism, and Gender in Modern Austrian Literature and Culture*, edited by Clemens Ruthner and Raleigh Whitinger, 249–58. New York: Peter Lang, 2011.

Spengler, Oswald. *Der Untergang des Abendlandes: Umrisse einer Morphologie der Weltgeschichte*. Munich: C. H. Beck'sche Verlagsbuchhandlung, 1923.

Stählin, Gustav. "Nun." In *Theologisches Wörterbuch zum Neuen Testament*, vol. 4, edited by Gerhard Kittel, 1099–1117. Stuttgart: Kohlhammer, 1942.

Stillwell, John. *Sources of Hyperbolic Geometry*. Providence, RI: American Mathematical Society, 1996.

Streim, Gregor. "'Krisis des Historismus' und geschichtliche Gestalt: Zu einem ästhetischen Geschichtskonzept der Zwischenkriegszeit." In *Literatur und*

Geschichte, edited by Daniel Fulda and Silvia Serena Tschopp, 463–88. Berlin: De Gruyter, 2002.
Taubes, Jacob. *Abendländische Eschatologie*. Bern: A. Francke, 1947.
——. *Die politische Theologie des Paulus: Vorträge, gehalten an der Forschungsstätte der evangelischen Studiengemeinschaft in Heidelberg, 23.–27. Februar 1987*. Edited by Aleida Assmann and Jan Assmann. Munich: Fink, 1993.
——. *Der Preis des Messianismus: Briefe von Jacob Taubes an Gershom Scholem und andere Materialien*. Edited by Elettra Stimilli. Würzburg: Königshausen & Neumann, 2006.
——. "Theodicy and Theology: A Philosophical Analysis of Karl Barth's Dialectical Theology." In *From Cult to Culture: Fragments toward a Critique of Historical Reason*, edited by Elisheva Fonrobert and Amir Engel, 177–94. Stanford, CA: Stanford University Press, 2010.
Tessitore, Fulvio. *Kritischer Historismus: Gesammelte Aufsätze*. Cologne: Böhlau, 2005.
Thiher, Allen. *Understanding Robert Musil*. Columbia: University of South Carolina Press, 2009.
Thums, Barbara. "Kracauer und die Detektive: Denk-Räume einer 'Theologie im Profanen.'" *Deutsche Vierteljahrsschrift für Literaturwissenschaft und Geistesgeschichte* 84, no. 3 (2010): 390–406.
Tillich, Paul. *Kairos: Zur Geisteslage und Geisteswendung*. Darmstadt: Reichl, 1926.
Tohline, Andrew M. "Towards a History and Aesthetics of Reverse Motion." PhD diss., College of Fine Arts of Ohio University, 2015.
Tolstaya, Katya. "Literary Mystification: Hermeneutical Questions of the Early Dialectical Theology." *Neue Zeitschrift für Systematische Theologie und Religionsphilosophie* 54, no. 3 (2012): 312–31.
Treml, Martin, and Daniel Weidner, eds. *Nachleben der Religionen: Kulturwissenschaftliche Untersuchungen zur Dialektik der Säkularisierung*. Munich: Fink, 2007.
——. "Zur Aktualität der Religionen: Einleitung." In *Nachleben der Religionen: Kulturwissenschaftliche Untersuchungen zur Dialektik der Säkularisierung*, edited by Martin Treml and Daniel Weidner, 7–22. Munich: Fink, 2007.
Troeltsch, Ernst. "Der Historismus und seine Probleme (1922)." In *Gesammelte Schriften*, vol. 3. Aalen: Scientia, 1977.
——. "Die Krisis des Historismus." *Die Neue Rundschau* 33, no. 6 (1922): 572–90.
Voegelin, Eric. *The New Science of Politics*. Chicago: University of Chicago Press, 1952.
Vollrath, Hans-Joachim. "Mathematische Bilder in Karl Barths Römerbrief." *Mathematica Didactica* 11 (1988): 3–10.
Von Moltke, Johannes. *The Curious Humanist: Siegfried Kracauer in America*. Oakland: University of California Press, 2016.
Wachter, David. *Konstruktionen im Übergang: Krise und Utopie bei Musil, Kracauer und Benn*. Freiburg im Breisgau: Rombach, 2013.

Warburg, Aby. "Italian Art and International Astrology in the Palazzo Schifanoia Ferrara." In *The Renewal of Pagan Antiquity: Contributions to the Cultural History of the European Renaissance*, translated by David Britt, 563–91. Los Angeles: Getty Research Institute for the History of Art and the Humanities, 1999.

———. "Italienische Kunst und internationale Astrologie im Palazzo Schifanoja zu Ferrara (1912)." In *Die Erneuerung der heidnischen Antike: Kulturwissenschaftliche Beiträge zur Geschichte der europäischen Renaissance*, edited by Horst Bredekamp and Michael Diers, 459–81. Berlin: Akademie, 1998.

Ward, Graham. "Barth, Modernity, and Postmodernity." In *The Cambridge Companion to Karl Barth*, edited by John Webster, 274–95. Cambridge: Cambridge University Press, 2000.

Weidner, Daniel. "Einleitung: Walter Benjamin, die Religion und die Gegenwart." In *Profanes Leben: Walter Benjamins Dialektik der Säkularisierung*, edited by Daniel Weidner, 7–35. Berlin: Suhrkamp, 2010.

———. "Thinking beyond Secularization: Walter Benjamin, the 'Religious Turn,' and the Poetics of Theory." *New German Critique* 37, no. 3 (2010): 131–48.

Weigel, Sigrid. "Warburg's 'Goddess in Exile': The 'Nymph' Fragment between Letter and Taxonomy, Read with Heinrich Heine." *Critical Horizons* 14, no. 3 (2013): 271–95.

Wiehl, Reiner. *Zeitwelten: Philosophisches Denken an den Rändern von Natur und Geschichte*. Frankfurt am Main: Suhrkamp, 1998.

Wittekind, Folkart. "Zwischen Deutung und Wirklichkeit: Überlegungen zum Bildcharakter eschatologischer Aussagen." In *Die Gegenwart der Zukunft: Geschichte und Eschatologie*, edited by Ulrich H. J. Körtner, 55–84. Neukirchen-Vluyn: Neukirchener Verlag, 2008.

INDEX

Page numbers followed by letter *f* refer to figures.

Abbott, Edwin A., 49–50
actuality (*Realität*), vs. reality (*Wirklichkeit*), Kracauer on, 197–98
Adorno, Theodor W.: as anti-Hegelian, 116n69; and metropolitan miniature, 3; *Minima Moralia*, 294–97; religious subtext of work of, 16; on standpoint of redemption, 295, 296–97
"The Affliction and Promise of the Christian Proclamation" (Barth), 108
afterlife (*Nachleben*): of photograph, Kracauer on, 209; of pictorial representations, Warburg's concept of, 70, 70n108; of religion, in modernism, 14–16, 17, 284

Agamben, Giorgio, 19–20
The Ages of the World (Schelling), 141–42
Ahasuerus (Wandering Jew), figure of: Baumann on, 229n92; in Kracauer's theory of history, 229–33, 258, 291, 293; von Moltke on, 230n93
Albrecht, Andrea, 236n3, 244n24, 246, 247
alienation: Adorno on, 295–96; Barth on, 105, 107; Kracauer on, 213, 217, 217n72, 221
allegory, theory of, 113
Altmann, Alexander, 139, 152n42
anatomy, metaphors from, in Rosenzweig's dual-covenant theology, 169–70

Anfang (beginning), end of history as: Barth on, 97, 114, 120–21; Rosenzweig on, 158–59, 180–81, 290. See also *Urgeschichte*

anteroom area, history as, Kracauer on, 182, 218, 227, 227n84, 228

antinomy of time: aesthetic reconciliation of, in Proust's novel, 231, 232; Kracauer's concept of, 8, 224–25, 229, 230n93, 231–32

arc, image of, in Rosenzweig's conception of history, 130, 148–51, 157, 170–71, 181, 253, 290

The Arcades Project (Benjamin), 71, 110

Archimedes, 90

Arendt, Hannah, 16

art: constructivist, 9, 11, 30, 36–39, 45–47, 79; as eschatological figuration of new reality, Barth on, 126–27; experimentation with spatiality of time in, 53–54; modernist, re-envisioning of space in, 30, 34; non-Euclidean geometries and innovations in, 9, 11, 37, 38–39, 45–47, 52; Renaissance, depiction of space in, 33–34; and shaped times, Kubler on, 223; time as element of plastic configuration in, 52–54

"Art and Pangeometry" (Lissitzky), 38, 46–47, 52–53

Auerbach, Erich: Barth's theory of contemporaneity and, 113; religious subtext of work of, 16

Aufbau (structure/construction): in art, Lissitzky on, 45, 46; in contemporary culture, Kracauer on, 195–96; of history, Barth on, 125; of history, Kracauer on, 187, 207–8; of history, Troeltsch on, 57, 67–69, 74–75, 79, 117, 125; modernist mathematics and, 247

Balthasar, Hans Urs von, 84n6

Barnouw, Dagmar, 185n5

Barth, Karl, 28, 82; "The Affliction and Promise of the Christian Proclamation," 108; on artistic creation as figuration of new reality, 126–27; constructivist historiography compared to, 112, 116–17, 118, 127–28; context for work of, 29, 30, 36, 80, 118; convergence of *Urgeschichte* and *Endgeschichte* in theology of, 12, 84, 120–21, 180; critique of historicism, 30, 83, 93, 112–17, 290; Dialectical Theology of, 5, 82, 83, 94, 109–10; on dual natures of Jesus Christ, 85–88, 89; *The Epistle to the Romans*, 19, 30, 82, 84–90, 95–105; and eschatological constructivism, 112, 116–17, 118, 125, 127–28; on eschatological now, 7, 12, 102–5, 103n40, 110–11, 113, 118, 127, 291; on eschatology as limit phenomenon, 5, 12, 89–90, 93–95, 120–24, 291, 294; geometrical metaphors used by, 12, 13, 17, 29, 83, 84–90, 93–104, 118, 244, 253, 290; and *Gestaltung* (figuration) of eschatology, 125–27; on God as wholly other, 82; historical context of 1920s' Europe and, 94n27, 106–7, 106n45; influences on, 82, 83, 89, 111, 112; Kracauer's theory of history compared to, 228, 228n87, 292; on liminal relations of history and eschatology, 29, 87–88, 94–95, 101, 102–5, 103n40, 109, 290, 291, 293; modernist aesthetics and, 17, 28, 83–84, 94n27, 105; Musil's ideas compared to, 291; and non-Euclidean geometry, interest in, 98, 99, 101, 102; Ouspensky's theories compared to, 55, 93n25; on problem of theodicy, 98–101; *The Resurrection of the Dead*, 94–95, 123–24; Rosenzweig's theory of history compared to, 129–30, 133, 147, 150, 158, 158n48, 161, 164n55, 290; scholarship on, 18, 19; on "situation" (*Lage*) of humanity, 106, 107–8, 110–11; spatial forms in theology of, 10, 30, 56, 83, 84–85, 93, 104, 105, 128; theology of history of, 4, 112–15, 118, 125, 127; theory of contemporaneity (*Gleichzeitigkeit*), 109–14, 119, 125, 155, 224, 224n80; theory of the moment, 82–83, 84, 87, 102–5, 104n43, 110–11, 118, 124, 128; on "time of the now" (*die Zeit des Jetzt*), 19, 55, 102–3, 128; and topos of being lifted out of historicity, 152; and *Urgeschichte*, concept of, 119–21; and *zwischen den Zeiten* (between the times), concept of, 109, 121, 158. See also *The Epistle to the Romans* (Barth)

Baudelaire, Charles, 2

Bauhaus, 9, 36

Baumann, Stephanie, 216n70, 229n92

Baur, Ferdinand Christian, 89

"The Bay" (Kracauer), 189, 195, 225

Beckett, Samuel, 149

beginning (*Anfang*), end of history as: Barth on, 97, 114, 120–21; Rosenzweig on, 158–59, 180–81, 290. See also *Urgeschichte*

Being and Time (Heidegger), 107

Beintker, Michael, 104n43

Bell, John L., 92

Beltrami, Eugenio: analogies and projections used by, 172; on curved spaces, 245; disc model of, 137, 137f, 159; and modernist mathematics, 31, 40; projective geometry of, 130, 136, 137, 159–60, 181, 292

Benjamin, Walter: *The Arcades Project*, 71, 110; Barth's work and, 19n44, 113; *Berlin Childhood around 1900*, 8; expansive concept of historicism, 28; historicist paradigm challenged by, 29; and metropolitan miniature, 2; on "now time" (*Jetztzeit*), 19, 20, 110; "On the Concept of History," 20, 110; presentist historiography of, 65n95; on redemption, 185n5; religious subtext of work of, 16; theory of allegory, 72; Troeltsch's theory of history compared to, 71–72

Benn, Gottfried, 107

Berlin Childhood around 1900 (Benjamin), 8

Bernoulli, Carl Albrecht, 119

Bernstein, Michael André, 269

"between the times" (*zwischen den Zeiten*), Barth's concept of, 109, 121, 158

biblical text: historicization of, Barth's opposition to, 112. See also Paul, Saint

Bieberich, Ulrich, 152n42, 154n45

Bloch, Ernst: concept of "religion in inheritance," 185n5; Kracauer's correspondence with, 184, 213n63, 227n84; on noncontemporaneity of the contemporaneous, 111, 224, 224n80

Blumenberg, Hans: critique of secularization theory, 14–15; *Work on Myth*, 297–300

Blumhardt, Christoph, 89

Böhme, Harmut, 271n59, 283

Bohrer, Karl Heinz, 270n57, 278n69, 279n70, 282, 283, 284–85

Bolyai, János, 11

Borges, Jorge Luis, 149

Braiterman, Zachary, 21n48, 140, 155, 177, 178n68

Brandi, Karl, 66

Brentano, Franz, 249n35

bridge, metaphor of: in Königsburg bridge problem, 193–94, 193f; in Musil's work, 29, 32, 248, 250–53, 274; in Rosenzweig's dual aspect theory of history, 145, 146

Broch, Hermann, 13

The Brothers Karamazov (Dostoevsky), 99–101

Brunelleschi, Filippo, 33, 34

Brunner, Emil, 88

Bultmann, Rudolf, 19n44, 130n1, 238n13

Burckhardt, Jacob, 220, 221

Butterfield, Herbert, 238n13

calculus: development of, concept of limit and, 90, 92; and Rosenzweig's dual-covenant theology, 164, 167–68, 171; tangent used in, 91

Cantor, Georg, 244n24, 246

Cassirer, Ernst, 24
Celan, Paul, 16
chance encounters, history as series of, Musil on, 234
Christianity: and Judaism, in Rosenzweig's dual-covenant theology, 129, 139, 143–44, 151–53, 156–70, 157n47; and philosophy of history, 6; rational numbers as metaphor for, Rosenzweig on, 165–66, 292; Rosenzweig's theory of, 143–44, 156–61, 157n47, 160–61, 162; understanding of redemption in, 20. *See also* Jesus Christ; Protestant liberal theology
Christianity and Culture (Overbeck), 119–20
chronological time, Kracauer on: as empty vessel, 1, 219, 220, 223, 225; historicist concept of, distortions associated with, 224; shaped times in relation to, 7, 223, 231–32, 291
circle: broken, in Rosenzweig's eschatological account of history, 149; deformation in projective geometry, 136; deformation in topology, 183, 194; in Hegelian model of history, Rosenzweig's rejection of, 147–48, 150n40, 157; squaring of, Barth on, 98. *See also* tangent
Clark, T. J., 37, 79
Cliver, Gwyneth E., 241n19, 243
Cohen, Hermann, 13, 164, 164n55, 171, 178; and messianic concept of history, 178–79, 181; and Rosenzweig, 178, 178n69; Rosenzweig's philosophy contrasted with, 179–81
Collingwood, R. G., 220, 221, 238n13
Comte, Auguste, 219–20
configuration. *See* Gestaltung
The Confusions of Young Törless (Musil), 46, 235, 239–46, 250–52, 253; bridge metaphor in, 251–52, 253; crisis of mathematical representation explored in, 237, 239, 244–46, 250–51, 253; "other condition" in, 235, 240–45, 250, 252, 254; relationship between mathematics and theological problems in, 241–43, 253
construction. *See* Aufbau
constructivism: in art, 9, 11, 30, 36–39, 45–47, 79; eschatological, Barth and, 112, 116–17, 118, 125, 127–28; in historiography, 9, 56–80, 109, 116–17, 118; and Kracauer's approach to history, 187, 188; non-Euclidean geometries and, 11, 37–38, 39, 45–47, 80; and Rosenzweig's geometrical imagination of history, 130. *See also* Russian Constructivism
contemporaneity (*Gleichzeitigkeit*): Barth's theory of, 109–14, 119, 125, 155, 224, 224n80; Bloch's view of, 111, 224, 224n80; Kierkegaard's concept of, 111, 224n80; Kracauer on, 230
Craver, Harry T., 185n5
Cremer, Douglas, 94n27
The Crisis of Historicism (Heussi), 107
"The Crisis of Intuition" (Hahn), 51–52
Critique of Pure Reason (Kant), on unitary nature of space, 34, 35
Croce, Benedetto, 220
Cubism, 9, 11, 34, 38; non-Euclidean geometries and, 45; reimagining of space in, 38, 39
current. *See* Strom
curved space(s): in art, 45; and Einstein's theory of relativity, 52; Helmholtz on, 44; insights from, and new shapes of time, 3–4, 10; and modernist imagination of history, 12, 36; of non-Euclidean geometries, 9–10, 11, 44–45, 53, 183; Riemann on, 9, 44, 172; and topology, 183. *See also* shape(s) of time/shaped times
Cusanus, Nicolaus, 149

d'Alembert, Jean le Rond, 91
The Decline of the West (Spengler), 76

deformation(s): in Heussi's theory of historiography, 66–67; in Kracauer's geometries of urban space, 184, 192; in Kracauer's theory of history, 188, 199, 216, 233, 291–92; in projective geometry, 136; in topology, 183–84, 192–93, 194

Derrida, Jacques, 16

destructive moment, of materialist historiography: Benjamin on, 71–72; Troeltsch on, 72–73

The Detective Novel (Kracauer), 229n90

development, principle of (*Entwicklung/Weiterentwicklung*): collapse of, 82; constructivist historiography and rejection of, 69, 75; in Dilthey's understanding of history, 69; in Hegel's philosophy of history, 6, 25; in Herder's understanding of history, 27; in historicism, 26–28; in Ranke's understanding of history, 27; Spengler's criticism of, 73–74

de Vries, Hent, 83n5, 112n60, 116n69

Dialectical Theology, Barth's, 5, 82, 83; constructivist historiography and, 112, 116–17, 118; geometrical language used by, 30–31; present perspective in, 94; Weimar intellectual thought and, 94n27

Diderot, Denis, 91

differential calculus, and Rosenzweig's dual-covenant theology, 164, 167–68, 171

differential geometry, Riemann's, 172–73

Dilthey, Wilhelm: Barth's constructivist eschatology compared to, 127; *The Formation of the Historical World in the Human Sciences*, 67; on history of development (*Entwicklungsgeschichte*), 69; on "philosophy of life" (*Lebensphilosophie*), 57

"discontinuous series" (*diskontinuierliche Reihe*), Rosenzweig on history as, 135–36, 138

disintegration (*Zerfall*), Kracauer on: and configuration (*Gestaltung*), dialectic of, 195–96, 197–99, 211, 232; history subject to, 216; of photographs, 199, 200, 202, 209, 211, 212–13, 212n60, 213–14, 218; and reconfiguration (*Neugestaltung*), eschatological dynamic of, 215

dividing line (*Scheidelinie*), image of, in Musil's work, 242–43

Dostoevsky, Fyodor, 99–101

Droysen, Johann Gustav: and historicism, 7, 25, 26, 56; on teleological orientation of history, 27–28

Drügh, Heinz, 113n63

dual-aspect theories of history, 7; Barth's concept of eschatological now and, 7, 12, 102–5, 103n40, 110–11, 113, 118, 127, 291; Kracauer's, 7, 223, 231–32, 291; Rosenzweig's, 7, 31, 130, 144–47, 181, 290, 291

Duchamp, Marcel, 45

Dürer, Albrecht, 33, 34

Eagleton, Terry, 284n78

Earth, interrupted homolosine projection of, 137–38, 138f

"Education of Mankind" (Lessing), 75

Edwards, C. H., 90

Ehrenberg, Hans, Rosenzweig's correspondence with, 132, 139, 140, 164–65

Ehrenfels, Christian von, 131

Einstein, Albert, on space-time continuum, 45, 52, 53

Elements (Euclid), 39; definition of tangent in, 91. *See also* Euclidean geometry

Elements of the Philosophy of Right (Hegel), 22, 23–24

Eliot, T. S., 81

ellipse, image of: in Barth's theology, 97; in projective geometry, 136; in topology, 183, 194

empirical space, as only one of many possible spaces, 35, 41–42, 44, 47, 194

Encyclopedia of the Philosophical Sciences in Basic Outline (Hegel), 24
Encyclopédie (Diderot), 91
"The End of All Things" (Kant), 4
end of time/history: Barth on, 5, 12, 84, 94–95; Heussi on, 67; Kant on, 4, 5; proximity to present moment, modernist thinkers on, 12, 32, 238, 238n13; reimagining in nonteleological terms, 5, 21; Rosenzweig on, 12, 158–59, 180–81. *See also* eschatology
Engelhardt, Nina, 12, 13
Entwicklung. See development, principle of
The Epistle to the Romans (Barth), 19, 30; language of mathematics in, 84–90, 93–104; problem of theodicy in, 98–101; rhetoric of crisis in, 84; Rosenzweig's *Star of Redemption* compared to, 129; on "situation" (*Lage*) of humanity, 106, 107–8, 110–11; theory of the moment in, 82–83, 128; on "time of the now" (*die Zeit des Jetzt*), 19, 55, 102–3, 128
eschatological now, Barth's concept of, 7, 12, 102–5, 103n40, 110–11, 113, 118, 127, 291; Musil's "other condition" compared to, 235, 291; Rosenzweig's concept of "today" compared to, 144–45, 154–55, 179; unrepresentability of, non-Euclidean space as metaphor for, 12, 90
eschatology: Adorno's concept of, 296–97; Barth's concept of, 5, 12, 89–90, 93–95, 117–24, 294; decoupled from teleological model of historical development, modernism and, 8, 237–38, 282, 285, 294, 300; etymology of term, 294n1; Heidegger on representation of, 124; in Heussi's constructivist historiography, 67; historicization/secularization of, 6, 7, 17–18, 21–24, 156, 282; Judaism's spatial proximity to, Rosenzweig on, 153–54; Kracauer's concept of, 31–32, 182–83, 186, 196n30, 197–98, 209–11, 213, 215, 226–33; as limit phenomenon, 8, 124–25, 294; vs. messianism, Agamben on, 20; modernist concept of, 4, 8, 17–21, 32, 215, 282, 284, 285, 294; modernist concept of, enduring legacy of, 294–95, 300; Musil's concept of, 29, 32, 235, 237–39, 251–52, 253, 268–69, 282, 287–88, 291; of myth, aesthetic experimentation with, 297–300; non-Euclidean geometry and modernist concept of, 11, 12, 14, 117, 120; reversal of traditional schema of, in Kafka's "Prometheus," 298; Rosenzweig's concept of, 12, 31, 131, 148–51, 162, 180–81, 180n75, 294; spatial understanding of, 4–5, 8, 287–88, 294, 300; temporal understanding of, and concept of history, 5–6, 8; temporal vs. spatial understanding of, 5, 178, 180, 282; Tillich's construction of, 117–18; traditional/teleological understanding of, 4–5, 20, 180. *See also* history and eschatology, liminal relation of; limit phenomenon, eschatology as
"Essay on the Interpretation of Non-Euclidean Geometry" (Beltrami), 136
eternal figures (*ewigen Gestalten*), Rosenzweig on, 143–44, 147, 177
eternal people (*das ewige Volk*), Rosenzweig on, 156, 162
eternal way (*der ewige Weg*), Rosenzweig on, 156–61, 162
eternity: Judaism and Christianity in relation to, Rosenzweig's theory of, 143–44, 156–61, 157n47, 160–61, 162; perspective of, needed to comprehend spatiality of time, 54; present containing moment of, Rosenzweig on, 130; represented as vector, 118, 125, 147; Rosenzweig's concept of, 130, 152–61. *See also* time and eternity, liminal relation of

Euclid: *Elements*, 39, 91; parallel postulate of, 39, 40
Euclidean geometry, 39; historicist concept of history compared to, 146; metaphysical foundation for, Kant and, 34; and Renaissance perspective, 34, 38; space in, 9, 34, 41–42, 44; space in, as only one of many possible spaces, 35, 41–42, 44, 47, 194; tangent in, 91, 93
Euler, Leonhard, 183; Königsburg bridge problem, 193–94, 193*f*
exile, historian's position of, Kracauer on, 198, 216–17, 220–21

Fanta, Walter, 275n64
Fenves, Peter, 21n48
"Figure and Disintegration" (Kracauer), 195–98, 209, 216, 232
figures: eternal (*ewigen Gestalten*), Rosenzweig on, 143–44, 147, 177; supra-mathematical, 170–71, 173–74, 292. *See also* deformation(s); geometrical figures; *Gestalten*; spatial form(s)
film: critical potential of, Kracauer on, 205, 205n50, 211n58, 215; technique of reverse motion in, and Musil's thought experiment, 266
Flatland: A Romance of Many Dimensions (Abbott), 49–50
Floris, Joachim von, 75
"The Flypaper" (Musil), 287–88
The Formation of the Historical World in the Human Sciences (Dilthey), 67
Frank, Joseph, on spatial form, 81–82, 83, 105, 255n39
Fuchs, Anne, 3, 7n14, 29n82
Fulda, Daniel, 23
Funkenstein, Amos, 175n63
Futurism, 9, 52, 53

Gauchy, Augustin-Louis, 92
Gauss, Carl Friedrich: on curved spaces, 9, 172; and modernist mathematics, 9, 10, 34–35, 40, 47, 193, 194

General Theory of Relativity: and art, innovations in, 45; and space-time continuum, 52
geometrical figures: instability in non-Euclidean geometries, 9, 10; and Heussi's theory of historiography, 66–67. *See also* deformation(s); *Gestalten*; spatial form(s)
geometry: algebraic approaches to, 194; decoupling from intuition, 52. *See also* Euclidean geometry; non-Euclidean geometries
German Idealism: notion of historical development in, 6; philosophical eschatology in, Taubes on, 75; vs. Rosenzweig's eschatological reading of history, 147
Gestalten (figures): historical motifs and formations treated as, 58–61; and Kracauer's theory of history, 207–8, 291; and Rosenzweig's theory of history, 130, 131, 143–44, 147, 173–74, 176–78, 181, 291; and Spengler's theory of history, 74, 75, 126; term, development of, 130–31
Gestalt psychology, 131; *Kippfiguren* (multistable figures) developed by, 2; and Mannheim's concept of history, 59–60
Gestaltung (figuration/configuration): in art, Lissitzky on, 38, 45, 46, 47, 52–53; Barth's theological constructivism and, 125–27; in constructivist historiographies, 57–58, 66, 74, 75, 125–26, 127; plastic (*plastische Gestaltung*), time as element of, 52–54; Tillich's rhetoric of, 127n86; and *Zerfall* (disintegration), Kracauer on dialectic of, 195–96, 197–99, 211, 232
Giles, Steve, 212n59
Gleichzeitigkeit. *See* contemporaneity
"Globe: Studies on the World-Historical Theory of Space" (Rosenzweig), 140

326   Index

goal (*Ziel*), of history: as beginning (*Anfang*), Rosenzweig on, 180–81; Jewish people's awareness of, Rosenzweig on, 153; Spengler's rejection of, 76; in teleological models of history, 18, 25, 27, 56–57

God: alienation from, Barth on, 107; unrepresentability of, non-Euclidean space as metaphor for, 12; as wholly other, Barth on, 82

God and humanity, relationship of: Barth on, 107–8, 290; language of mathematics used to express, 84, 85–87, 99–101, 290; in Rosenzweig's star of redemption, 172–73, 174f. *See also* history and eschatology

Goethe, Johann Wolfgang von, theory of morphology: applied to historical forms and figures, 57–58, 77; and *Gestalten*, concept of, 131

Gogarten, Friedrich, 109n53, 228n87

Goode, John Paul, interrupted homolosine projection of Earth, 137–38, 138f

Gordon, Peter, 150n40, 175n63, 180n75

grace: geometrical figure used to represent, Barth and, 122–23; and sin, as incommensurable magnitudes, Barth on, 101–2, 292

Gray, Jeremy, 10, 13, 39, 171, 247

Grenze. *See* limit

Gropius, Walter, 36

Gualtieri, Elena, 217n72

Günzel, Stephan, 194

Hahn, Hans, 51–52

Handelman, Matthew, 13, 164n55, 183n1, 194, 195n28

Hansen, Miriam, 184n4, 186n8, 196n30, 201n44, 202n45, 204n48, 205n50, 206n52, 212n60, 213, 213n61, 214n64, 227n86

haptic space, 42

Harnack, Adolf, 115n66

Harvey, David, 2, 3, 13

Hausdorff, Felix, 246

Hausdorff space, 35

Hegel, Georg Wilhelm Friedrich: *Elements of the Philosophy of Right*, 22, 23–24; *Encyclopedia of the Philosophical Sciences in Basic Outline*, 24; on freedom as goal of human history, 25, 27; and historicization/secularization of eschatology, 6, 7, 17, 18, 22, 23–24, 156; influence on Barth, 82; *Lectures on the Philosophy of History*, 24–25; philosophy of history of, 6, 21, 22, 23–25, 27, 56, 75, 114, 116, 116n69, 147; on redemptive function of history, 134; Rosenzweig's critique of, 130, 132, 133, 134, 135, 141, 147–48, 150n40, 157

Hegel and the State (Rosenzweig), 134

Heidegger, Martin: *Being and Time*, 107; on eschatological representation, 124

Heine, Heinrich, 3n6

Helmholtz, Hermann von: analogies and projections used by, 172; on construction of space, 43–44; on limits of representability, 48–49; "On the Origin and Meaning of Geometric Axioms," 40–41, 48–49; on parallel lines meeting at infinity, 40–41, 101; and popularization of non-Euclidean geometry, 40–41, 42, 101, 121

"Helpless Europe, or a Journey from the Hundredth to the Thousandth" (Musil), 263–65

Henderson, Linda D., 9, 13, 40, 45

Herder, Johann Gottfried: influence on Kracauer, 224n80; philosophy of history of, 6, 25, 27, 75; and *Strom* (current), metaphor of, 144, 144n3; *This Too a Philosophy of History*, 25, 144n31

Hering, Ewald, 41

Heussi, Karl: Barth's concept of history compared to, 116–17, 118; constructivist historiography of, 9, 56, 57, 58, 63–67, 78, 109; *The Crisis of*

Historicism, 107; eschatological perspective of, 67; on historical situation, 107; on history as sublime current, 63, 66, 79; and spatial concepts of time and history, 30, 36, 57, 66–67

Heute. See today

Hilbert, David, 167n57, 236, 236n7, 245; foundational crisis of mathematics and, 246; on mathematics as language of symbols, 248, 248n34

Hilbert space, 35

historian: perspectival standpoint of, Mannheim's theory of, 57, 61–63, 67; position of exile, Kracauer on, 198, 216–17, 220–21

historical object: as counterpart with no fixed shape, 66–67; as fragment, Benjamin on, 71–72

historical time: and symbolic time, in Rosenzweig's dual aspect theory of history, 146–47. *See also* chronological time

historicism, 7, 25–28; Adorno's rejection of, 296; Barth's critique of, 30, 83, 93, 112–17, 290; collapse of, 82; vs. constructivist historiography, 56, 68, 73, 78; Euclidean geometry compared to, 145, 146; Kracauer's critique of, 182, 199, 204–5, 207, 216, 217, 218–19, 218n75, 224; vs. Mannheim's model of perspectivism, 61; Musil's critique of, 234, 263–65; paradigm of time in, vs. modernist concepts, 29, 182, 186, 224; philosophy of history compared to, 25–26, 27, 28; and photography, Kracauer's analogy of, 199, 204–5; present moment in, 145, 146; Ranke and, 7, 25, 26, 56, 61, 65n95; Rosenzweig in relation to, 130, 132, 140–47, 180n75, 290; temporal understanding of eschatology and, 5–6, 8

"Historicism" (Mannheim), 58–59

historiography: 20th-century theorists of, 26; alternative models of, 8; Barth's antihistoricist approach to, 116–17; constructivist, 9, 56–80, 109, 116; materialist, destructive moment of, 71–73

history: 19th-century concepts of, 21–28; Barth's theology of, 4, 112–15, 118, 125, 127; constructivism and new models of, 9, 56–80; developmental model of, 6, 25–28; elemental forces of, Troeltsch on retrieval of, 69–70, 79; eschatology and understanding of, in early 20th century, 5–6, 8, 17–21, 124–25; Kracauer's theory of, 4, 185–88, 196n30, 197–99, 207–8, 213n63, 215, 217–33, 291, 292–93; modernism and new models of, 4, 7–9, 28–29, 289–90, 294–95, 300; modernist mathematics and concepts of, 3, 10, 12, 14, 29, 290; morphology of, concept of, 73; Musil's concept of, 7, 12, 234, 258–67, 269, 270, 281, 282, 291, 293; present-interest theory of, 220, 221; recognition of unhistorical in, Barth and, 114–15; redemptive function of, Hegel on, 134; Rosenzweig's theory of, 4, 131–32, 134–36, 138–40; spatial approach to, 30, 57, 66–67, 139–41, 147, 177–78, 181, 292–93; as spatial construction, Kracauer on, 207–8, 216; Spengler's morphology of, 57–58, 77–78; temporal exterritoriality of, Kracauer on, 182, 220–21, 226, 230n93. *See also* dual-aspect theories of history; history and eschatology; philosophy of history; teleological concepts of history

history and eschatology, liminal relation of, 8, 10, 29, 31, 80; Adorno on, 296–97; Barth on, 29, 87–88, 94–95, 101, 102–5, 103n40, 109, 290, 291, 293; enduring legacy of concept of, 300; Kracauer on, 215, 228; language of mathematics used to express, 5, 8, 10, 11–12, 31, 84,

history and eschatology (*continued*)
253, 290, 293; Musil on, 253, 273, 291; Rosenzweig on, 130, 147, 148–51, 154–56, 290; shaped times and, 4–5, 12, 28, 289, 290, 293–94
"History as Figuration" (Brandi), 66
History of the Revolt of the Netherlands (Schiller), 23
The History of the Thirty Years' War (Schiller), 23
History: The Last Things before the Last (Kracauer), 1, 182, 188, 216–32; and "Photography," continuities between, 217–18, 217n73
Hollander, Dana, 154n45, 180n73, 181n76
hollow spaces (*Hohlräume*): in Barth's theory of history, 5, 83, 88, 96, 292, 295; in Kracauer's geometry of urban spaces, 189, 225; in Kracauer's theory of history, 184, 199, 215, 224–25, 226, 253, 273, 292, 296; photograph revealing, Kracauer on, 199, 204
Honold, Alexander, 236n8, 239, 263n46, 269, 283
"The Hotel Lobby" (Kracauer), 227
humanity: as "between the times" (*zwischen den Zeiten*), Barth's concept of, 109, 121, 158; dual nature of, Barth on, 123; in Rosenzweig's star of redemption, 172–73, 174f. *See also* God and humanity, relationship of
Husserl, Edmund, 62
Huyssen, Andreas, 2–3, 194, 195n28
hyperbola, mathematical concept of, 169; Barth's use of, 96; Rosenzweig's use of, 150–51, 169

"Idea for a Universal History with a Cosmopolitan Aim" (Kant), 22–23
"The Idea of the Messiah" (Cohen), 178–79
imaginary numbers: in *The Confusions of Young Törless* (Musil), 244, 250–51; Lissitzky on, 46, 47; nonintuitive nature of, 245

Impressionism, 34; reimagining of space in, 38, 39
infinity: Christianity's vs. Judaism's relationship to, Rosenzweig on, 165–66; debates on, in history of mathematics, 244n24; experience of, in *The Confusions of Young Törless* (Musil), 243–44. *See also* parallel lines, intersecting at infinity
Innes, Keith, 95n30
intuition: breaking with, foundational crisis of mathematics and, 247–48; decoupling of geometry from, non-Euclidean spaces and, 9, 50–52; space and time as pure forms of, Kant on, 2, 52, 100
irrational numbers, in Rosenzweig's dual-covenant theology, 164, 165, 166–67, 181, 292
Itkin, Alan, 221n77, 230n93, 290n94

Jameson, Frederic, 13
Jaspers, Karl, 106
Jennings, Michael W., 19n44
Jesus Christ: dual nature of, Barth on, 85–88, 89, 93; historicization of figure of, Schleiermacher and, 88–89; as paradox, Kierkegaard on, 89; as point of intersection between time and eternity, Barth on, 109
Jewish Gnosticism: and Kracauer's account of disintegration, 196n30. *See also* Judaism
Joyce, James: and modernist aesthetics, 105; and spatial form, 81
Judaism: and Christianity, in Rosenzweig's dual-covenant theology, 129, 139, 143–44, 151–53, 156–70, 157n47; irrational numbers as metaphor for, Rosenzweig on, 165, 166–67, 292; as outside of history, Rosenzweig on, 139, 152–53, 179–80; and philosophy of history, 6; Rosenzweig's theory of, 139, 143–44, 152–53, 156, 157n47, 160, 161, 179–80; understanding of redemption in, 20

Kafka, Franz: historicist paradigm challenged by, 29; and metropolitan miniature, 2; and modernist aesthetics, 105; "Prometheus," 297–300; religious subtext of work of, 16

Kandinsky, Wassily, 36, 178n68

Kant, Immanuel: categorical schema of, and Heussi's figuration of history, 66; concept of the sublime, 64, 243, 243n23; *Critique of Pure Reason*, 34; "The End of All Things," 4; on end of history, 4, 5; "Idea for a Universal History with a Cosmopolitan Aim," 22–23; influence on Barth, 82; philosophy of history of, 56, 75, 179; on space and time as forms of intuition, 2, 52, 100; on space as a priori form, 34, 35, 59

Kantorowicz, Ernst, 85, 86

Kaplan, Gregory, 154n45, 157n47, 163, 180n74

Kermode, Frank, 238, 238n15

Kessler, Michael, 185n5, 229n91

Kierkegaard, Søren: concept of Christ as paradox, 89; and contemporaneity (*Gleichzeitigkeit*), concept of, 111, 224n80; influence on Barth, 82, 83, 87, 89, 111, 224n80; and topos of being lifted out of historicity, 152

King Lear (Shakespeare), 238

The King's Two Bodies (Kantorowicz), 85, 86

Kippfiguren (multistable figures), 2

Klee, Paul, 36, 140, 178n68

Klein, Felix: on curved spaces, 245; disc model of, 137, 137f; projective geometry of, 31, 130, 136, 137, 159–60, 181, 292; scale of elliptically equidistant points, 159, 160f

Koch, Gertrud, 185n5, 232n96

Koffka, Kurt, 59

Köhler, Wolfgang, 59

Königsburg bridge problem, 193–94, 193f

Koselleck, Reinhart, 6, 62–63

Koshar, Rudy, 115n66

Kracauer, Siegfried, 28; on actuality (*Realität*) vs. reality (*Wirklichkeit*), 197–98; on antinomy of time, 8, 224–25, 229, 230n93, 231–32; Barth's theory of history compared to, 228, 228n87, 292; "The Bay," 189, 195, 225; on chronological time and shaped time, relationship of, 7, 223, 231–32, 291; concept of redemption, 185n5, 210–11, 211n58, 213, 217, 227, 227n86; on configuration (*Gestaltung*) and disintegration (*Zerfall*), dialectic of, 195–96, 197–99, 211, 232; constructivist dimension of approach to history, 187, 188; context for work of, 29, 30, 36, 80; correspondence with Bloch, 184, 213n63, 227n84; critique of historicism, 182, 199, 204–5, 207, 216, 217, 218–19, 218n75; on deformation of figures, 184, 188, 192, 199, 233, 291–92; *The Detective Novel*, 229n90; eschatological view of history, 196n30, 197–98, 226–33; eschatology from perspective of, 29, 31–32, 182–83, 186, 209–11, 213, 215; "Figure and Disintegration," 195–98, 209, 216, 232; on geometries of urban space, 184, 186, 188–95, 196, 225, 292; on history as spatial construction, 207–8, 291; *History: The Last Things before the Last*, 1, 182, 188, 216–32; "The Hotel Lobby," 227; interest in mathematics, 183n1; language of mathematics/spatial forms and, 10, 12, 14, 29, 56; *The Mass Ornament*, 188, 194–95; and modernist imagination of history, 28; Musil's concept of history compared to, 258, 291; "Photography," 199–215, 216; on religious vacuum, movements responding to, 196–97; and Rosenzweig, 158n48, 228n87; Rosenzweig's theory of history compared to, 228, 291; scholarship

Kracauer (*continued*)
on, 185n5, 186, 194–95, 195n28; on shapes of time/shaped times, 1, 12, 31, 182–83, 186, 216, 223–24, 225, 233, 291, 292–93; on "simultaneity of the asynchronous," 223–24; *Sociology as Science*, 187; *Streets in Berlin and Elsewhere*, 190–91; teleology of history rejected by, 29; on temporal exterritoriality of history, 182, 220–21, 226; on theology in the profane, 16, 184–85, 185n5; theory of history, 4, 185–88, 196n30, 197–99, 207–8, 213n63, 215, 217–33, 291, 292–93; "Those Who Wait," 196; "Time and History," 216; topology and theories of, 12, 31, 183–84; training as architect, 186, 188; "Two Forms of Communication," 184; "Two Planes," 188, 189–90, 192, 194, 225; "View from a Window," 190–91, 192

Kraus, Justice, 235n3, 245, 247n27, 250

Kreienbrock, Jörg, 106

Kristeller, Paul Oskar, 216n69

Kubler, George, 222–23

Lage. See "situation" of humanity

Lash, Scott, 214, 214n65

last judgment. See *Weltgericht*

Latour, Bruno, 226

Lectures on the Philosophy of History (Hegel), 24–25

Lefebvre, Henri, 13

Leibniz, Gottfried Wilhelm: development of calculus by, 90; and origins of topology, 193

Leslie, Esther, 194, 195n28

Lessing, Gotthold Ephraim: "Education of Mankind," 75; on pictorial space vs. narrative time, 81

Levinas, Emmanuel, 16, 150n40

Leyden, W. von, 224n80

liminal relations. See history and eschatology; time and eternity

limit (*Grenze*), mathematical concept of, 8, 90; Barth's use of, 31, 93, 95, 95n30, 104, 120, 244, 292; and calculus, development of, 90, 92; definitions of, 91–92; Musil's use of, 235, 242–44; Newton and, 90, 91, 92*f*

limit phenomenon, eschatology as, 8, 124–25, 294; Adorno on, 296–97; Barth on, 5, 12, 89–90, 93–95, 120–24, 291, 294; Musil on, 235, 251–52, 253, 287–88, 291

Lindemann, Ferdinand von, 98

linear development, history conceived as: Spengler's rejection of, 73–74. *See also* development, principle of; teleological concepts of history

lines, conceived as curves: in Riemann's differential geometry, 172; in Rosenzweig's philosophy of religion, 172–73, 181. *See also* curved space(s); dividing line; parallel lines; polyline (*Linienzug*)

Lissitzky, El, 30, 36; approach to art, constructivist historiography compared to, 79; "Art and Pangeometry," 38, 46–47, 52–53; and new concept of space, 37, 45–46; on plastic configuration, 38, 52–53; *Proun* paintings/essay of, 45–46

Listing, Johann Benedict, 183, 193

literature, modernist (literary modernism), 11; modernist eschatology and, 237–39; modernist mathematics and, 12–13; non-Euclidean geometry and, 99, 100; nonteleological narrative form of, 234–35, 238, 239n15, 254–55; spatial form in, Frank on, 81–82, 83, 105, 255n39; spatial imagination of time and, 2–3, 8–9

Lobachevsky, Nikolai, 9, 11, 40, 47, 101

Lowenthal, David, 221n76

Löwith, Karl, 6, 14, 17–18

Luther, Martin, on "time of the now," 103n41

Mach, Ernst, 41–42

magnitude, mathematical concept of, 65

Malachi, Prophet, 135
Malevich, Kazimir, 30, 36; approach to art, constructivist historiography compared to, 79; and new concept of space, 37
Man, Paul de, 284n78
Mann, Thomas, 29
Mannheim, Karl: Barth's concept of history compared to, 112, 116–17, 127; constructivist historiography of, 9, 56, 57, 58–63, 78, 109; Gestalt psychology and, 59–60; "Historicism," 58–59; presentist historiography of, 58, 61, 63, 76, 79; and spatial approach to history, 30, 57; theory of perspectival standpoint of historian, 57, 61–63, 67
The Man without Qualities (Musil), 254–63, 265–81; beginning of, 254, 263n46, 286; concept of history in, 234, 258–63; destabilization of empirical space in, 236; dialectic of reality and possibility in, 265; eschatological metaphors in, 270–78, 270n57, 283–85; essayistic sections of, 254, 255, 268; fractal-like structure of, 269; literary form of, 239n15; modernist eschatology and, 268–69, 282; narrative impasse of, 12, 236n8, 237, 256–57, 258, 265–66, 267–70, 279–81, 282, 286; nonteleological narrative form of, 12, 234–35, 238–39, 254–55; "other condition"/suspension of time in, 5, 235, 238, 239, 255, 256, 270–80, 282, 283; structural paradox of, 286; thought-images in, 260, 262, 266–67; World War I and horizon of, 265–66
Marc, Franz, 178n68
Marion, Jean-Luc, 16
Marquardt, Friedrich-Wilhelm, 106n45
Marx, Karl: philosophy of history, Kracauer on, 219–20, 221; and secularization of eschatology, 17, 18
The Mass Ornament (Kracauer), 188, 194–95

"The Mathematical Human Being" (Musil), 246–47
mathematics, modernist, 10–11; and aesthetic modernism, 237, 247; and art, 9, 11, 37, 38–39; Barth's use of language of, 12, 13, 17, 29, 83, 84–90, 93–104; concept of magnitude in, disappearance of, 65; and constructivism, 11, 37–38, 39; creative freedom associated with, 35; and creative potential, Musil on, 237, 247, 250; foundational crisis of, 237, 246, 247–48, 249; as language of symbols, 248; liminal relations of history and eschatology expressed through language of, 5, 8, 10, 11–12, 31, 84, 253, 290, 293; and literature, scholarship on, 12–13; metaphorical function for Rosenzweig, 175n63, 176; and multiple concepts of space, 34, 35; and Musil's works, 235–37; and new concepts of history and eschatology, 3, 10, 12, 14; nonintuitive character of, 50–52; as "organon of thought," Rosenzweig on, 164, 175–76; and relationship between God and humanity, reflections on, 84, 85–87, 99–101; and religion, reflections on mysteries of, 95–104, 95n31, 241–43; and theological perspective on history, 21, 29, 130, 290; and visualization of nonintuitive, 95–104. *See also* calculus; non-Euclidean geometries
McCormack, Bruce, 86, 106n45, 116n67
McDowell, John C., 106n44
Mehrtens, Herbert, 10, 11, 13, 34, 35, 65
Meinecke, Friedrich, 132, 141, 290
Mendes-Flohr, Paul, 136
messianism: Cohen's teleological understanding of, 178–79, 181; vs. eschatology, Agamben on, 20; and Kracauer's thought, 186n8; philosophy of history as, questioning of, 180

metropolitan miniature, 2–3
Millennial Kingdom, eschatological motif of, 270, 271, 277–78, 285
"Mind and Experience: Notes for Readers Who Have Escaped the Decline of the West" (Musil), 248
Minima Moralia (Adorno), 294–97
modernism: and art, 30, 34; and eschatology, revised concept of, 8, 237–38, 282, 285, 294, 300; and history, new models of, 4, 7–9, 28–29, 181, 289–90, 294–95, 300; and music, 11; and religion, persistence/afterlives of, 14–16, 17; and representation of space, shift in, 34; and time, new concepts of, 2–3, 7. *See also* literature, modernist; mathematics, modernist
Moholy-Nagy, László, 36
moment: Barth's theory of, 82–83, 84, 87, 102–5, 104n43, 110–11, 118, 124, 128; eschatology as equidistant to, Rosenzweig on, 12, 31, 131; Musil's utopia of, 282, 283, 284–85; Ouspensky's expansive concept of, 54; as spatial form, 82
morphology of history, 73; Spengler's, 57–58, 73, 77–78, 79
Mosès, Stéphane, 134, 138, 139, 145, 146, 166
Mülder-Bach, Inka, 184n4, 186n8, 217n71, 231n95, 259, 262, 265
Müller, Johannes, 41
music, modernism in, 11
Musil, Robert, 28–29; bridge metaphor in works of, 248, 250–53, 274; concept of eschatology, 251–52, 253, 287–88; concept of history, 7, 12, 234, 258–67, 269, 270, 281, 282, 291, 293; *The Confusions of Young Törless*, 46, 235, 239–46, 250–52, 253; context for work of, 29, 30, 36, 80; critique of historicism, 234, 263–65; eschatological metaphors used by, 270–78, 270n57, 283–85; "The Flypaper"/"Roman Summer," 287–88; "Helpless Europe, or a Journey from the Hundredth to the Thousandth," 263–65; and language of mathematics/spatial forms, 4, 10, 12, 13, 29, 32, 56, 234, 239, 241n19; *The Man without Qualities*, 4, 12, 234–35, 238–39, 254–63, 265–81; "The Mathematical Human Being," 246–47; and metropolitan miniature, 2; "Mind and Experience: Notes for Readers Who Have Escaped the Decline of the West," 248; modernist aesthetics and, 105, 237, 247; modernist eschatology and, 29, 237–39, 268–69, 282; modernist mathematics and, 12, 13, 235–37, 241n19, 246–49; nonnarrative approach of, 12, 234–35, 236n7, 237, 238–39, 291; "other condition" in works of, 5, 16–17, 29, 32, 235–45, 250, 252, 270–80, 293, 299; *Posthumous Papers of a Living Author*, 286, 286n82, 287; religious and theological subtext of work of, 16, 237–38, 241–42; "Sketch of the Knowledge of the Poet," 249–50; teleology of history rejected by, 29, 32, 234; "Toward a New Aesthetic: Observations on a Dramaturgy of Film," 252–53; and "utopia of the moment," 282, 283, 284–85
Myers, David N., 21n48, 152n43, 154n45
myth, eschatology of, aesthetic experimentation with, 297–300

Nachleben: concept of, 70n108. *See also* afterlife
narrative continuum: absence in Musil's works, 12, 234–35, 236n7, 237, 238–39, 291; modernist eschatology and rejection of, 237–39
narrative philosophy, Rosenzweig's, 141–43, 290
natural history, Kracauer on, 219–20
negation: in Adorno's theory of history, 295–96; in Barth's theory of history, 88, 89–90, 96–97, 103–4, 105, 124,

292; in Kracauer's theory of history, 183, 184, 292; in Rosenzweig's theory of history, 133
The New Thinking (Rosenzweig), 142, 143
Newton, Isaac: and concept of limit, 90, 91; development of calculus by, 90. *See also* calculus
Nicholas of Cusa, 242
Nietzsche, Friedrich: concept of *Urgeschichte*, 119; on falsification of reality, 264n47; influence on Barth, 82, 83, 112; influence on Musil, 264; influence on Spengler, 73; Kracauer on, 218n75; and topos of being lifted out of historicity, 152
Nobel, Nehemias Anton, 228n87
non-Euclidean geometries: and art, innovations in, 9, 37, 38–39, 45–47, 52; Barth's interest in, 98–99, 101, 102, 290; basic problem of, 159; construction of space in, 44; and constructivism, 11, 37–38, 39, 45–47, 80; curved spaces of, 9–10, 11, 44–45, 53, 183; Dostoevsky's use of images from, 99, 100; and eschatology, representation of, 11, 12, 14, 117; Helmholtz on, 40–41, 42, 101, 121; instability of figures in, 9, 10; and Kracauer's images of urban spaces, 189, 191; Lobachevsky and, 9, 11, 40, 47, 101; vs. multidimensional spaces, Helmholtz on, 48; Musil's interest in, 247, 248–49; nonintuitive character of, 9, 50–52; parallel lines in, intersection of, 9, 11–12, 39–41; popularization of, 40–41, 42–43; projective geometry and, 136; and Rosenzweig's theory of history, 17, 130, 131, 146–47, 164, 172–73, 181; and shape of time, 3–4, 10, 12; space in, 9–10, 30, 35, 42, 46–50, 52; and space-time continuum, concept of, 45; topology compared to, 194. *See also* space(s), in non-Euclidean geometries

nontemporal constructivism, Troeltsch and, 68
The Notebooks of Malte Laurids Brigge (Rilke), 8
novel, modernist form of, 239n15, 254–55
now: Ouspensky's concept of, 54. *See also* eschatological now; "now time"; present; today (*Heute*)
"now time" (*Jetztzeit*): Barth on, 110; Benjamin on, 19, 20, 110. *See also* "time of the now" (*die Zeit des Jetzt*)

"On the Concept of History" (Benjamin), 20, 110
"On the Origin and Meaning of Geometric Axioms" (Helmholtz), 40–41, 48–49
"On the So-Called Non-Euclidean Geometry" (Klein), 136
Oppenheim, Gertrud, 144
"other condition," Musil on, 5, 235–39; anarchic quality of, 283; in *The Confusions of Young Törless*, 235, 240–45, 250, 252, 254; dialectic of drive and stasis in, 299; eschatological metaphors used to describe, 270–78, 270n57, 283–85; Kant's concept of the sublime compared to, 243, 243n23; as limit phenomenon, 235, 253; in *The Man without Qualities*, 5, 235, 238, 239, 255, 256, 270–80; mathematical/spatial metaphors used to represent, 29, 32, 240, 248–53, 271–73; as new shape of time, 12, 235, 239, 293; as "profane religiosity," 16–17, 284, 284n80
Ouspensky, P. D.: hyperspace philosophy of, 53; *Tertium Organum*, 53–56; on time and eternity, relationship of, 93n25
Overbeck, Franz: *Christianity and Culture*, 119–20; and concept of *Urgeschichte*, 119, 120, 121; influence on Barth, 82, 89

parallel lines, in Euclidean geometry, 39, 40; and Kracauer's images of urban spaces, 191–92, 195; nonapplicability in curved spaces, 44

parallel lines, intersecting at infinity: in Barth's theory of history, 98–99, 290; Helmholtz on, 40–41, 101; as metaphor for liminal relation of time and eternity/history and eschatology, 11–12, 31; and Musil's "other condition," 32, 240, 245–46; in non-Euclidean geometries, 9, 11–12, 39–41; problem of theodicy explored through, 98–101; in Rosenzweig's dual-aspect theory of history, 146–47, 181, 290

Pascal, Blaise, 149

past: as constantly changing, constructivist historiography and, 61, 67; contemporaneity with present, Barth on, 109–14, 119, 125, 155, 224, 224n80; elemental forces of, Troeltsch on retrieval of, 69–70, 79; objective and mimetic representation of, historicism and, 26; and present, dialectic in photographs, Kracauer on, 211

Paul, Saint: Epistle to the Corinthians, Barth's interpretation of, 94; Epistle to the Romans, Agamben's reading of, 19, 19n44; Epistle to the Thessalonians, Heidegger on, 124; and immanence of eschatological end, idea of, 238. *See also The Epistle to the Romans* (Barth); "time of the now" (*die Zeit des Jetzt*)

Pecora, Vincent P., 233n97

perceptual space, 41–42; space of non-Euclidean geometry compared to, 42

perspectivism: Heussi's model of, 66; Mannheim's model of, 57, 61–63, 67. *See also* Renaissance perspective

philosophy of history: vs. constructivist historiography, 73, 75; Hegel's, 6, 21, 22, 23–25, 27, 56, 75, 114, 116, 116n69, 147; Herder's, 6, 25, 27, 75; historicism compared to, 25–26, 27, 28; Kant's, 56, 75, 179; Kracauer on, 219; Schiller's, 6, 21–23, 27

"philosophy of life" (*Lebensphilosophie*), 57

photograph(s)/photography, Kracauer on: afterlife of, 209; as death mask, 201–2, 204, 204n48, 212; dialectic of past and present represented in, 211; disintegration of, 199, 200, 202, 209, 211, 212–13, 212n60, 213–14, 218; and eschatology, 209–11, 213, 215; and historicism, analogy between, 199, 204–5; hollow spaces of time and history revealed in, 199, 200, 202, 208–9; vs. memory image, 205–7, 208–9; redemptive qualities of, 210–11, 211n58, 213; spatial and temporal aspects of, 200–201, 204, 206; and theory of history, 217–18, 227; and time, reflections on relationship of, 201–3; topological objects compared to, 214

"Photography" (Kracauer), 199–215, 216; and *History: The Last Things before the Last*, continuities between, 217–18, 217n73

Picasso, Pablo, 38

Pike, Burton, 269

plastic configuration (*plastische Gestaltung*): Lissitzky on, 38, 52–53; time as element of, 52–54

Poincaré, Henri: analogies and projections used by, 172; foundational crisis of mathematics and, 246; projective geometry of, 31, 130, 159–60, 181, 292; on representative space, 42–43; and topology, discipline of, 183

point (*Punkt*), mathematical concept of: Barth's use of, 87–88, 103–4, 108–9, 118, 128; in Kracauer's theory of history, 187–88; Spengler

on, 104n42; Tillich's use of, 117–18, 122. *See also* standpoint
Pointillism, 34
polyline (*Linienzug*), 208n54; compression of history into, Kracauer on, 207–8
Posthumous Papers of a Living Author (Musil), 286, 286n82, 287
Pound, Ezra, 81
present: contemporaneity of past with, Barth on, 109–14, 119, 125, 155, 224, 224n80; dual aspect of, Rosenzweig on, 7, 31, 130, 144–47, 152–56, 181; eschatological limit determining, 94; in historicist concept of history, 145, 146; history constructed from vantage point of, theories of, 9, 57, 58, 61, 63, 67–69, 75, 76–77, 79, 94, 117–18; Judaism living in, Rosenzweig on, 152–54; and past, dialectic in photographs, Kracauer on, 211; proximity of end to each moment of, modernist thinkers on, 12, 32, 238, 238n13; Schleiermacher on, 109–10; self-understanding in, insight into the past oriented toward, 68. *See also* now; today (*Heute*)
present-interest theory of history, 220; Kracauer's critique of, 220, 221
primal history. *See Urgeschichte*
Principia Mathematica (Newton), 90
progress, ideology of: collapse of, 82; Herder and, 27; Spengler's criticism of, 75
projective geometry, 136–38, 137*f*, 138*f*, 159–60; and Rosenzweig's ideas of history, 31, 130, 137–39, 160, 181, 292
"Prometheus" (Kafka), 297–300
Protestant liberal theology, 82; view of history, 133
Protestant "theology of crisis," and Rosenzweig's rethinking of history, 130n1
Proun (Lissitzky), 45–46

Proust, Marcel: and antinomy of time, aesthetic reconciliation of, 231, 232; and spatial form, 81
pseudosphere: geometry of, in Kracauer's images of urban spaces, 189, 191; Helmholtz on, 44, 48; history conceived as, 12; in projective geometry, 137–38, 137*f*, 138*f*
Punkt. *See* point, mathematical concept of

"The Quadrangle" (Kracauer), 189–90

Ranke, Leopold von: historicism of, 7, 25, 26, 56, 61, 65n95; on progress in human history, 27; and Rosenzweig's concept of history, 130, 132, 140–41, 142, 143; and *Strom* (current), metaphor of, 144, 144n3
rational numbers, in Rosenzweig's dual-covenant theology, 165–66, 292
reality (*Wirklichkeit*), vs. actuality (*Realität*), Kracauer on, 197–98
reconfiguration (*Neugestaltung*): disintegration (*Zerfall*) and, Kracauer on eschatological dynamic of, 215. See also *Gestaltung*
redemption: Adorno on, 295, 296–97; Benjamin on, 185n5; Christian vs. Judaic understanding of, 20; Kracauer's concept of, 185n5, 210–11, 211n58, 213, 217, 227, 227n86; in Rosenzweig's star of redemption, 173, 174*f*, 293
relativity. *See* General Theory of Relativity
religion: absence of, movements responding to vacuum of, Kracauer on, 196–97; afterlives in modernity/persistence in post-secular societies, 14–16, 17, 21, 32; with antihistoricist undertones, Rosenzweig and concept of, 134; and concept of history in 20th century, 17–20; and history, Rosenzweig on relationship of, 139; mathematics used to visualize mysteries of, 95–104, 95n31, 241–43;

religion (*continued*)
  in modernist imagination, scholarship on, 16; and modernist mathematics, 20th-century rethinking of history through, 21, 29, 290. *See also* Christianity; Judaism; theology
Renaissance perspective, 33–34, 38; art forms of breaking with, 38, 79; vs. Mannheim's model of perspectivism, 61; space of negative curvature supplanting, 189
Renan, Ernest, 89
representability, limits of: Abbott on, 49–50; Helmholtz on, 48–49; Lissitzky on, 46–47
representative space, vs. geometric space, 42–43
"Resignation" (Schiller), 21–22
resurrection, as limit phenomenon, in Barth's theology, 93–94, 121–24, 291
The Resurrection of the Dead (Barth), 94–95, 123–24
revelation: every moment in time as potential moment of, Barth on, 87; in Rosenzweig's star of redemption, 173, 174f
reverse motion: film technique of, 266n49; Musil's thought experiment involving, 266–67
Richter, Gerhard, 295
Richter, Silvia, 150n40
Riemann, Bernhard: on curved spaces, 9, 44, 172–73, 245; and modernist mathematics, 9, 10, 40, 44, 47, 194, 236; and topology, discipline of, 183, 194
Rilke, Rainer Maria, 2; *The Notebooks of Malte Laurids Brigge*, 8
"Roman Summer" (Musil), 287–88
Rosenstock-Huessy, Eugen, Rosenzweig's correspondence with, 129, 135, 139–40, 142, 164, 167–70, 176
Rosenzweig, Franz, 28, 129; antihistoricism of, 139; Barth's theory of history compared to, 129–30, 133, 147, 150, 158, 158n48, 161, 164n55, 290; Cohen and, 178, 178n69; vs. Cohen's teleological narrative of history, 179–81; concept of eschatological now, 144–45, 154–55, 179; concept of eternity, 130, 152–61; concept of today (*Heute*), 7, 31, 130, 144–47, 154–55, 180, 293; context for work of, 29, 30, 36, 80; correspondence with Ehrenberg, 132, 139, 140, 164–65; correspondence with Kracauer, 158n48; correspondence with Rosenstock-Huessy, 129, 135, 139–40, 142, 164, 167–70, 176; critique of Hegel's philosophy of history, 130, 132, 133, 134, 135, 141, 147–48, 150n40, 157; on curved spaces of history, 12, 131, 141; dual-aspect theory of history, 7, 31, 130, 144–47, 181, 290, 291; dual-covenant theology of, 129, 139, 143–44, 151–53, 156–70, 157n47; eschatology from perspective of, 12, 31, 131, 148–51, 162, 180–81, 180n75, 294; geometrical metaphors used by, 29, 31, 130, 131, 148–51, 156, 164, 169, 176, 181; *Gestalten* (figures) and theory of history of, 130, 131, 143–44, 147, 173–74, 176–78; "Globe: Studies on the World-Historical Theory of Space," 140; on goal (*Ziel*) of history as beginning (*Anfang*), 180–81; *Hegel and the State*, 134; historicism and, 130, 132, 140–47, 180n75, 290; on history as "discontinuous series" (*diskontinuierliche Reihe*), 135–36, 138; and image of arc of history, 130, 148–51, 157, 170–71, 181, 253, 290; influences on, 130, 132–33, 141–42; and Kracauer, 158n48, 228n87; Kracauer's theory of history compared to, 228, 291; language of mathematics/spatial forms and, 10, 12, 13, 14, 56, 130, 164; and liminal concept of eschatology, 29; on mathematics as "organon

of thought," 164, 175–76; and Meinecke, 132, 141, 290; modernist aesthetics influencing, 17, 177–78, 178n68; and modernist imagination of history, 28; narrative philosophy of, 141–43, 290; *The New Thinking*, 142, 143; non-Euclidean geometry and theories of, 17, 130, 131, 146–47, 164, 172–73, 181; projective geometry and ideas of history, 31, 130, 137–39, 160; Protestant "theology of crisis" and, 130n1; Ranke's historicism and, 130, 132, 140–41, 142–43; rethinking of history in spatial terms, 139–40; scholarship on, 18–19, 154n45; on shapes of time/shaped times, 129–30, 291; spatial approach to history/eschatology, 139–41, 147, 177–78; and star of redemption, figure of, 151, 162, 164, 172–74, 174f; *The Star of Redemption*, 31, 129–31, 147–64, 170–77, 180; supra-mathematical figures used by, 170; teleology of history rejected by, 29; theory of history, 4, 131–32, 134–36, 138–40; on today as present moment and springboard to eternity, 7, 31, 130, 144–47, 293

Rothacker, Erich, 107

rupture, of historical continuum: Jesus Christ and, Barth on, 86–87; as moment of nonpresence, Barth on, 103–4

Russian Constructivism, 9, 36, 52, 53; constructivist historiography compared to, 79; Lissitzky's *Proun* essay on, 45–46; reimagining of space in, 36–37, 38, 39, 45–47

Scheidelinie (dividing line), image of, in Musil's work, 242–43

Schelling, Friedrich Wilhelm Joseph: *The Ages of the World*, 141–42; and Rosenzweig's narrative philosophy, 142, 156, 157; and secularization of eschatology, 17, 18

Schiller, Friedrich: and historicization of eschatology, 6, 7, 17, 21–22; *History of the Revolt of the Netherlands*, 23; *The History of the Thirty Years' War*, 23; philosophy of history of, 6, 21–23, 27; "Resignation," 21–22; "What Is, and to What End Do We Study, Universal History?", 22–23

Schleiermacher, Friedrich: present according to, 109–10; Rosenzweig's critique of, 132, 133; theology of, 82, 88–89

Schmitt, Carl, 14

Scholem, Gershom, 13

Schröter, Michael, 210

Schweitzer, Albert, 19n44, 89

The Sense of an Ending (Kermode), 238

Serres, Michel, 226

set theory, mathematical uncertainties surrounding, 246, 249

Shakespeare, William, *King Lear*, 238

shape(s) of time/shaped times, 35–36; Barth on, 111, 129; chronological time in relation to, 7, 223, 231–32, 289, 291; curved space of modernist mathematics and, 3–4, 10; Kracauer on, 1, 12, 31, 182–83, 186, 216, 223–24, 225, 233, 291, 292–93; Kubler on, 222–23; and modernist imagination of history, 4–5, 12, 28, 178, 289, 290, 293–94; Musil on ("other condition"), 12, 235, 239, 293; Rosenzweig on, 129–30, 131, 141, 181, 291; topology and, 226

The Shape of Time: Remarks on the History of Things (Kubler), 222–23

Simmel, Georg: and *Gestalten*, concept of, 131; historicist paradigm challenged by, 29

Simonis, Annette, 130

Simons, Oliver, 2, 9

"simultaneity of the asynchronous" (*Gleichzeitigkeit des ungleichzeitigen*): Kracauer on, 223–24. *See also* contemporaneity (*Gleichzeitigkeit*)

sin, and grace, as incommensurable magnitudes, Barth on, 101–2, 292
"situation" (*Lage*) of humanity: Barth on, 106, 107–8, 110–11; in German discourses of 1920s-1930s, 106–7
"Sketch of the Knowledge of the Poet" (Musil), 249–50
Smith, John H., 13, 164n55, 171
social theory, on privileging of time over space in modernity, 3, 3n6
sociology, Kracauer's account of, 187–88
Sociology as Science (Kracauer), 187
Soja, Edward W., 13
space(s): as a priori form, Kant on, 34, 35, 59; alternative concepts of, Musil and, 247, 248–49; as basic category of our perceptual apparatus, 2; as conceptual bridge between mathematical symbols and profane reality, Musil on, 248; construction of, modernist mathematics on, 35, 43–44, 247; eschatology conceived in terms of, 5, 8, 294, 300; in Euclidean geometry, 9, 34, 41–42, 44; as only one of many possible spaces, 35, 41–42, 44, 47, 194; of history, end of history conceived as, 181; modernist art and re-envisioning of, 30, 34, 36–37, 38, 39, 45–47; multidimensional, 46, 47, 51; multiple concepts of, modernist mathematics and, 34, 35, 40–42, 44, 47, 194; new theory of, non-Euclidean geometry and, 9–10, 30; nonintuitive/counterintuitive, modernist mathematics and study of, 236; as object of construction, 35–36; of "other condition," Musil on, 5, 32, 235, 236–39; perceptual/visual, 41–42; privileging of time over, modernity and, 3, 3n6; pseudospherical, 44, 48; as pure form of intuition, Kant on, 2; Renaissance perspective on, 33–34, 38; representative, 42–43; spherical, 44; and time, conceived as separate and opposite categories, 2; time as fourth dimension of, theories of, 53–54; time as motion in, idea of, 52–53; topological objects as, 183, 214, 214n65; of touch (haptic space), 42. *See also* curved space(s); space(s), in non-Euclidean geometries; spatial form(s)

space(s), in non-Euclidean geometries, 9–10, 30; Helmholtz on, 48–49; Lissitzky on, 46–47; nonrepresentability of, 46–50; perceptual/visual space compared to, 42; Spengler on, 50–51

space-time continuum, Einstein's concept of, 45, 52, 53; and constructivist art, 46

spatial form(s): in Barth's theory of history, 10, 30, 56, 83, 84–85, 93, 104, 105, 128; compression of history into, Kracauer on, 207–8; language of, nonintuitive relations of time and eternity represented in, 5, 10, 56; in modernist literature, Frank on, 81–82, 83, 105, 255n39; moment as, 82; in Musil's works, 10, 56, 234, 239; and religious thought in modernity, 17; in Rosenzweig's theory of history, 10, 56, 130, 164, 173–74; time as, modernist interest in, 82

"Spatial Form in Modern Literature" (Frank), 81–82

Spengler, Oswald: Barth's concept of history compared to, 116–17, 118, 125–26, 127; constructivist historiography of, 9, 56, 57–58, 73–78; *The Decline of the West*, 76; Kracauer's theory of history compared to, 188; and morphology of history, 57–58, 73, 77–78, 79; Musil's essay on, 248; on non-Euclidean spaces, 50–51; on plurality of historical cultures, 74, 75; on presentist historiography, critique of, 76–77; on simultaneity of historical distant figures, 112n57;

and spatial approach to history, 30, 57, 74, 75; on status of *Punkt* in modern mathematics, 104n42
sphere: Being as, Hegelian concept of, 148; in *Flatland: A Romance of Many Dimensions* (Abbott), 49–50; in projective geometry, 136; in Rosenzweig's eschatological account of history, 149; surface of, and Helmholtz's visualization of non-Euclidean geometry, 40–41, 44, 48–49, 101, 121. *See also* pseudosphere
springboard, metaphor of, in Rosenzweig's dual aspect theory of history, 7, 31, 130, 144–47, 145, 146
Stählin, Gustav, 103n40
standpoint: of eschatology, Tillich on, 117–18, 122, 296; of redemption, Adorno on, 295, 296–97
star of redemption, Rosenzweig's figure of, 151, 162, 164, 172–74, 174*f*, 181, 291; as demathematicized figure, 175; first sketch of, 176
The Star of Redemption (Rosenzweig), 31, 129–31, 147–64, 170–77, 180; central idea of, 139; on Christianity as "eternal way" (der *ewige Weg*), 156–61; coordinates of history in, 138–39; on eternal figures (*ewigen Gestalten*), 143–44; historicism and, 143; on Judaism standing outside of history, 139, 152–53; non-Euclidean geometry and, 164; postscript to, 142; rhetoric of *Gestalten* (figures) in, 130, 131, 143–44, 176–77; spatial images of history in, 139–41, 147
Stillwell, John, 136
Strauß, David Friedrich, 89
Strauss, Leo, 221
Streets in Berlin and Elsewhere (Kracauer), 190–91
Strom (current/stream), image of: and Heussi's understanding of history, 63, 66, 79; in historicist understanding of time, 1, 144, 144n3; in Musil's description of "other condition," 276, 278; Schiller on, 23

sublime, Kant's concept of: and Heussi's theory of history, 64; and Musil's "other condition," 243, 243n23
supra-mathematical figure(s): Rosenzweig's interest in, 170–71, 292; star of redemption as, 173–74
Suprematism, 9, 36, 52, 53
Surrealism, 9
symbolic time, and historical time, in Rosenzweig's dual aspect theory of history, 146–47

tangent, mathematical concept of: in Barth's understanding of eschatology, 89–90, 93, 95, 253, 290; Euclidean understanding of, 91, 93; in modern calculus, 91
Tatlin, Vladimir, 38, 53
Taubes, Jacob, 18–19; on expressionism of Barth's theology, 84n6; on philosophical eschatology in German Idealism, 75; on Rosenzweig's dual-covenant theology, 163
teleological concepts of history, 21–25, 27–28, 56–57; historicism and, 27–28; vs. history as object of construction, 56–57, 75–76; modernist eschatology decoupled from, 8, 237–38, 282, 285, 294, 300; modernist thinkers rejecting, 29, 31–32, 56–57, 75–76, 79, 178, 179–81, 234; Musil's critique of, 234, 260–62; philosophy of history and, 21–25, 75; shapes of time as alternative to, 178, 293–94
temporal exterritoriality of history, Kracauer on, 182, 220–21, 226, 230n93
Tertium Organum (Ouspensky), 53–56
theodicy, problem of: Barth on, 98–101; Dostoevsky on, 100
theology: afterlife in modernity, 14–16, 17, 284; constructivist form of, Barth's, 89–90, 93–94, 116–17, 117–18, 127–28; dual-covenant, Rosenzweig's, 129, 139, 143–44,

theology (*continued*)
151–53, 156–70, 157n47; and modernist mathematics, perspective on history based on, 130; in the profane, Kracauer on, 184–85, 185n5; Schleiermacher's, 82, 88–89. *See also* Protestant liberal theology

Thiher, Allen, 241n18

This Too a Philosophy of History (Herder), 25, 144n31

"Those Who Wait" (Kracauer), 196

Thums, Barbara, 185n5

Thurneysen, Eduard, 99, 109n53, 120, 228n87

Tillich, Paul: rhetoric of *Gestaltung* (figuration), 127n86; on standpoint of eschatology, 117–18, 122, 296; on Troeltsch's historiography, 117–18

time(s): antinomy of, Kracauer's concept of, 8, 224–25, 229, 230n93, 231–32; concept of, modernism and transformation of, 1–3, 7; curvature of, concepts of, 52; as element of plastic configuration, Lissitzky on, 52; fold in, Serres's concept of, 226; as fourth dimension of space, theories of, 53–54; historical and symbolic, in Rosenzweig's dual aspect theory of history, 146–47; historicist paradigm of, vs. modernist concepts, 29, 145, 182, 186, 224, 289; interruption and rupture of, geometrical language used to represent, 31; as motion in space, idea of, 52–53; non-Euclidean geometries and new possibilities for thinking about, 3–4, 10, 12, 52; photography and, Kracauer's reflections on relationship of, 201–3; privileging over space, 3, 3n6; as pure form of intuition, Kant on, 2; spatial representation of, 2–3, 8–9, 52–54, 82; standstill of, Musil's articulation of, 5, 275–77, 293. *See also* chronological time; shape(s) of time/shaped times; time and eternity

time and eternity (time of history and eschatological time), liminal relation of, 8, 10; Adorno on, 296–97; Barth on, 29, 87–88, 94–95, 101, 102–5, 103n40, 109, 290, 291, 293; enduring legacy of concept of, 300; Kracauer on, 228; mathematical language used to represent, 5, 8, 10, 11–12, 31, 84, 253, 293; Musil on, 253, 273, 282, 291; Ouspensky on, 93n25; Rosenzweig on, 130, 147, 148–51, 154–56, 290; shaped times and, 4, 12, 28, 289, 290, 293–94

"Time and History" (Kracauer), 216

"time of the now" (*die Zeit des Jetzt*), Paul's concept of, 30, 83; Barth's interpretation of, 19, 55, 102–3, 128; Luther's interpretation of, 103n41; and Musil's "other condition," 235

time-sense, Ouspensky on, 55

today (*Heute*): historicism and concept of, 145, 146; Rosenzweig on, 7, 31, 130, 144–47, 154–55, 180, 293

topology, mathematical discipline of, 31, 192–94, 292; deformation of object-spaces in, 183–84, 192–93, 194; Königsburg bridge problem in, 193–94, 193f; and Kracauer's reflections on photography, 214; and Kracauer's theory of history, 12, 31, 183, 225, 226, 233, 292–93; non-Euclidean geometry compared to, 194; objects in, as spaces, 183, 214, 214n65

touch, space of, 42

"Toward a New Aesthetic: Observations on a Dramaturgy of Film" (Musil), 252–53

tragedy, modern, rise of, 238

Treml, Martin, 15

Troeltsch, Ernst: on *Aufbau* (structure/construction) of history, 57, 67–69, 74–75, 79, 125; Barth's concept of history compared to, 112, 116–17, 118, 127; Benjamin's theory of history compared to, 71–72;

constructivist historiography of, 9,
  56, 57, 58, 67–73, 74–75, 78, 117;
  on elemental forces of history,
  retrieval of, 69–70, 79; Kracauer's
  theory of history compared to, 188;
  and spatial concepts of time and
  history, 30, 36, 57; Tillich's critique
  of, 117–18
"Two Forms of Communication"
  (Kracauer), 184
"Two Planes" (Kracauer), 188,
  189–90, 192, 194, 225

universal history of human development, idea of, 7
urban space, geometries of, Kracauer
  on, 184, 186, 188–95, 196, 225, 292
Urgeschichte, concept of: Barth on,
  119–21; Nietzsche on, 119;
  Overbeck on, 119, 120, 121
Urgeschichte and *Endgeschichte*,
  convergence of: in Barth's theory of
  history, 12, 84, 120–21, 180, 290; in
  Kafka's "Prometheus," 299; Musil's
  bridge metaphor compared to, 251;
  in Rosenzweig's theory of history,
  158–59, 180–81, 290

vector, representation of eternity as,
  118, 125, 147
Vico, Giambattista, 220
"View from a Window" (Kracauer),
  190–91, 192
visual space, 41–42; space of non-
  Euclidean geometry compared to, 42
Voegelin, Eric, 17
voids. *See* hollow spaces (*Hohlräume*)
von Moltke, Johannes, 230n93

Wandering Jew. *See* Ahasuerus
Warburg, Aby: on afterlife (*Nachleben*)
  of pictorial representations, 70;
  Barth's theory of contemporaneity
  and, 113

wave interference, history compared
  to, 225
Weber, Max, 14
Weidner, Daniel, 15, 19n44
Weigel, Sigrid, 70n108
Weiterentwicklung. *See* development,
  principle of
Weltgericht (last judgment), historicization of eschatology as, 6, 21–22;
  Hegel on, 22, 24; Schiller on, 6, 17,
  21–22
Wertheimer, Max, 59
"What Is, and to What End Do We
  Study, Universal History?" (Schiller),
  22–23
Wiehl, Reiner, 154n45
Wittekind, Folkart, 94n28
Wittgenstein, Ludwig, 2
Woolf, Virginia, 105
Work on Myth (Blumenberg), 297–300
World War I: and contingency of
  history, Musil's idea of, 264; cultural
  crisis in aftermath of, 106, 134;
  impact on Rosenzweig, 134,
  158n48; and *The Man without
  Qualities* (Musil), horizon of,
  265–66

Zerfall. *See* disintegration
Zermelo, Ernst, 246
Ziel (goal), of history: as beginning
  (*Anfang*), Rosenzweig on, 180–81;
  Jewish people's awareness of,
  Rosenzweig on, 153; Spengler's
  rejection of, 76; in teleological
  models of history, 18, 25, 27, 56–57
Zweck (purpose), of history: Droysen
  on, 27–28; Hegel on, 24, 25; Schiller
  on, 23
Zwischen den Zeiten (journal), 108,
  109n53
zwischen den Zeiten (between the
  times), Barth's concept of, 109, 121,
  158

www.ingramcontent.com/pod-product-compliance
Lightning Source LLC
Chambersburg PA
CBHW032359230426
43672CB00007B/755